Probability

An Introduction with Statistical Applications

Probability

An Introduction with Statistical Applications

▶ **John J. Kinney**

Rose–Hulman Institute of Technology

John Wiley & Sons, Inc.

New York • Chichester • Brisbane • Toronto • Singapore • Weinheim

ACQUISITION EDITOR	Brad Wiley II
MARKETING MANAGER	Jay Kirsch
PRODUCTION EDITOR	Ken Santor
DESIGNER	Nancy Field
MANUFACTURING MANAGER	Dorothy Sinclair
ILLUSTRATION EDITOR	Sigmund Malinowski

This book was set in 10/12 Times Roman by Eigentype Compositors and printed and bound by R.R. Donnelley/Crawfordsville. The cover was printed by Lehigh Press.

Recognizing the importance of preserving what has been written, it is a policy of John Wiley & Sons, Inc. to have books of enduring value published in the United States printed on acid-free paper, and we exert our best efforts to that end.

The paper on this book was manufactured by a mill whose forest management programs include sustained yield harvesting of its timberland. Sustained yield harvesting principles ensure that the number of trees cut each year does not exceed the amount of new growth.

Library of Congress Cataloging-in-Publication Data:
Kinney, John J.
    Probability : an introduction with statistical applications / John J. Kinney
        p.   cm
    Includes bibliographical references (p.   –   ).
    ISBN 0-471-12210-6 (cloth : alk. paper)
    1. Probabilities.   I. Title.
QA273.K493 1997
519.2 – dc20
                                                                96-15306
                                                                CIP

Printed in the United States of America

10 9 8 7 6 5 4

▶ *Preface*

▌ Historical Note

The theory of probability is concerned with events that occur when randomness or chance influences the result. While the data from a sample survey or the occurrence of extreme weather patterns are common enough examples of situations where randomness is involved, we have come to presume that many models of the physical world contain elements of randomness as well. Scientists now commonly suppose that their models contain random components as well as deterministic ones. Randomness, of course, does not involve any new physical forces; rather than measure all the forces involved and thus predict the exact outcome of an experiment, we choose to combine all these forces and call the result random. The study of random events is the subject of this book.

It is impossible to chronicle the first interest in events involving randomness or chance, but we do know of a correspondence between Blaise Pascal and Pierre de Fermat in the middle of the seventeenth century regarding questions arising in gambling games. Appropriate mathematical tools for the analysis of such situations were not available at that time, but interest continued among some mathematicians. For a long time the subject was connected only to gambling games and its development was considerably restricted by the situations arising from such considerations. Mathematical techniques suitable for problems involving randomness have produced a theory not only applicable to gambling situations, but to more practical situations as well. It has not been until recent years, however, that scientists and engineers have become increasingly aware of the presence of random factors in their experiments and manufacturing processes and have become interested in measuring or controlling these factors.

It is the realization that the statistical analysis of experimental data, based on the theory of probability, is of great importance to experimenters that has brought the theory to the forefront of applicable mathematics. The history of probability and the statistical analysis it makes possible illustrate a prime example of seemingly useless mathematical research that now has an incredibly wide range of practical application. Mathematical models for experimental situations now commonly involve both deterministic and random terms. It is perhaps a simplification to say that science, while interested in deterministic models to explain the physical world, now is interested as well in separating deterministic factors from random factors and measuring their relative importance.

There are two facts that strike me as most remarkable about the theory of probability. One is the apparent contradiction that random events are in reality well-behaved and that there are laws of probability. The outcome on one toss of a coin cannot be predicted, but given 10,000 tosses of the same coin, many events can be predicted with a high degree of accuracy. The second fact, which the reader will soon perceive, is the pervasiveness of a probability distribution known as the normal distribution. This distribution, which will be defined and discussed at some length, arises in situations which at first glance have little in common; the normal distribution is an essential tool in statistical modeling and is perhaps the single most important concept in statistical inference.

There are reasons for this and it is my purpose to explain these in this book.

About the Text

From the author's perspective, the characteristics of this text which most clearly differentiate it from others currently available include:

- Applications to a variety of scientific fields, including engineering, appear in every chapter.
- Integration of computer algebra systems such as Mathematica provides insight into both the structure and results of problems in probability.
- A great variety of problems at varying levels of difficulty provides a desirable flexibility in assignments.
- Topics in statistics appear throughout the text so that professors can include or omit these as the nature of their course warrants.
- Some problems are structured and solved using recursions since computers and computer algebra systems facilitate this.
- Significant and practical topics in quality control and quality production are introduced.

It has been my purpose to write a book that is readable by students who have some background in multivariable calculus. Mathematical ideas are often easily understood until one sees formal definitions which frequently obscure such understanding. Examples allow us to explore ideas without the burden of language. Therefore I often begin with examples and follow with the ideas motivated first by them; this is quite purposeful on my part, since language often obstructs understanding of otherwise simply perceived notions.

I have attempted to give examples that are interesting and often practical in order to show the wide-spread applicability of the subject. I have sometimes sacrificed exact mathematical precision for the sake of readability; readers who seek a more advanced explication of the subject will have no trouble in finding suitable sources. I have proceeded in the belief that beginning students want most to know what the subject encompasses and for what it may be useful. More theoretical courses may then be chosen as time and opportunity allow. For those interested, the bibliography contains a number of current references.

An author has considerable control over the reader by selecting the material, its order of presentation, and the explication. I am hopeful that I have executed these duties with due regard for the reader. While the author may not be described with any sort of precision as the holder of a tightrope, I have been guided by the admonition: "It's not healthy for the tightrope walker to be misunderstood by the person who's holding the rope."[1]

The book makes free use of the now widely available computer algebra systems. I have used Mathematica, Maple, and Derive for various problems and examples in the book and I hope the reader has access to one of these marvelous mathematical aids. These systems allow us the incredible opportunity to see graphs and surfaces easily, which otherwise would be very difficult and time-consuming to produce. Computer algebra systems make some parts of mathematics visual and thereby add immensely to our understanding. Derivatives, integrals, series expansions, numerical computation, and the solution of recursions are used throughout the book, but the reader will find that only the results are included; in my opinion there is no longer any reason to dwell on calculation of either a numeric or algebraic sort. We can now concentrate on the meaning of the results without being restrained by the often mechanical effort in achieving them; hence our concentration on the structure of the problem and the insight the solution gives. Graphs are freely drawn and, when appropriate, a geometric view of the problem is given so that the solution and the problem can be visualized. Numerical approximations are given when exact solutions are not feasible. The reader without a computer algebra system can still do the problems; the reader with such a system can reproduce every graph in the book exactly as it appears. I have included a fairly extensive appendix in which computer commands in Mathematica are given for many of the examples in which Mathematica was used; this should also ease the translation to other computer algebra systems. The reader with access to a computer algebra system should refer to Appendix 1 fairly frequently. Occasionally there are problems and examples which cannot easily be done without such a system or for which such a system is very useful; these problems are marked by the symbol ⌨.

Although I hope the book is readable and as completely explanatory as a probability text may be, I know that students often do not read the text, but proceed directly to the problems. There is nothing wrong with this; after all, if the ability to solve practical problems is the goal, then the student who can do this without reading the text is to be admired. Readers are warned, however, that probability problems are rarely repetitive; the solution of one problem does not necessarily give even any sort of hint as to the solution of the next problem. I have included over 840 problems so that a reader who solves the problems can be reasonably assured that the concepts involving them are understood.

The problem sections begin with the easiest problems and gradually work their way up to some reasonably difficult problems while remaining within the scope and level of the book. In discussing a forthcoming examination with my students, I summarize the material and give some suggestions for practice problems, so I have followed each chapter by a Chapter Summary, some suggestions for review problems, and finally some Supplementary Problems.

---

[1] *Smilla's Sense of Snow*, by Peter Hoeg (Farrar, Straus and Giroux: New York, 1993).

For the Instructor

Texts on probability often use generating functions and recursions in the solution of many complex problems; with our use of computer algebra systems we can determine generating functions, and often their power series expansions, with ease. The structure of generating functions is also used to explain limiting behavior in many situations. Many interesting problems can be best described in terms of recursions and since computer algebra systems allow us to solve such recursions, some discussion of recursive functions is given. Proofs are often given using recursions, a novel feature of the book. Occasionally the more traditional proofs are given in the exercises.

While numerous applications of the theory are given in the text and in the problems, the text by no means exhausts the applications of the theory of probability. In addition to solving many practical and varied problems, the theory of probability also provides the basis for the theory of statistical inference and the analysis of data. Statistical analysis is combined with the theory of probability throughout the book. Hypothesis testing, confidence intervals, acceptance sampling, and control charts are considered at various points in the text. The order in which these topics are to be considered is entirely up to the instructor; the book is quite flexible in allowing sections to be skipped or delayed, resulting in rearrangement of the material. This book will serve as a first introduction to statistics, but the reader who intends to apply statistics should also elect a course in applied statistics. In my opinion, statistics will be the centerpiece of applied mathematics in the twenty-first century.

Acknowledgments

I have been able to do what I have done because of where I am and because of my students and my colleagues. I am grateful to Rose-Hulman Institute of Technology for granting me a sabbatical leave so that I could begin this book. Many of my colleagues gave me considerable encouragement and advice. Among them I would especially note Dr. Robert Lopez, Dr. Nacer Abrouk, Professor Alfred Schmidt, and Dr. Yosi Shibberu. I am greatly indebted to Dr. Ralph Grimaldi who read every word of the text and problems; his counsel added to the readability of the book. I have been privileged to work with some remarkable students throughout my career; I often have learned much from them. Don Jenkins, Chris Rolenc, Mike Ley, Patrick Swickard, and Gabe Ferland were especially open with comments for me when I used a preliminary edition of the book in Mathematics 311, but in a sense I owe a debt to every student who took that course from me over the years. I am grateful to Mary Lou McCullough for her secretarial assistance.

It has been a great privilege to work with my editor, Brad Wiley II, as well as Mary O'Sullivan and Ken Santor of John Wiley & Sons. They have all contributed to the book and eased both the creation and production processes.

I owe a lot to my daughter, Kaylyn, who encouraged me and who inspired some of the problems. My greatest debt, however, is to my wife, Cherry, who, in addition to providing encouragement, endured my long hours in the study over a course of two years with great understanding.

John J. Kinney
Terre Haute, Indiana
October 1996

This book is for Cherry and Kaylyn.

▶ *Contents*

Chapter 4 ▶ **Functions of Random Variables, Generating Functions, and Statistical Applications** 237

► *Chapter 1*

Sample Spaces and Random Variables

▌1.1 ▶ Discrete Sample Spaces

Probability theory deals with situations in which there is an element of randomness or chance. Some models of the physical world are *deterministic*; that is, they predict exactly what will happen under certain circumstances. For example, if an object is dropped from a height and given no initial velocity, its distance, s, from the starting point is given by $s = \frac{1}{2} \cdot g \cdot t^2$ where g is the acceleration due to gravity and t is time. If one tried to apply the formula in a practical situation, one would not find very satisfactory results. The problem is that the formula applies only in a vacuum and ignores the shape of the object and the resistance of the air as well as other factors. While some of these factors can be determined, we generally combine them and say that the result has a random or chance component. Our model then becomes $s = \frac{1}{2} \cdot g \cdot t^2 + \varepsilon$ where ε denotes the random component of the model. In contrast with the deterministic model, this model is *stochastic*.

Science often considers stochastic models; in formulating new models, the scientist may try to determine the contributions of both the deterministic and the random components of the model in predicting accurate results.

The mathematical theory of probability arose in consideration of games of chance, but, as the above example shows, it is now widely used in far more practical and applied situations. We encounter other circumstances frequently in everyday life in which we presume that some random factors are at work. Here are some simple examples. What is the chance I will find that all eight traffic lights I pass through on my way to work are green? What are my chances for winning a lottery? I have a ten-volume encyclopedia which I have packed in separate boxes. If the boxes become mixed up and I draw the volumes out at random, what is the chance that my encyclopedia will be in order? My desk lamp has a bulb that is "guaranteed" to last 5000 hours. It has been used for 3000 hours. What is the chance that I must replace it before 2000 more hours are used? Each of these situations involves a random event whose specific outcome is unpredictable in advance.

Probability theory has become important due to the wide variety of practical problems it solves and because of its role in science. It is also the basis of the statistical analysis of data that is widely used in industry and in experimentation. Consider some examples. A manufacturer of television sets may know that 1% of the television sets manufactured have defects of some kind. What is the chance that a shipment of 200 sets a dealer has received contains 2% defective sets? Solving problems such as these has become important to manufacturers anxious to produce high-quality products, and indeed such considerations play a central role in what has become known in manufacturing as *statistical process control*. Sample surveys, in which only a portion of a population or reference set is investigated, have become commonplace. A recent survey, for example, showed that two-thirds of welfare recipients in the United States were not old enough to vote. But surely we do not know that exactly two-thirds of all welfare recipients were not old enough to vote; there is some uncertainty, largely dependent upon the size of the sample investigated as well as the manner in which the survey was conducted, connected with this result. How is this uncertainty calculated?

As a final example, consider a scientific investigation into, say, the relationship between temperature, a catalyst, and pressure in creating a chemical compound. A scientist can carry out only a few experiments in which several combinations of temperatures, amount of catalyst, and level of pressure are investigated. Furthermore, there is an element of randomness (due largely to other, unmeasured factors) that influence the amount of compound produced. How is the scientist to determine which combination of factors maximizes the amount of chemical compound? We will encounter many of these examples in this book.

In some situations we could measure all the forces involved and predict the outcome precisely but very often choose not to do so. In the traffic light example, we could, by knowledge of the timing of the lights, my speed, and the traffic pattern, predict precisely the color of each light as I approach it. While this is possible, it is probably not worth the effort, so we combine all the forces involved and call the result "chance." So "chance" as we use it does not imply any new or unknown physical forces; it is simply an umbrella under which we put forces we choose not to measure.

How then do we measure the probability of events such as those described above? How do we determine how likely such events are? Such probability problems may be puzzling to us because we lack a framework in which to solve them. We lack a strategy for dealing with the randomness involved in these situations. A sensible way to begin is to consider all the possibilities that could occur. Such a list, or set, is called a *sample space*.

We begin here with some situations that are admittedly much simpler than some of those described above; more complex problems will also be encountered in this book.

We will consider situations that we call *experiments*. These are situations that can be repeated under identical circumstances. Those of interest to us will involve some randomness so that the outcomes cannot be precisely predicted in advance. As examples, consider the following:

- Choose two people at random from a group of five people.
- Choose one of two brands of breakfast cereal at random.
- Throw two fair dice.
- Take an actuarial examination until it is passed for the first time.
- Perform any laboratory experiment.

Clearly the first four of these experiments involve random factors. Laboratory experiments involve random factors as well, and we would probably choose not to measure all the factors necessary to be able to predict the exact outcome in advance.

Once the conditions for the experiment are set, and we are assured that these conditions can be repeated exactly, we can form the *sample space*, that we define as follows:

Definition: A *sample space* is a set of all the possible outcomes from an experiment.

Example 1.1.1

The sample spaces for the first four experiments mentioned above are:

    a. (Choose two people at random from a group of five people.)
        Denoting the five people as A, B, C, D, and E, we find, if we

disregard the order in which the persons are chosen, that there are ten possible samples of two people:

$$S = \{AB, AC, AD, AE, BC, BD, BE, CD, CE, DE\}.$$

This set, S, comprises the sample space for the experiment.

If we regard the choice of people as random, we might expect that each of these ten samples occurs about 10% of the time. Further, we see that any particular person, say B, occurs in exactly four of the samples, so we say the *probability* that any particular person is in the sample is $\frac{4}{10} = \frac{2}{5}$. The reader may be interested to show that if three people were selected from a group of five people then the probability that a particular person is in the sample is $\frac{3}{5}$. There is a pattern here that we can establish with some results to be developed later in this chapter.

b. (Choose one of two brands of breakfast cereal at random.) Denote the brands as K and P. We take the sample space as

$$S = \{K, P\}$$

where the set S contains each of the *elementary* outcomes K and P .

c. (Toss two fair dice.) Unlike the first two examples, we might consider several different sample spaces. Suppose first that we distinguish the two dice by color; say one is red and the other green. Then we could write the result of a toss as an ordered pair indicating the outcome on each die, giving say the result on the red die first and the result on the green die second. Let a sample space be

$$S_1 = \{(1, 1), (1, 2), \ldots, (1, 6), (2, 1), (2, 2), \ldots, (2, 6), \ldots, (6, 6)\}.$$

It is useful to see this sample space as a geometric space as in Figure 1.1.

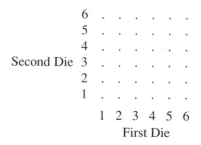

Figure 1.1   Sample space for tossing two dice.

Note that the 36 dots represent the only possible outcomes from the experiment. The sample space is not continuous in any sense in this case and may differ from our usual notions of a geometric space.

We could also describe all the possible outcomes from the experiment by the set

$$S_2 = \{2, 3, 4, 5, 6, 7, 8, 9, 10, 11, 12\}$$

since one of these *sums* must occur when the two dice are thrown.

Which sample space should be chosen? Note that each point in S_2 represents at least one point in S_1. So, while we might consider each of the 36 points in S_1 to occur with equal frequency if we threw the dice a large number of times, we would not consider that to be true if we chose sample space S_2. A sum of 7, for example, occurs on six of the points in S_1, while a sum of 2 occurs at only one point in S_1.The choice of sample space is largely dependent on what sort of outcomes are of interest when the experiment is performed. It is not uncommon for an experiment to admit more than one sample space. We generally select the sample space most convenient for the analysis of the probabilities involved in the problem.

d. (Take an actuarial examination until it is passed for the first time.) Letting P and F denote passing and failing the examination respectively, we note that the sample space here is infinite:

$$S = \{P, FP, FFP, FFFP, \ldots\}.$$

However, S here is a *countably infinite* sample space because its elements can be counted in the sense that they can be placed in a one-to-one correspondence with the set of natural numbers $\{1, 2, 3, 4, \ldots\}$ as follows:

$$P \leftrightarrow 1$$
$$FP \leftrightarrow 2$$
$$FFP \leftrightarrow 3$$
$$\cdot$$
$$\cdot$$
$$\cdot$$

The rule for the one-to-one correspondence is: Given an entry in the left column, the corresponding entry in the right column is the number of the attempt on which the examination is passed; given an entry in the right column, say n, consider $n - 1$ F's followed by P to construct the corresponding entry in the left column. Hence the correspondence with the set of natural numbers is one-to-one. Such sets are called *countable* or *denumerable.* We will consider countably infinite sets in much the same way that we will consider finite sets. In the next chapter we will encounter infinite sets that are not countable.

e. Sample spaces for laboratory experiments are usually difficult to enumerate and may involve a combination of finite and infinite factors.   ◄

Example 1.1.2

As a more difficult example, consider observing single births in a hospital until two girls are born consecutively. The sample space now is a bit more challenging to write down than the sample spaces for the situations considered in Example 1.1.1. For convenience, we write the points, showing the births in order and grouped by the total number of births:

Number of Births	Sample Points	Number of Sample Points
2	GG	1
3	BGG	1
4	$BBGG$	2
	$GBGG$	
5	$BBBGG$	3
	$BGBGG$	
	$GBBGG$	
6	$BBBBGG$	5
	$BBGBGG$	
	$BGBBGG$	
	$GBBBGG$	
	$GBGBGG$	
⋮	⋮	⋮

and so on. We note that the number of sample points as we have grouped them follows the sequence $1, 1, 2, 3, 5, \ldots$, which we recognize as the beginning of the Fibonacci sequence. The Fibonacci sequence is found by starting with the sequence 1, 1. Subsequent entries are found by adding the two immediately preceding entries. However, we only have evidence that the Fibonacci sequence applies to a few of the groups of points in the sample space. We will have to establish the general pattern in this example before concluding that the Fibonacci sequence does indeed give the number of sample points in the sample space. The reader may wish to do that before reading the next paragraphs!

Here's the reason the Fibonacci sequence occurs. Consider a sequence of B's and G's in which GG occurs for the first time at the nth birth. Let a_n denote the number of ways in which this can occur. If GG occurs for the first time on the *n*th birth, there are two possibilities for the beginning of the sequence. These possibilities are mutually exclusive; that is, they cannot occur together.

One possibility is that the sequence begins with a *B* and is followed for the first time by the occurrence of *GG* in $n - 1$ births. Since we are requiring the sequence *GG* to occur for the first time at the $n - 1$ birth, this can occur in a_{n-1} ways.

The other possibility for the beginning of the sequence is that the sequence begins with *G*, which then must be followed by *B* (else the pattern *GG* will occur in two births), and then the pattern *GG* occurs in $n - 2$ births. This can occur in a_{n-2} ways. Since the sequence begins with either *B* or *G*, it follows that

$$a_n = a_{n-1} + a_{n-2}, \ n \geq 4,$$
$$\text{where} \quad a_2 = a_3 = 1,$$

(1.1)

which describes the Fibonacci sequence.

The sequences for which *GG* occurs for the first time in seven births can then be found by writing *B* followed by the sequences for six births and by writing *GB* followed by *GG* in five births:

$$B|BBBBGG$$
$$B|BBGBGG$$
$$B|BGBBGG$$
$$B|GBBBGG$$
$$B|GBGBGG$$

$$GB|BBBGG$$
$$GB|BGBGG$$
$$GB|GBBGG$$

Formulas such as Equation 1.1 often describe a problem in a very succinct fashion. They are called *recursions* because they describe one value of a function, here a_n, in terms of other values of the same function; in addition, they are easily programmed. Computer algebra systems are especially helpful in giving large numbers of terms determined by recursions. One can find, for example, that there are 46,368 ways for the sequence *GG* to occur for the first time on the 25th birth. It is difficult to imagine determining this number without the use of a computer.   ◀

Exercises 1.1

1. An experiment consists of drawing 2 numbered balls from a box of balls numbered from 1 to 9. Describe the sample space if

    a. the first ball is not replaced before the second is drawn.

    b. the first ball is replaced before the second is drawn.

2. In the following diagram, A, B, and C are switches that may be closed (current flows through the switch) or open (current cannot flow through the switch). Show the sample space indicating all the possible positions of the switches in the circuit. Then indicate the sample points for which current flows through the circuit.

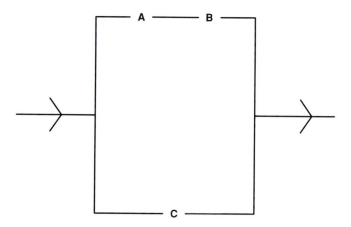

3. Items being produced on an assembly line can be good (G) or not meeting specifications (N). Show the sample space for the next 5 items produced by the assembly line.

4. A student decides to take an actuarial examination until it is passed, but will attempt the test at most 5 times. Show the sample space.

5. In the World Series, games are played until one of the teams has won 4 games. Show all the points in the sample space in which the American League (A) wins the series over the National League (N) in at most 6 games.

6. We are interested in the sequence of male and female births in 5-child families. Show the sample space.

7. Twelve chips numbered 1 through 12 are mixed in a bowl. Two chips are drawn successively and without replacement. Show the sample space for the experiment.

8. An assembly line is observed until items of both types – good (G) items and items not meeting specification (N) – are observed. Show the sample space.

9. Two numbers are chosen without replacement from the set {2, 3, 4, 5, 6, 7} , with the additional restriction that the second number chosen must be smaller than the first. Describe an appropriate sample space for the experiment.

10. Computer chips coming off an assembly line are marked defective (D) or non-defective (N). The chips are tested and their condition listed. This is continued until 2 consecutive defectives are produced or until 4 chips have been tested, whichever occurs first. Show a sample space for the experiment.

11. A coin is tossed 5 times and a running count of the heads and tails is kept (so the number of heads and the number of tails tossed so far is recorded at each toss). Show all the sample points where the heads count *always* exceeds the tails count.

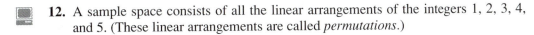 **12.** A sample space consists of all the linear arrangements of the integers 1, 2, 3, 4, and 5. (These linear arrangements are called *permutations*.)

   a. Use your computer algebra system to list all the sample points.
   b. If the sample points are equally likely, what is the probability that the number 3 is in the third position?
   c. What is the probability that none of the integers occupies its natural position?

▌ 1.2 ▶ Events; Axioms of Probability

After establishing a sample space, we are often interested in particular points, or sets of points, in that sample space. Consider the following examples:

   a. An item is selected at random from a production line. We are interested in the selection of a good item.
   b. Two dice are tossed. We are interested in the occurrence of a sum of 5.
   c. Births are observed until a girl is born. We are interested in this occurring in an even number of births.

Let's begin by defining an *event*.

Definition: An *event* is a subset of a sample space.

Events then contain one or more elementary outcomes in the sample space.

In the examples above, "a good item is selected," "the sum is 5," and "an even number of births was observed" can be described by subsets of the appropriate sample space and are, therefore, events.

We say that an event *occurs* if any of the elementary outcomes contained in the event occurs.

We will be interested in the *relative frequency* with which these events occur. In example **a**, we would most likely say, if 99% of the items produced in the production line are good, a good item will be selected about 99% of the time the experiment is performed, but we would expect some variation from this figure. In example **b**, such a calculation is more complex because the event "the sum of the spots showing on the dice is 5" is comprised of several more elementary events. If the sample space is

$$S = \{(1, 1), (1, 2), \ldots, (1, 6), (2, 1), \ldots, (6, 6)\},$$

then the points where the sum is 5 are

$$(1, 4), (2, 3), (3, 2), (4, 1).$$

If the dice are fair, then each of the 36 points in S occurs about $\frac{1}{36}$ of the time, so we conclude that the sum of the spots showing is 5 occurs about $4 \cdot \frac{1}{36} = \frac{1}{9}$ of the time.

In example **c**, observing births until a girl is born, the event "an even number of births is observed" is much more complex than examples **a** and **b** because there is an infinity of possibilities. How are we to judge the frequency of occurrence of each one? We can't answer this question at this time, but we will consider it later.

Now we consider a structure so that we can deal with such questions, as well as many others far more complex than those considered so far. We start with some assumptions about any sample space.

Axioms of Probability

We consider the *long-range relative frequency* or *probability* of an event in a sample space. If we perform an experiment 120 times and an event, A, occurs 30 times, then we say that the *relative frequency* of A is $\frac{30}{120} = \frac{1}{4}$. In general, if in n trials an event A occurs $n(A)$ times, then we say that the *relative frequency* of A is

$$\frac{n(A)}{n}.$$

Of course if we perform the experiment another n times, we do not expect A to occur exactly the same number of times as before, giving another relative frequency for the event A. We do expect these variable ratios representing relative frequencies to settle down in some manner as n grows large. If A is an event, we denote this limiting relative frequency by the *probability* of A and denote this by $P(A)$.

Definition: If A is an event then the *probability* of A is

$$P(A) = \lim_{n \to \infty} \frac{n(A)}{n}.$$

We assume at this point that the limit exists. We will return to this point in Chapter 4.

In considering events, it is most convenient to use the language and notation of sets where the following notations are common:

The *union* of sets A and B is denoted by $A \cup B$ where
$$A \cup B = \{x | x \in A \text{ or } x \in B\},$$

where the word "or" is used in the inclusive sense; that is, an element in both sets A and B is included in the union of the sets.

The *intersection* of sets A and B is denoted by $A \cap B$ where
$$A \cap B = \{x | x \in A \text{ and } x \in B\}.$$

We will consider the following as axiomatic, or self-evident:

1. $P(A) \geq 0$, where A is an event,
2. $P(S) = 1$, where S is the sample space, and
3. $P(A \text{ or } B) = P(A \cup B) = P(A) + P(B)$ if A and B are *disjoint*, or *mutually exclusive*; that is, they have no sample points in common.

Axioms of probability of course should reflect our common intuition about the occurrence of events. Because an event cannot occur with a negative relative frequency, **1** is evident. Because *something* must occur when the experiment is done, and because S denotes the entire sample space, S must occur with relative frequency 1, hence assumption **2**. Now suppose A and B are events with no sample points in common. We can illustrate events in a graphic manner by drawing a rectangle that represents all the points in S; events are subsets of this sample space. A diagram showing the event A – that is, the set of all elements of S that are in the event A – is shown in Figure 1.2. Illustrations of sets and their relationships to each other are called *Venn diagrams*.

The event A *or* B consists of the all points in A or in B and so its relative frequency is the sum of the relative frequencies of A or B. This is assumption **3**. Figure 1.3 shows a Venn diagram illustrating the disjoint events A and B.

No further axioms will be needed in our development of probability theory. We now consider some consequences of these assumptions.

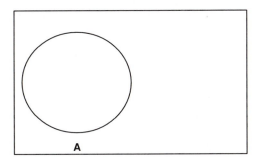

Figure 1.2   Venn diagram showing event A.

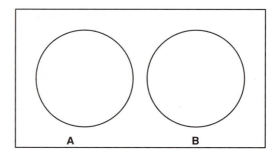

Figure 1.3   Venn diagram showing disjoint events A and B.

1.3 ▶ Probability Theorems

In example **b** above we considered the event that the sum was 5 when two dice were thrown. This event in turn was comprised of elementary events

$$(1, 4), (2, 3), (3, 2), (4, 1)$$

each of which had probability $\frac{1}{36}$. Since the events $(1, 4)$, $(2, 3)$, $(3, 2)$, and $(4, 1)$ are disjoint, axiom **3** shows that the probability of the event that the sum is 5 is the sum of the probabilities of these four elementary events or

$$\frac{1}{36} + \frac{1}{36} + \frac{1}{36} + \frac{1}{36} = \frac{4}{36} = \frac{1}{9}.$$

Assumption **3** shows that if A is an event that is comprised of elementary *disjoint* events $a_1, a_2, a_3, \ldots, a_n$, then

THEOREM 1: $P(A) = \sum_{i=1}^{n} P(a_i).$ ■

This fact is often used in the establishment of the theorems we consider in this section. Although we will not do so, all of them can be explained using Theorem 1.

What can we say about $P(A \cup B)$ if A and B have sample points in common? If we find

$$P(A) + P(B),$$

we will have counted the points in the intersection $A \cap B$ twice, as the diagram in Figure 1.4 shows. So the intersection must be subtracted once, giving

THEOREM 2: $P(A \cup B) = P(A) + P(B) - P(A \cap B).$ ■

We call this the *addition theorem* (for two events).

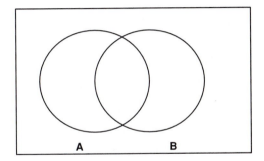

Figure 1.4 Venn diagram showing arbitrary events A and B.

Example 1.3.1

Choose a card from a well-shuffled deck of cards. Let A be the event, "the selected card is a heart" and let B be the event, "the selected card is a face card." Let the sample space S consist of one point for each of the 52 cards. If the deck is really well-shuffled, each point in S can be presumed to have probability $\frac{1}{52}$. The event A contains 13 points and the event B contains 12 points so $P(A) = \frac{13}{52}$ and $P(B) = \frac{12}{52}$. But the events A and B have three sample points in common, those for the king, queen, and jack of hearts. The event $A \cup B$ is then the event "the selected card is a heart or a face card," and its probability is

$$P(A \cup B) = P(A) + P(B) - P(A \cap B)$$
$$= \frac{13}{52} + \frac{12}{52} - \frac{3}{52} = \frac{22}{52} = \frac{11}{26}.$$

It is also easy to see by direct counting that the event "the selected card is a heart or a face card" contains exactly 22 points in the sample space of 52 points. ◄

How can the addition theorem for two events be extended to three or more events? First consider events A, B, and C in a sample space S. By adding and subtracting probabilities, the reader may be able to see that

THEOREM 3:

$$P(A \cup B \cup C) = P(A) + P(B) + P(C) - P(A \cap B)$$
$$- P(A \cap C) - P(B \cap C) + P(A \cap B \cap C) \qquad ■$$

We offer another proof as well. This proof will be based on the fact that a correct expression for $P(A \cup B \cup C)$ must count each sample point in the event $A \cup B \cup C$

once and only once. The Venn diagram in Figure 1.5 shows that S is comprised of eight disjoint regions labeled as follows:

0: points outside $A \cup B \cup C$ (1 region)
1: points in A, B, or C alone (3 regions)
2: points in exactly two of the events (3 regions)
3: points in $A \cap B \cap C$ (1 region).

Now we show that the right side of Theorem 3 counts each point in the event $A \cup B \cup C$ once and only once. By symmetry, we can consider only four cases.

Case 1: Suppose a point is in event A only. Then its probability is counted only once, in $P(A)$, on the right side of Theorem 3.

Case 2: Suppose a point is in $A \cap B$ only. Then its probability is counted in $P(A)$, $P(B)$, and $P(A \cap B)$, a net count of one on the right side in Theorem 3.

Case 3: Suppose a point is in $A \cap B \cap C$. Then its probability is counted in each term on the right side of Theorem 3, yielding a net count of one.

Case 4: If a point is outside $A \cup B \cup C$, it is not counted on the right side in Theorem 3.

So Theorem 3 must be correct because it counts each point in $A \cup B \cup C$ exactly once and never counts any point outside the event $A \cup B \cup C$. This proof uses a combinatorial principle, that of *inclusion and exclusion,* a principle used in other ways as well in the field of combinatorics. We will make some use of this principle in the remainder of the book. Theorem 2 is of course a special case of Theorem 3.

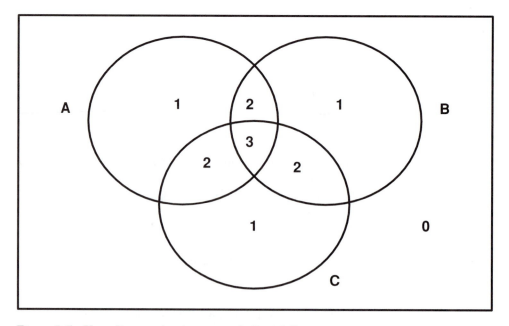

Figure 1.5   Venn diagram showing events A, B, and C.

We would like to extend Theorem 3 to n events, but this requires some combinatorial facts that will be developed later, so we postpone this extension until they are established.

Example 1.3.2

A card is again drawn from a well-shuffled deck. Consider the events
  A : the card shows an even number (2, 4, 6, 8, or 10),
  B : the card is a heart, and
  C : the card is black.
  We use a sample space containing one point for each of the 52 cards in the deck.
  Then $P(A) = \frac{20}{52}$, $P(B) = \frac{13}{52}$, $P(C) = \frac{26}{52}$, $P(A \cap B) = \frac{5}{52}$, $P(A \cap C) = \frac{10}{52}$, $P(B \cap C) = 0$, and $P(A \cap B \cap C) = 0$,
so, by Theorem 3,

$$P(A \cup B \cup C) = \frac{20}{52} + \frac{13}{52} + \frac{26}{52} - \frac{5}{52} - \frac{10}{52} = \frac{44}{52} = \frac{11}{13}. \qquad \blacktriangleleft$$

We will show one more fact in this section. Consider S and an event, A, in S. Denoting the set of points where the event A does not occur by $\overline{A}$, it is clear that the events $\overline{A}$ and A are disjoint. So, by Theorem 2, $P(A \cup \overline{A}) = P(A) + P(\overline{A}) = 1$, which is most often written as

THEOREM 4: $P(\overline{A}) = 1 - P(A).$ $\blacksquare$

Example 1.3.3

Throw a pair of fair dice. What is the probability that the dice show different numbers? Here it is convenient to let A be the event, "the dice show different numbers." Referring to the sample space shown in Figure 1.1, we compute $P(\overline{A})$, since

$$P(\overline{A}) = P \text{ (the dice show the same numbers)} = \frac{6}{36} = \frac{1}{6}.$$

So $P(A) = 1 - \frac{6}{36} = \frac{5}{6}.$

  This is easier than counting the 30 sample points out of 36 for which the dice show different numbers. $\blacktriangleleft$

  The theorems we have developed so far appear to be fairly simple; the difficulty arises in applying them.

Exercises 1.3

1. A fair coin is tossed 5 times. Find the probability of obtaining

   a. exactly 3 heads.
   b. at most 3 heads.

2. A recall of pickup trucks is required for the repair of possible defects in the steering column and defects in the brake linings. Dealers have been notified that 3% of the trucks have defective steering only and that 6% of the trucks have defective brake linings only. If 87% of the trucks have neither defect, what percentage of the trucks have both defects?

3. A hat contains tags numbered 1, 2, 3, 4, and 5. A tag is drawn from the hat and replaced, then a second tag is drawn. Assume that the points in the sample space are equally likely.

   a. Show the sample space.
   b. Find the probability that the number on the second tag exceeds the number on the first tag.
   c. Find the probability that the first tag has a prime number and the second tag has an even number. The number 1 is not considered to be a prime number.

4. A fair coin is tossed 4 times.

   a. Show a sample space for the experiment, showing each possible sequence of tosses.
   b. Suppose the sample points are equally likely and a running count is made of the number of heads and the number of tails tossed. What is the probability the heads count always exceeds the tails count?
   c. If the last toss is a tail, what is the probability that an even number of heads was tossed?

5. In a sample space is it possible to have $P(A) = \frac{1}{2}$, $P(A \cap B) = \frac{1}{3}$, and $P(B) = \frac{1}{4}$?

6. If A and B are events in a sample space, explain why $P(A \cap \overline{B}) \geq P(A) - P(B)$.

7. In testing the water supply for various cities for two kinds of impurities commonly found in water, it was found that 20% of the water supplies had neither sort of impurity, 40% had an impurity of type A, and 50% had an impurity of type B. If a city is chosen at random, what is the probability that its water supply has exactly one type of impurity?

8. A die is loaded so the probability that a face turns up is proportional to the number on that face. If the die is thrown, what is the probability that an even number occurs?

9. Show that $P(\overline{A} \cap B) = P(B) - P(A \cap B)$.

10. **a.** Explain why $P(A \cup B) \le P(A) + P(B)$.
    b. Explain why $P(A \cup B \cup C) \le P(A) + P(B) + P(C)$.

11. Find a formula for $P(A \text{ or } B)$ using the word "or" in an exclusive sense: that is, A or B means that event A occurs or event B occurs, but not both.

12. The entering class in an engineering college has 34% who intend to major in mechanical engineering, 33% who indicate an interest in taking advanced courses in mathematics as part of their major field of study, and 28% who intend to major in electrical engineering, while 23% have other interests. In addition, 59% are known to major in mechanical engineering or take advanced mathematics, while 51% intend to major in electrical engineering or take advanced mathematics. Assuming that a student can major in only one field, what percent of the class intends to major in mechanical engineering or in electrical engineering, but shows no interest in advanced mathematics?

1.4 ▶ Conditional Probability and Independence

Example 1.4.1

Suppose a card is drawn from a well-shuffled deck of 52 cards. What is the probability the card is a jack? If the sample space consists of a point for each card in the deck, the answer to the question is $\frac{4}{52}$ because there are four jacks in the deck.

Now suppose the person choosing the card gives us some additional information. Specifically, suppose we are told that the drawn card is a face card. Now what is the probability the card is a jack? An appropriate sample space for the experiment becomes the set of 12 points consisting of all the possible face cards that could be selected:

$$\{jh, qh, kh, jd, qd, kd, js, qs, ks, jc, qc, kc\}.$$

Considering each of these 12 outcomes to be equally likely, the probability the chosen card is a jack is now $\frac{4}{12}$. The given additional information that the card is a face card has altered the probability of the event in question. Such additional information, or *conditions,* generally has the effect of changing the probability of an event as the conditions change. Specifically, the conditions often reduce the sample space and hence alter the probabilities on those points that satisfy the conditions.

Let us denote by

A: the event "the chosen card is a jack"
  and
B: the event "the chosen card is a face card."

Further, we will use the notation $P(A|B)$ to denote the probability of the event A given that the event B has occurred. We call $P(A|B)$ the *conditional probability of* A *given* B. In this example we see that $P(A|B) = \frac{4}{12}$.

Now we can establish a general result by reasoning as follows. Suppose the event B has occurred. While this reduces the sample space to those points in B, we cannot presume that the probability of the set of points in B is 1. However, if the probability of each point in B is divided by $P(B)$, then the set of points in B has probability 1 and can therefore serve as a sample space. This division by a constant also preserves the *relative* probabilities of the points in the original sample space; if one point in the original sample space was k times as probable as another, it is still k times as probable as the other point in the new sample space. Clearly $P(A|B)$ accounts for the points in $A \cap B$ in the new sample space. We have found that

$$P(A|B) = \frac{P(A \cap B)}{P(B)}$$

where we have presumed of course that $P(B) \neq 0$.

In the above example, $P(A \cap B) = \frac{4}{52}$ and $P(B) = \frac{12}{52}$, so $P(A|B) = \frac{4}{12}$ as before. In this example, $P(A \cap B)$ reduces to $P(A)$, but this will not always be the case.

We can also write this result as

$$P(A \cap B) = P(B) \cdot P(A|B), \text{ or, interchanging } A \text{ and } B,$$
$$P(A \cap B) = P(A) \cdot P(B|A).$$

We call this result the *multiplication theorem*.   ◄

Example 1.4.2

A box of transistors has four good transistors mixed up with two bad transistors. A production worker, in order to sample the product, chooses two transistors at random, the first chosen transistor not being replaced before the second transistor is chosen. What is the probability that both transistors are good?

If the events are

    A: the first transistor chosen is good

      and

    B: the second transistor chosen is good,

then we want $P(A \cap B)$.

Now $P(A) = \frac{4}{6}$ while $P(B|A) = \frac{3}{5}$ because the box, after the first good transistor is drawn, contains five transistors, three of which are good ones. So the probability both chosen transistors are good is

$$P(A \cap B) = P(A) \cdot P(B|A)$$

$$P(A \cap B) = \frac{4}{6} \cdot \frac{3}{5} = \frac{2}{5}$$

by the multiplication theorem. ◄

Example 1.4.3

In the context of the example above, what is the probability that the second transistor chosen is good?

We need $P(B)$. Now B can occur in two mutually exclusive ways: the first transistor is good and the second transistor is good also, or the first transistor is bad and the second transistor is good. So,

$$P(B) = P[(A \cap B) \cup (\overline{A} \cap B)]$$

$$= P(A) \cdot P(B|A) + P(\overline{A}) \cdot P(B|\overline{A})$$

$$P(B) = \frac{4}{6} \cdot \frac{3}{5} + \frac{2}{6} \cdot \frac{4}{5} = \frac{2}{3}. \qquad ◄$$

In this example, we used the fact that

$$P(B) = P(A) \cdot P(B|A) + P(\overline{A}) \cdot P(B|\overline{A})$$

since B occurs when either A or $\overline{A}$ occurs.

This result can be generalized. Suppose the sample space consists of disjoint events so that

$$S = A_1 \cup A_2 \cup \ldots \cup A_n$$

where A_i and A_j have no sample points in common if $i \neq j$, $i, j = 1, 2, \ldots, n$.

Then if B is an event,

$$P(B) = P[(A_1 \cap B) \cup (A_2 \cap B) \cup \ldots \cup (A_n \cap B)]$$
$$= P(A_1 \cap B) + P(A_2 \cap B) + \cdots + P(A_n \cap B)$$
$$= P(A_1) \cdot P(B|A_1) + P(A_2) \cdot P(B|A_2) + \cdots + P(A_n) \cdot P(B|A_n).$$

We have then

THEOREM 1:   (Law of Total Probability): If $S = A_1 \cup A_2 \cup \ldots \cup A_n$ where A_i and A_j have no sample points in common if $i \neq j$, $i, j = 1, 2, \ldots, n$, then, if B is an event,

$$P(B) = P(A_1) \cdot P(B|A_1) + P(A_2) \cdot P(B|A_2) + \cdots + P(A_n) \cdot P(B|A_n)$$

$$\text{or } P(B) = \sum_{i=1}^{n} P(A_i) \cdot P(B|A_i). \qquad \blacksquare$$

Example 1.4.4

A supplier purchases 10% of its parts from factory A, 20% of its parts from factory B, and the remainder of its parts from factory C. Three percent of A's parts are defective, 2% of B's parts are defective, and 0.5% of C's parts are defective. What is the probability that a randomly selected part is defective?

Let $P(A)$ denote the probability that the part is from factory A and define $P(B)$ and $P(C)$ similarly. Let $P(D)$ denote the probability that an item is defective. Then, from the Law of Total Probability,

$$P(D) = P(A) \cdot P(D|A) + P(B) \cdot P(D|B) + P(C) \cdot P(D|C).$$

so

$$P(D) = (0.10) \cdot (0.03) + (0.20) \cdot (0.02) + (0.70) \cdot (0.005) = 0.0105.$$

So 1.05% of the items are defective.

We will encounter other uses of the Law of Total Probability in the following examples.   ◄

Example 1.4.5

Suppose, in the context of the previous example, we are given that the second chosen transistor is good. What is the probability that the first was also good?

Using the events A and B in the previous example, we want to find $P(A|B)$. This is

$$P(A|B) = \frac{P(A \cap B)}{P(B)}.$$

From the previous example, $P(A \cap B) = \frac{4}{6} \cdot \frac{3}{5} = \frac{2}{5}$, and we found in Example 1.4.3 that $P(B) = \frac{2}{3}$, so

$$P(A|B) = \frac{3}{5}. \qquad \blacktriangleleft$$

When the above results are combined we see that

$$P(A|B) = \frac{P(A \cap B)}{P(B)} = \frac{P(A) \cdot P(B|A)}{P(A) \cdot P(B|A) + P(\overline{A}) \cdot P(B|\overline{A})}. \qquad (1.2)$$

This result is sometimes known as *Bayes' Theorem*. The theorem can easily be extended to three or more mutually disjoint events.

THEOREM 6:   (Bayes' Theorem): If $S = A_1 \cup A_2 \cup \ldots \cup A_n$ where A_i and A_j have no sample points in common if $i \neq j$ then, if B is an event,

$$P(A_i|B) = \frac{P(A_i \cap B)}{P(B)}$$

$$P(A_i|B) = \frac{P(A_i) \cdot P(B|A_i)}{\begin{array}{c} P(A_1) \cdot P(B|A_1) + P(A_2) \cdot P(B|A_2) + \cdots \\ + P(A_n) \cdot P(B|A_n) \end{array}}$$

and

$$P(A_i|B) = \frac{P(A_i) \cdot P(B|A_i)}{\sum_{j=1}^{n} P(A_j) \cdot P(B|A_j)}. \qquad \blacksquare$$

Rather than remember this result, it is useful to look at Bayes' Theorem in a geometric way; it is not nearly as difficult as it may appear. This will be illustrated first using the current example.

Draw a square of side 1; as shown in Figure 1.6, divide the horizontal axis proportional to $P(A)$ and $P(\overline{A})$, in this case (returning to the context of Example 1.4.5) in the proportions $\frac{4}{6}$ to $\frac{2}{6}$. Along the vertical axis the conditional probabilities are shown. The vertical axis shows $P(B|A) = \frac{3}{5}$ and $P(B|\overline{A}) = \frac{4}{5}$ respectively.

The shaded area above $P(A)$ shows $P(A) \cdot P(B|A)$. The total shaded area shows $P(B) = \frac{4}{6} \cdot \frac{3}{5} + \frac{2}{6} \cdot \frac{4}{5} = \frac{2}{3}$. The more darkly shaded region is the proportion of the shaded area arising from the occurrence of A, which is $P(A|B)$. We see that this is

$$\frac{\frac{4}{6} \cdot \frac{3}{5}}{\frac{4}{6} \cdot \frac{3}{5} + \frac{2}{6} \cdot \frac{4}{5}} = \frac{3}{5}$$

yielding the same result found using Bayes' Theorem.

Figure 1.7 shows a geometric view of the general situation.

Bayes' Theorem, then, simply involves the calculation of areas of rectangles.

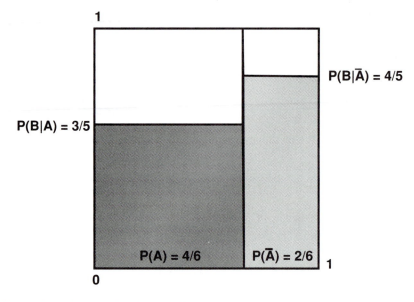

Figure 1.6   Diagram for Example 1.4.5.

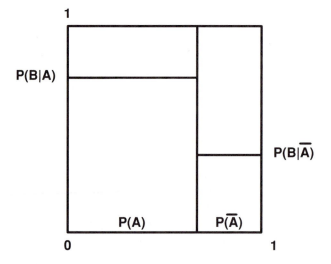

Figure 1.7   A geometric view of Bayes' Theorem.

Example 1.4.6

According to the *New York Times* (September 5, 1987), a test for the presence of human immunodeficiency virus (HIV) exists that gives a positive result (indicating the virus) with certainty if a patient actually has the virus. However, associated with this test, as with most tests, there is a *false positive rate*; that is, the test will sometimes indicate the presence of the virus in patients actually free of the virus. This test has a false positive rate of 1 in 20,000. So the test would appear to be very sensitive. Assuming now that 1 person in 10,000 is actually HIV positive, what proportion of patients for whom the test indicates HIV actually have the virus? The answer may be surprising.

A picture (greatly exaggerated so that the relevant areas can be seen) is shown in Figure 1.8.

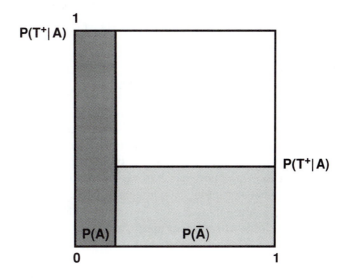

Figure 1.8   HIV example.

Define the events as

A: patient has HIV
  and
T^+: test indicates patient has HIV.

Then $P(A) = 0.0001$; $P(T^+|A) = 1$; $P(T^+|\overline{A}) = \frac{1}{20,000}$ from the data given. We are interested in $P(A|T^+)$. So, from Figure 1.8, we see that

$$P(A|T^+) = \frac{(0.0001) \cdot 1}{(0.0001) \cdot 1 + (0.9999) \cdot \frac{1}{20000}}$$

or

$$P(A|T^+) = \frac{20000}{29999}.$$

We could also apply Bayes' Theorem to find that

$$P(A|T^+) = \frac{P(A \cap T^+)}{P(T^+)} = \frac{P(A \cap T^+)}{P[(A \cap T^+) \cup (\overline{A} \cap T^+)]}$$

$$= \frac{P(A) \cdot P(T^+|A)}{P(A) \cdot P(T^+|A) + P(\overline{A}) \cdot P(T^+|\overline{A})}$$

$$= \frac{(0.0001) \cdot 1}{(0.0001) \cdot 1 + (0.9999) \cdot \frac{1}{20000}}$$

$$= \frac{20000}{29999},$$

giving the same result as that found using simple geometry.

At first glance, the test would appear to be very sensitive due to its small false positive rate, but only $\frac{2}{3}$ of those people testing positive would actually have the virus, showing that wide-spread use of the test, while detecting many cases of HIV, would also falsely detect the virus in about $\frac{1}{3}$ of the population who test positive. This risk may be unacceptably high.

A graph of $P(A|T^+)$ (shown in Figure 1.9) shows that this probability is highly dependent upon $P(A)$.

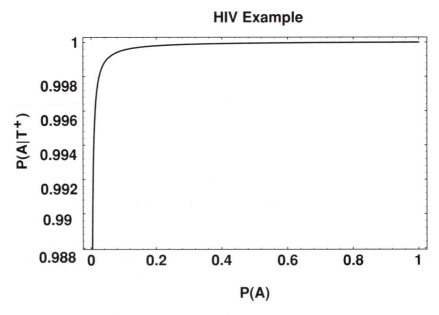

Figure 1.9 $P(A|T^+)$ as a function of $P(A)$.

The graph shows that $P(A|T^+)$ increases as $P(A)$ increases and that $P(A|T^+)$ is very large even for small values of $P(A)$. For example, if we desire $P(A|T^+)$ to be ≥ 0.9, then we must have $P(A) \geq 0.0045$.

The sensitivity of the test may incorrectly be associated with $P(T^+|A)$. The patient, however, is concerned with $P(A|T^+)$. This example shows how easy it is to confuse $P(A|T^+)$ with $P(T^+|A)$.   ◀

Example 1.4.6 (Continued)

Let's generalize the HIV example to a more general medical test in this way. Assume a test has a probability p of indicating a disease among patients actually having the disease. Assume also that the test indicates the presence of the disease with probability $1 - p$ among patients not having the disease. Finally, suppose the incidence rate of the disease is r.

If T^+ denotes that the test indicates the disease, and if A denotes the occurrence of the disease, then

$$P(A|T^+) = \frac{r \cdot p}{r \cdot p + (1 - r) \cdot (1 - p)}.$$

For example, if $p = 0.95$ and $r = 0.005$ (indicating that the test is 95% accurate on both those who have the disease and those who don't, and that 5 patients out of 1000 actually have the disease), then $P(A|T^+) = 0.087156$. Since $P(\overline{A}|T^+) = 0.912844$, a positive result on the test appears to indicate the absence and not the presence of the disease!

This odd result is actually due to the small incidence rate of the disease. Figure 1.10 shows $P(A|T^+)$ as a function of r assuming that $p = 0.95$. We see that $P(A|T^+)$ becomes quite large (≥ 0.8) for $r \geq 0.21$.

It is also interesting to see how r and p, varied together, affect $P(A|T^+)$. The surface is shown in Figure 1.11. The surface shows that $P(A|T^+)$ is large when the test is sensitive; that is, when $P(T^+|A)$ is large, or when the incidence rate $r = P(A)$ is large. But there are also combinations of these values that give large values of $P(A|T^+)$: one of these is $r = 0.2$ and $P(T^+|A) = 0.8$, for then $P(A|T^+) = 0.5$.   ◀

Example 1.4.7

A game show contestant is shown three doors, one of which conceals a valuable prize, while the other two are empty. The contestant is allowed to choose one door. Regardless of the choice made, at least one (that is, exactly one or perhaps both) of the remaining doors is empty. The show host opens one door to show it empty. The contestant is now given the opportunity to switch doors. Should the contestant switch?

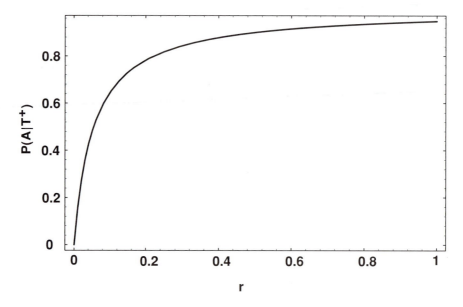

Figure 1.10 $P(A|T^+)$ as a function of the incidence rate, r, if $p = 0.95$.

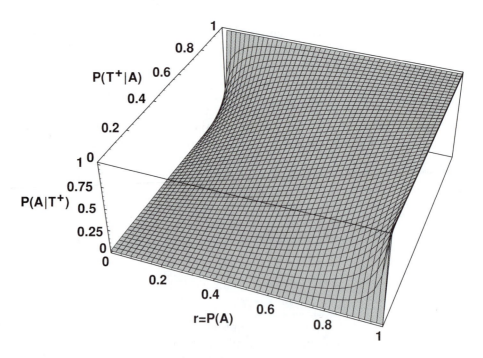

Figure 1.11 $P(A|T^+)$ as a function of r, the incidence rate, and $P(T^+|A)$

The problem is often called the Monty Hall problem due to its origin on the television show "Let's Make A Deal." It has been written about extensively, possibly because of its non-intuitive answer and perhaps because people unwittingly change the problem in the course of thinking about it.

The contestant clearly has a probability of $\frac{1}{3}$ of choosing the prize if a random choice of the door is made. So a probability of $\frac{2}{3}$ rests with the remaining two doors. The fact that one door is opened and revealed empty does not change these probabilities; hence the contestant should switch and will gain the prize with probability $\frac{2}{3}$.

When the empty door is opened the problem does not suddenly become a choice between two doors (one of which conceals the prize). This change in the problem ignores the fact that the game show host sometimes has a choice of one door to open and sometimes two. Persons changing the problem in this manner may think, incorrectly, that the probability of choosing the prize is now $\frac{1}{2}$, indicating that switching may have no effect in the long run; the strategy in reality has a great effect on the probability of choosing the prize.

To analyze the situation, suppose the contestant chooses the first door and the host opens the second door. Other possibilities are handled here by symmetry. Let A_i, $i = 1, 2, 3$ denote the event "the prize is behind door i" and D denote the event "door 2 is opened." The condition here is then D; we now calculate the conditional probability that the prize is behind door 3, that is, the probability that the contestant will win if he switches.

We assume that $P(A_1) = P(A_2) = P(A_3) = \frac{1}{3}$. Then $P(D|A_1) = \frac{1}{2}$, $P(D|A_2) = 0$ and $P(D|A_3) = 1$. The situation is shown in Figure 1.12.

It is clear from the shaded area in Figure 1.12 that the probability that the contestant wins if the first choice is switched to door 3 is

$$P(A_3|D) = \frac{\frac{1}{3} \cdot 1}{\frac{1}{3} \cdot \frac{1}{2} + \frac{1}{3} \cdot 0 + \frac{1}{3} \cdot 1} = \frac{2}{3}$$

which verifies our previous analysis.

This example illustrates that some events are highly dependent upon others. We now turn our attention to events for which this is not so. ◄

Independence

We have found that $P(A \cap B) = P(A) \cdot P(B|A)$. Occasionally, the occurrence of A has no effect on the occurrence of B so that $P(B|A) = P(B)$. If this is the case, we call A and B *independent* events. When A and B are independent, we have $P(A \cap B) = P(A) \cdot P(B)$.

Definition: Events A and B are called *independent* events if $P(A \cap B) = P(A) \cdot P(B)$.

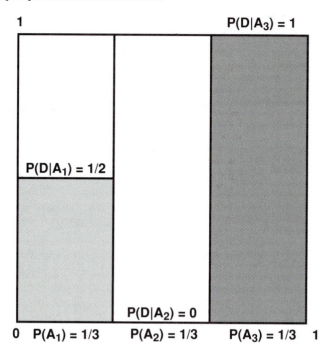

Figure 1.12   Diagram for the Monty Hall problem.

If we draw cards from a deck, replacing each drawn card before the next card is drawn, then the events denoting the cards drawn are clearly independent because the deck is full before each drawing and each drawing occurs under exactly the same conditions. If the cards are not replaced, however, then the events are not independent.

For three events, say A, B, and C, we define the events as independent if

$$P(A \cap B) = P(A) \cdot P(B),$$
$$P(A \cap C) = P(A) \cdot P(C),$$
$$P(B \cap C) = P(B) \cdot P(C), \quad \text{and} \tag{1.3}$$
$$P(A \cap B \cap C) = P(A) \cdot P(B) \cdot P(C).$$

The first three of these conditions establishes that the events are independent in pairs, so we call events satisfying these three conditions *pairwise independent*. Example 1.4.8 will show that events satisfying these three conditions may not satisfy the fourth condition so pairwise independence does not determine independence.

We also note that there is some confusion between *independent* events and *mutually exclusive* events. Often people speak of these as "having no effect on each other," but that is not a precise characterization in either case. Note that while mutually exclusive events cannot occur together, independent events must be able to occur together. To be specific, suppose that neither $P(A)$ nor $P(B)$ is 0 and that A and B are mutually exclusive. Then $P(A \cap B) = 0 \neq P(A) \cdot P(B)$. Hence A and B cannot be independent. So if A

and B are mutually exclusive, then they cannot be independent. This is equivalent to the statement that if A and B are independent, then they cannot be mutually exclusive, but the reader may enjoy establishing this from first principles as well.

Example 1.4.8

This example shows that pairwise independent events are not necessarily independent.

A fair coin is tossed four times. Consider the events A: the first coin shows a head, B: the third coin shows a tail, and C: there are equal numbers of heads and tails. Are these events independent?

Suppose the sample space consists of the 16 points showing the tosses of the coins in order. The sample space, indicating the events that occur at each point, is shown here.

Point	Event
HHHH	*A*
HHHT	*A*
HHTH	*A, B*
THHH	
HTHH	*A*
HHTT	*A, B, C*
HTHT	*A, C*
THHT	*C*
THTH	*B, C*
HTTH	*A, B, C*
TTHH	*C*
TTTH	*B*
TTHT	
THTT	*B*
HTTT	*A, B*
TTTT	*B*

Then $P(A) = \frac{1}{2}$ and $P(B) = \frac{1}{2}$ while C consists of the 6 points with exactly two heads and two tails, so $P(C) = \frac{6}{16} = \frac{3}{8}$.

Now $P(A \cap B) = \frac{4}{16} = \frac{1}{4} = P(A) \cdot P(B)$; $\quad P(A \cap C) = \frac{3}{16}$ $= P(A) \cdot P(C)$; and $P(B \cap C) = \frac{3}{16} = P(B) \cdot P(C)$, so the events A, B, and C are pairwise independent.

Now $A \cap B \cap C$ consists of the two points HTTH and HHTT with probability $\frac{2}{16} = \frac{1}{8}$.

Hence $P(\ A \cap B \cap C\ ) \neq P(A) \cdot P(B) \cdot P(C)$, so A, B, and C are not independent.

Formulas 1.3 also show that establishing only that $P(A \cap B \cap C) = P(A) \cdot P(B) \cdot P(C)$ is not sufficient to establish the independence of events A, B, and C. ◄

Exercises 1.4

1. Box I contains 4 green and 5 brown marbles. Box II contains 6 green and 8 brown marbles. A marble is chosen from box I and placed in box II, then a marble is drawn from box II.

  a. What is the probability the second marble chosen is green?

  b. If the second marble chosen is green, what is the probability a brown marble was transferred?

2. A football team wins its weekly game with probability 0.7. Suppose the outcomes of games on 3 successive weekends are independent. What is the probability the number of wins exceeds the number of losses?

3. Three manufacturers of floppy disks, A, B, and C, produce 15%, 25%, and 60% respectively of all floppy disks made. Manufacturer A produces 5% defective disks, manufacturer B produces 7% defective disks, and manufacturer C produces 4% defective disks.

  a. What proportion of floppy disks are defective?

  b. If a floppy disk is found to be defective, what is the probability it came from manufacturer B?

4. A chest has 3 drawers, each containing a coin. One coin is silver on both sides, one is gold on both sides, and the third is silver on one side and gold on the other side. A drawer is chosen at random and one face of the coin is shown to be silver. What is the probability the other side is silver also?

5. If A and B are independent events in a sample space, show that

$$P(A \cup B) = P(B) + P(A) \cdot P(\overline{B}) = P(A) + P(\overline{A}) \cdot P(B).$$

6. In a sample space, events A and B are independent, events B and C are mutually exclusive, and A and C are independent. If $P(A \cup B \cup C) = .9$, $P(B) = 0.5$, and $P(C) = 0.3$, find $P(A)$.

7. If $P(A \cup B) = 0.4$ and $P(A) = 0.3$, find $P(B)$ if

  a. A and B are independent.

  b. A and B are mutually exclusive.

8. A coin, loaded so the probability that it shows heads when tossed is $\frac{3}{4}$, is tossed twice. Let the events A, B, and C be, respectively, "first toss is heads," "second toss is heads," and "tosses show the same face."

  a. Are the events A and B independent?

  b. Are the events A and $B \cup C$ independent?

  c. Are the events A, B, and C independent?

9. Three missiles, whose probabilities of hitting a target are 0.7, 0.8, and 0.9, respectively, are fired at a target. Assuming independence, what is the probability the target is hit?

10. A student takes a driving test until it is passed. If the probability that the test is passed on any attempt is 4/7, and if the attempts are independent, what is the probability that the test is taken an even number of times?

11. **a.** Let p be the probability of obtaining a five at least once in n independent tosses of a die. What is the least value of n so that $p \geq \frac{1}{2}$?

  b. Generalize the result in part **a**. Suppose an event has probability p of occurring at any one of n independent trials of an experiment. What is the least value of n so the probability that the event occurs at least once is $\geq r$?

  c. Graph the surface in part b], showing n as a function of p and r.

12. Box I contains 7 red and 3 black balls; Box II contains 4 red and 5 black balls. After a randomly selected ball is transferred from Box I to Box II, 2 balls are drawn from Box II without replacement. Given that the 2 balls are red, what is the probability a black ball was transferred?

13. In rolling a fair die, what is the probability of rolling a one before rolling an even number?

14. **a.** There is a 50–50 chance that firm A will bid for the construction of a bridge. Firm B submits a bid and the probability that it will get the job is $\frac{2}{3}$, provided firm A does not bid; if firm A submits a bid, the probability that firm B gets the job is 1/5. Firm B is awarded the job; what is the probability that firm A did not bid?

  b. In part **a**, suppose now that the probability firm B gets the job if firm A bids on the job is p. Graph the probability that firm A did not bid given that B gets the job as a function of p.

  c. Generalize parts **a** and **b** further and suppose that the probability that B gets the job given that firm A does not bid on the job is r. Graph the surface showing the probability that firm A did not bid given that firm B gets the job as a function of p and r.

15. In a sample space, events A and B have probabilities $P(A) = P(B) = \frac{1}{2}$, and $P(A \cup B) = \frac{2}{3}$.

  a. Are A and B mutually exclusive?

  b. Are A and B independent?

  c. Calculate $P(\overline{A} \cap B)$.

  d. Calculate $P(\overline{A} \cap \overline{B})$.

16. Suppose that events A, B, and C are independent with $P(A) = \frac{1}{4}$, $P(B) = \frac{1}{2}$, and $P(A \cup B \cup C) = \frac{3}{4}$. Find $P(C)$.

17. A fair coin is tossed until the same face occurs twice in a row, but it is tossed no more than 4 times. If the experiment is over no later than the third toss, what is the probability that it was over by the second toss?

18. A collection of 65 coins contains one with two heads; the remainder of the coins are fair. If a coin, selected at random from the collection, turns up heads 6 times in 6 tosses, what is the probability it is the two-headed coin?

19. Three distinct methods, A, B, and C, are available for teaching a certain industrial skill. The failure rates are 30%, 20%, and 10%, respectively. However, due to costs, A is used twice as frequently as B, which is used twice as frequently as C.

    a. What is the overall failure rate in teaching the skill?

    b. A worker is taught the skill, but fails to learn it correctly. What is the probability he was taught by method A?

20. Sixty percent of new drivers have had driver education. During their first year of driving, drivers without driver education have a 0.08 probability of having an accident, but new drivers with driver education have a 0.05 probability of having an accident. What is the probability that a new driver with no accidents during the first year had driver education?

21. Events A, B, and C have $P(A) = 0.3$, $P(B) = 0.2$, and $P(C) = 0.4$. Also A and B are mutually exclusive, A and C are independent, and B and C are independent. Find the probability that *exactly one* of the events A, B, or C occurs.

22. A set consists of the 6 possible arrangements of the letters a, b, and c, as well as the points (a, a, a), (b, b, b) and (c, c, c). Let A_k be the event "letter a is in position k" for $k = 1, 2, 3$. Show that the events A_k are pairwise independent, but not independent.

23. Assume that the probability that a first-born child is a boy is p, and that the sex of subsequent children follows a chance mechanism so that the probability the next child is the same sex as the previous one is r.

    a. Let P_n denote the probability that the nth child is a boy. Find P_i, $i = 1, 2, 3$, in terms of p and r.

    b. Are the events A_i : "the ith child is a boy," $i = 1, 2, 3$ independent?

    c. Find a value for r so that A_1 and A_2 are independent.

24. A message is coded into the binary symbols 0 and 1 and sent over a communication channel. The probability a 0 is sent is 0.4 and the probability a 1 is sent is 0.6. The channel, however, has a random error that changes a 1 to a 0 with probability 0.2, and changes a 0 to a 1 with probability 0.1.

    a. What is the probability a 0 is received?

    b. If a 1 is received, what is the probability a 0 was sent?

25. a. Hospital patients with a certain disease are known to recover with probability $\frac{1}{2}$ if they do not receive a certain drug. The probability of recovery is $\frac{3}{4}$ if the drug is used. Of 100 patients, 10 are selected to receive the drug. If a patient recovers, what is the probability that the drug was used?

  b. In part **a**, let the probability that the drug was used be p. Graph the probability that the drug was used, given that the patient recovers, as a function of p.

  c. Find p if the probability that the drug was used, given that the patient recovers, is $\frac{1}{2}$.

26. Two people each toss 4 fair coins. What is the probability that they each throw the same number of heads?

27. In sample surveys people may be asked questions they regard as sensitive and so they may or may not answer them truthfully. An example might be, "Are you using illegal drugs?" If it is important to discover the real proportion of illegal drug users in the population, the following procedure, often called a *randomized response* technique, may be used.

　　The respondent is asked to flip a fair coin and not reveal the result to the questioner. If the result is heads then the respondent answers the question, "Is your Social Security number even?" If the coin comes up tails the respondent answers the sensitive question. Clearly the questioner cannot tell whether a response of "yes" is a consequence of illegal drug use or of an even Social Security number. Explain, however, how the results of such a survey with a large number of respondents can be used to find accurately the percentage of the respondents who are users of illegal drugs.

28. a. The individual events in a series of independent events have probabilities $\frac{1}{2}, (\frac{1}{2})^2, (\frac{1}{2})^3, \ldots, (\frac{1}{2})^n$. Show the probability that at least one of the events occurs approaches 0.711 as $n \to \infty$.

  b. Show, if the probabilities of the events are $\frac{1}{3}, (\frac{1}{3})^2, (\frac{1}{3})^3, \ldots, (\frac{1}{3})^n$, the probability that at least one of the events occurs approaches 0.440 as $n \to \infty$.

  c. Show, if the probabilities of the events are $p, p^2, p^3, \ldots, p^n$, the probability that at least one of the events occurs can be very well approximated by the function $p + p^2 - p^5 - p^7$ for $\frac{1}{11} \le p \le \frac{1}{2}$.

29. a. If events A and B are independent, show that

  i. $\overline{A}$ and B are independent.
  ii. A and $\overline{B}$ are independent.
  iii. $\overline{A}$ and $\overline{B}$ are independent.

  b. (Huff [17]). Show that events A and B are independent if and only if $P(A|B) = P(A|\overline{B})$.

30. A lie detector is accurate $\frac{3}{4}$ of the time; that is, if a person is telling the truth, the lie detector indicates he or she is telling the truth with probability $\frac{3}{4}$, while if the person is lying, the lie detector indicates that he or she is lying with probability $\frac{3}{4}$. Assume that a person taking the lie detector test is unable to influence its results, and also assume that 95% of the people taking the test tell the truth. What is the probability that a person is lying if the lie detector indicates that he or she is lying?

∎ 1.5 ▶ Some Examples

We now show two examples of probability problems that have interesting results that may counter intuition.

Example 1.5.1 (The Birthday Problem)

This problem exists in many variations in the literature on probability and has been written about extensively. The basic problem is this: There are n people in a room; what is the probability that at least two of them have the same birthday?

Let A denote the event "at least two people have the same birthday"; we want to find $P(A)$. It is easier in this case to calculate $P(\overline{A})$ (the probability that the birthdays are all distinct) rather than $P(A)$. To find $P(\overline{A})$ note that the first person can have any day as a birthday. The birthday of the next person cannot match that of the first person; this has probability $\frac{364}{365}$; the birthday of the third person cannot match that of either of the first two people; this has probability $\frac{363}{365}$, and so on. Multiplying these conditional probabilities,

$$P(\overline{A}) = \frac{365}{365} \cdot \frac{364}{365} \cdot \frac{363}{365} \cdots \frac{365 - (n-1)}{365}.$$

It is easy with a computer algebra system to calculate exact values for $P(A) = 1 - P(\overline{A})$ for various values of n:

n	$P(A)$	n	$P(A)$	n	$P(A)$	n	$P(A)$
2	0.002740	12	0.167025	22	0.475695	32	0.753348
3	0.008204	13	0.194410	23	0.507297	33	0.774972
4	0.016356	14	0.223103	24	0.538344	34	0.795317
5	0.027136	15	0.252901	25	0.568700	35	0.814383
6	0.040462	16	0.283604	26	0.598241	36	0.832182
7	0.056236	17	0.315008	27	0.626859	37	0.848734
8	0.074335	18	0.346911	28	0.654461	38	0.864068
9	0.094624	19	0.379119	29	0.680969	39	0.878220
10	0.116948	20	0.411438	30	0.706316	40	0.891232
11	0.141141	21	0.443688	31	0.730455		

We see that $P(A)$ increases rather rapidly; it exceeds $\frac{1}{2}$ for $n = 23$, a fact that surprises many, most people guessing that the value of n to make

$P(A) \geq \frac{1}{2}$ is much larger. In thinking about this, note that the problem says that *any* two people in the room can share *any* birthday. If some specific date comes to mind, such as August 2, then, because the probability that a particular person's birthday is not August 2 is $\frac{364}{365}$, the probability that at least one person in a group of n people has that specific birthday is

$$1 - (\frac{364}{365})^{n}.$$

It is easy to solve this for some specific probability. We find, for example, that for this probability to equal $\frac{1}{2}$, $n = 253$ people are necessary.

We show a graph, in Figure 1.13, of $P(A)$ for $n = 1, 2, 3, \ldots, 40$. The graph indicates that $P(A)$ increases quite rapidly as n, the number of people, increases.

It would appear that $P(A)$ might be approximated by a polynomial function of n. To consider how such functions can be constructed would be a diversion now so we will not discuss it. For now we state that the approximating function found by applying a principle known as least squares is

$$f(n) = -6.44778 \cdot 10^{-3} - 4.54359 \cdot 10^{-5} \cdot n$$
$$+ 1.51787 \cdot 10^{-3} \cdot n^{2} - 2.40561 \cdot 10^{-5} \cdot n^{3}.$$

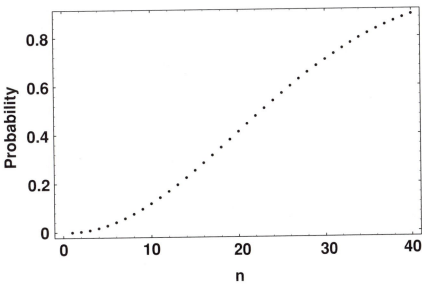

Figure 1.13   The birthday problem as a function of n, the number of people in the group.

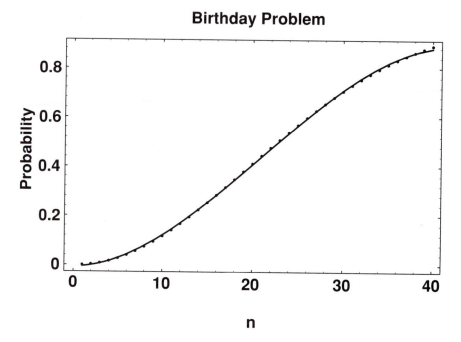

Birthday Problem

Figure 1.14   Polynomial approximation to the birthday data.

It can be shown that $f(n)$ fits $P(A)$ quite well in the range $2 \le n \le 40$. For example, if $n = 13$, $P(A) = 0.194410$ while $f(13) = 0.196630$; if $n = 27$, $P(A) = 0.626859$ while $f(27) = 0.625357$. A graph of $P(A)$ and the approximating function $f(n)$ is shown in Figure 1.14. The principle of least squares will be considered in Section 4.16.   ◀

Example 1.5.2

How many people must be in a group so that the probability that at least two of them have birthdays within at most one day of each other is at least $\frac{1}{2}$?

Suppose there are n people in the group, and that A represents the event "at least two people have birthdays within at most one day of each other." If a person's birthday is August 2, for example, then the second person's birthday must not fall on August 1, 2, or 3, giving 362 choices for the second person's birthday. The third person, however, has either 359 or 360 choices, depending on whether the second person's birthday is August 4 or July 31 or some other day that has not previously been excluded from the possibilities. We give then an approximate solution as

$$P(\overline{A}) = \frac{365 \cdot 362 \cdot 359 \cdots (368 - 3n)}{365 \cdot 365 \cdots 365}.$$

We seek $P(A) = 1 - P(\overline{A})$. It is easy to make a table of values of n and $P(A)$ with a computer algebra system:

n	$P(A)$	n	$P(A)$
2	0.008219	10	0.316058
3	0.024522	11	0.372273
4	0.048575	12	0.429026
5	0.079855	13	0.485341
6	0.117669	14	0.540332
7	0.161181	15	0.593226
8	0.209442	16	0.643376
9	0.261424		

So 14 people are sufficient to make the probability exceed $\frac{1}{2}$ that at least two of the birthdays differ by at most one day. In the previous example, we found that a group of 23 people was sufficient to make the probability exceed $\frac{1}{2}$ that at least two of them shared the same birthday. The probability is approximately 0.8915 that at least two of these people have birthdays that differ by at most one day. ◄

Example 1.5.3 (Mowing the Lawn)

Jack and his daughter, Kaylyn, choose who will mow the lawn by a random process: Jack has one green and two red marbles in his pocket; two are selected at random. If the colors match, Jack mows the lawn; otherwise, Kaylyn mows the lawn. Is the game fair?

The sample space here is most easily shown by a diagram with vertices as the colors of the marbles and the edges as the two marbles chosen. Assuming that the three possible samples are equally likely, then two of them lead to Kaylyn mowing the lawn, while Jack only mows it $\frac{1}{3}$ of the time. If we mean by the word "fair" that each mows the lawn with probability $\frac{1}{2}$, then the game is clearly unfair.

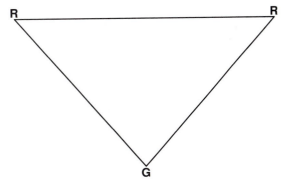

Three marbles in the lawn-mowing example.

If we are allowed to add marbles to Jack's pocket, can the game be made fair? The reader might want to think about this before proceeding.

What if a green marble is added? Then the sample space becomes all the sides and diagonals of a square.

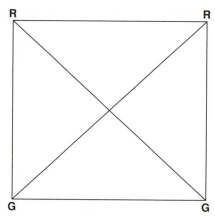

Four marbles in the lawn-mowing example.

While there are now six possible samples, four of them involve different colors while only two of them involve the same colors. So the probability that the colors differ is $\frac{4}{6} = \frac{2}{3}$; the addition of the green marble has not altered the game at all! The reader will easily verify that the addition of a *red* marble, rather than a green marble, will produce a fair game.

The problem of course is that, while the numbers of red and green marbles is important, the relevant information is the number of sides and diagonals of the figure produced since these represent the samples chosen. If we wish to find other compositions of marbles in Jack's pocket that make the game fair, we need to be able to count these sides and diagonals. We now show how to do this.

Consider a figure with n vertices (Figure 1.15)

In order to count the number of sides and diagonals, choose one of the n vertices. Now, to choose a side or diagonal, choose any of the other $n - 1$ vertices and join them. We have $n \cdot (n - 1)$ choices. Since it doesn't matter which vertex is chosen first, we have counted each side or diagonal twice. We conclude that there are

$$\frac{n \cdot (n - 1)}{2}$$

sides and diagonals. This is also called the number of *combinations of n distinct objects chosen two at a time* which we denote by the symbol $\binom{n}{2}$. So

$$\binom{n}{2} = \frac{n \cdot (n - 1)}{2}.$$

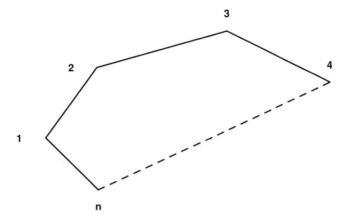

Figure 1.15 *n* marbles for the lawn-mowing problem.

If the game is to be fair, and if we have r red and g green marbles, then $\binom{r}{2}$ and $\binom{g}{2}$ represent the number of sides and diagonals connecting two red or two green marbles respectively. We want r and g so that the sum of these is $\frac{1}{2}$ of the total number of sides and diagonals; that is, we want r and g so that

$$\binom{r}{2} + \binom{g}{2} = \binom{1}{2} \cdot \binom{r+g}{2}.$$

The reader can verify that $r = 6$, $g = 3$ will satisfy the above equation as will $r = 10$, $g = 6$. The reader may also enjoy trying to find a general pattern for r and g before reading the following exercise. ◄

Exercises 1.5

1. Show that the "mowing the lawn" game is fair if, and only if, r and g, the numbers of red and green marbles respectively, are consecutive triangular numbers. (The first few triangular numbers are $1, 1 + 2 = 3, 1 + 2 + 3 = 6, \ldots$)

2. A fair coin is tossed until a head appears or until 6 tails have been obtained.

   a. What is the probability the experiment ends in an even number of tosses?
   b. Answer part **a** if the coin has been loaded so as to show heads with probability p.

3. Let P_r be the probability that among r people, at least two have the same birth month. Make a table of values of P_r for $r = 2, 3, \ldots, 12$. Plot a graph of P_r as a function of r.

4. Two defective transistors become mixed up with 2 good ones. The 4 transistors are tested one at a time, without replacement, until all the defectives are identified. Find P_r, the probability that the rth transistor tested will be the second defective, for $r = 2, 3, 4$.

5. A coin is tossed 4 times and the sequence of heads and tails is observed.

   a. What is the probability that heads and tails occur equally often if the coin is fair and the tosses are independent?
   b. Now suppose the coin is loaded so that $P(H) = \frac{1}{3}$ and $P(T) = \frac{2}{3}$ and that the tosses are independent. What is the probability that heads and tails occur equally often, given that the first toss is a head?

6. The following model is sometimes used to model the spread of a contagious disease. Suppose a box contains b black and r red marbles. A marble is drawn and c marbles of that color together with the drawn marble are placed in the box before the next marble is drawn, so that infected persons infect others while immunity to the disease may also increase.

   a. Find the probability that the first 3 marbles drawn are red.
   b. Show that the probability of drawing a black on the second draw is the same as the probability of drawing a black on the first draw.
   c. Show by induction that the probability the kth marble is black is the same as the probability of drawing a black on the first draw.

7. A set of 25 items contains 5 defective items. Items are sampled at random one at a time. What is the probability that the 3rd and 4th defectives occur at the 5th and 6th sample draws if

   a. the items are replaced after each is drawn?
   b. the items are not replaced after each is drawn?

 8. A biased coin has probability $\frac{3}{8}$ of coming up heads. A and B toss this coin with A tossing first.

   a. Show that the probability that A gets a head before B gets a tail is very close to $\frac{1}{2}$.
   b. How can the coin be loaded so as to make the probability in part **a** $\frac{1}{2}$?

1.6 ▶ Reliability of Systems

Mechanical and electrical systems are often composed of separate components which may or may not function independently. The space shuttle, for example, is comprised of hundreds of systems, each of which may have hundreds or thousands of components. The components are, of course, subject to possible failure, and these failures in turn may cause individual systems to fail, and ultimately cause the entire system to fail. We pause here to consider in some situations how the probability of failure of a component may influence the probability of failure of the system of which it is a part.

In general, we refer to the *reliability*, $R(t)$, of a component as the probability the component will function properly, or survive, for a given period of time. If we denote the event "the component survives at least t units of time" by $T > t$ then

$$R(t) = P(T > t)$$

where t is fixed.

The reliability of the system depends on two factors: the reliability of its component parts as well as the manner in which they are connected. We will consider some systems in this section that have few components and elementary patterns of connection.

We will presume that interest centers on the probability an entire system lasts a given period of time; we will calculate this as a function of the probabilities the components last for that amount of time. To do this, we repeatedly use the addition law and multiplication of probabilities.

Series Systems

If a system of two components functions only if both of the components function, then the components are connected in *series*. Such a system is shown in Figure 1.16.

Figure 1.16   A series system of two components

Let p_A and p_B denote the reliabilities of the components A and B; that is,

$$p_A = P(A \text{ survives at least } t \text{ units of time), and}$$
$$p_B = P(B \text{ survives at least } t \text{ units of time)}$$

for some fixed value t.

If the components function independently then the reliability of the system, say R, is the product of the individual reliabilities, so

$$R = P(A \text{ survives at least } t \text{ units of time and } B \text{ survives at least } t \text{ units of time)}$$
$$= P(A \text{ survives at least } t \text{ units of time}) \cdot P(B \text{ survives at least } t \text{ units of time)}$$

so

$$R = p_A \cdot p_B.$$

Parallel Systems

If a system of two components functions if either or both of the components function, then the components are connected in *parallel*. Such a system is shown in Figure 1.17.

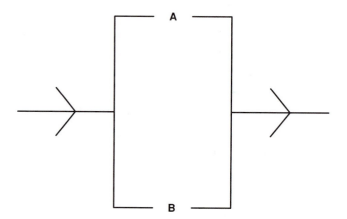

Figure 1.17   A parallel system of two components.

One way to calculate the reliability of the system depends on the fact that at least one of the components must function properly for the given period of time:

$$R = P(A \text{ or } B \text{ survives for a given period of time) so,}$$

by the addition law,

$$R = p_A + p_B - p_A \cdot p_B.$$

It is also clear, if the system is to function, that not both of the components can fail, so

$$R = 1 - (1 - p_A) \cdot (1 - p_B).$$

These two expressions for R are equivalent.

Figure 1.18 shows the reliability of both series and parallel systems as a function of p_A and p_B. The parallel system is always more reliable than the series system because, for the parallel system to function, at least one of the components must function, while the series system functions only if both components function simultaneously.

Series and parallel systems may be combined in fairly complex ways. We can calculate the reliability of a system from the formulas we have established.

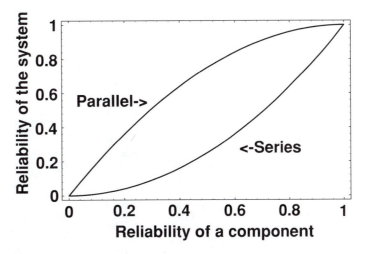

Figure 1.18 Reliability of series and parallel systems.

Example 1.6.1

The reliability of the system shown in Figure 1.19 can be calculated by using the addition law and multiplication of probabilities.

The connection of components A and B in the top section can be replaced by a single component with reliability $p_A \cdot p_B$. The parallel connection of switches C and D can be replaced by a single switch with reliability $1 - (1 - p_C) \cdot (1 - p_D)$. The reliability of the resulting parallel system is then

$$1 - (1 - p_A \cdot p_B) \cdot [1 - \{1 - (1 - p_C) \cdot (1 - p_D)\}].$$

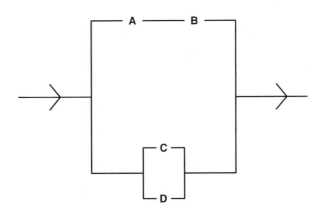

Figure 1.19 System for Example 1.6.1.

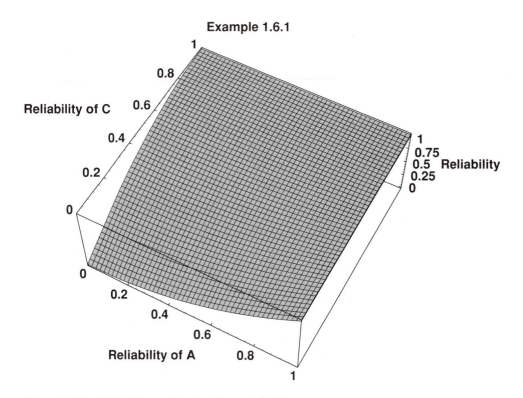

Figure 1.20   Reliability surface for Example 1.6.1.

A graph of the surface generated, assuming $p_A = p_B$ and $p_C = p_D$, is shown in Figure 1.20.

A contour plot of a surface shows values of p_A and p_C for which the reliability takes on particular values. Figure 1.21 shows a contour plot of the surface for Example 1.6.1, with contours specified at levels 0.80, 0.85, 0.90, 0.95, 0.99, and 0.995 for the reliability. The contour plot shows that if either p_A or p_C is 1, then the reliability is 1. The next contour shows choices of p_A and p_C giving reliability 0.995. The surface indicates that the system is highly reliable if either of the components is highly reliable; otherwise, the reliability declines rapidly.   ◄

Contour Plot

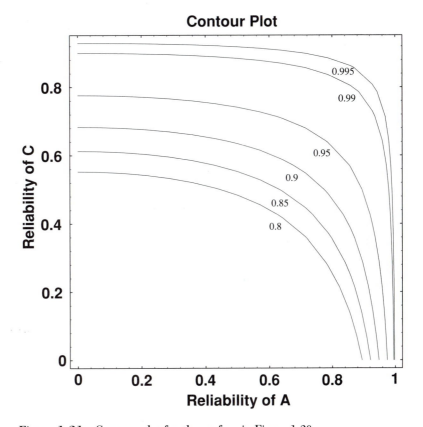

Figure 1.21   Contour plot for the surface in Figure 1.20.

Exercises 1.6

1. In the diagram following, let p_A, p_B, and p_C be the reliabilities of the individual switches. Determine the reliability of the system if

   a. at least one switch must function.
   b. at least two switches must function.

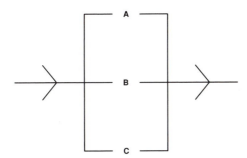

2. Determine the reliability of the following system if the reliability of any of the individual components is p.

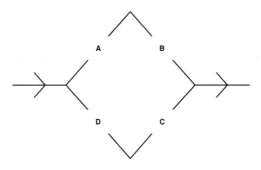

3. Find the reliability of the following system if $p_A = p_B$ and $p_C = p_D$. Then show the surface giving the reliability of the system as a function of p_A and p_C and draw a contour plot of the surface.

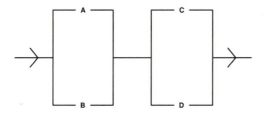

4. Find the reliability of the following system if each component has reliability 0.92.

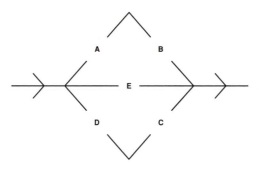

1.7 ▶ Counting Techniques

Occasionally sample spaces are encountered for which the sample points are equally likely. If this is the case, and if the sample space S contains n points, then, since the total probability in the sample space is 1, each point has probability $\frac{1}{n}$. If we denote the mutually exclusive points in A by a_i, $i = 1, 2, 3, \ldots, n$, then the probability of an event, A, is the sum of the probabilities of the sample points in A. That is,

$$P(A) = \sum_{a_i \varepsilon A} P(a_i) = \sum_{a_i \varepsilon A} \frac{1}{n}$$

so

$$P(A) = \frac{\text{Number of points in } A}{n} = \frac{\text{Number of points in } A}{\text{Number of points in } S}.$$

In order to consider problems leading to sample spaces with equally likely sample points, we pause to examine some techniques for counting sets of points. These techniques provide some challenging problems.

The reader is cautioned to beware of concluding that just because a sample space has n points, each point has probability $\frac{1}{n}$. For example, an airplane journey is either safely completed or not. One hopes these do not each have probability $\frac{1}{2}$!

The counting techniques considered here are based on two fundamental counting principles concerning mutually exclusive events A and B:

Principle 1:

If events A and B can occur in n and m ways respectively, then A *and* B can occur together in $n \cdot m$ ways.

Principle 2:

If events A and B can occur in n and m ways respectively, then A *or* B (but not both) can occur in $n + m$ ways.

Principle 1 is easily established because A can occur in n ways and then must be followed by each way in which B can occur. A tree diagram, shown in Figure 1.22, illustrates the result. Principle 2 simply uses the word "or" in an exclusive sense.

A linear arrangement of n distinct objects is called a *permutation*. For example, three distinct objects, say A, B, and C, can be arranged in six different ways: ABC, ACB, BAC, BCA, CAB, and CBA. So there are six permutations of three distinct objects. To count these permutations for n distinct objects, we use Principle 1. We have n choices for the object in the first position; that object chosen, we have $n - 1$ choices for the object in the second position. Principle 1 tells us that there are $n \cdot (n - 1)$ ways to fill the first two positions. Continuing, we have

$$n \cdot (n - 1) \cdot (n - 2) \cdots 3 \cdot 2 \cdot 1$$

ways to arrange all n of the items. We call this expression $n!$ and note, for example, that $3! = 3 \cdot 2 \cdot 1 = 6$, verifying the number of permutations of A, B, and C above.

The values of $n!$ increase very rapidly: $1! = 1, 2! = 2, 3! = 6, 4! = 24, 5! = 120$, and $10!$ is over 3 million. If we are interested in the number of permutations of even a small set we must be prepared to deal with immense quantities. For example, the cards in

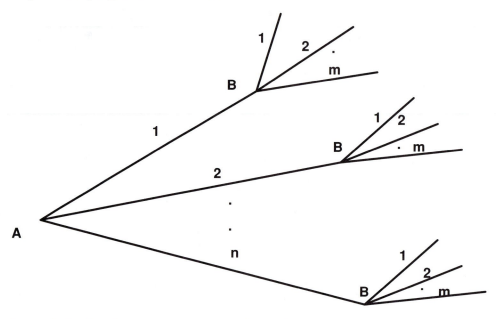

Figure 1.22   Tree diagram showing counting Principle 1.

a deck of 52 cards can be arranged in 52! = 80,658,175,170,943,878,571,660,636,856, 403,766,975,289,505,440,883,277,824,000,000,000,000 different ways. The reader may be surprised to find out how long it would take to enumerate these, even at a rate of 10,000 different permutations per second. This consideration may also persuade us that shuffling a deck so that each of these orders is equally likely is extremely unlikely.

A useful fact is that

$$n! = n \cdot (n-1)!$$

If we wish to permute only r, say, of the n distinct objects, this can be done in

$$n \cdot (n-1) \cdot (n-2) \cdots (n - (r-1)) \text{ ways.}$$

Multiplying and dividing by $(n-r)!$ shows that we can permute r of the n distinct objects in

$$\frac{n!}{(n-r)!} \text{ ways.}$$

So that this formula will work when $r = n$, we define $0! = 1$. If we wish to permute 5 cards chosen from a deck of 52, this can be done in

$$52 \cdot 51 \cdot 50 \cdot 49 \cdot 48 = 311,875,200 \text{ ways.}$$

We note, multiplying and dividing by 47!, that this can also be written as $\frac{52!}{47!}$, a fact that will be useful later.

If a list of permutations is desired, the reader is advised to do this using a computer algebra system. The $4! = 24$ permutations of the set $\{a, b, c, d\}$ is shown by a computer algebra system to be

a	b	c	d	c	a	b	d
a	b	d	c	c	a	d	b
a	c	b	d	c	b	a	d
a	c	d	b	c	b	d	a
a	d	b	c	c	d	a	b
a	d	c	b	c	d	b	a
b	a	c	d	d	a	b	c
b	a	d	c	d	a	c	b
b	c	a	d	d	b	a	c
b	c	d	a	d	b	c	a
b	d	a	c	d	c	a	b
b	d	c	a	d	c	b	a

If we regard these permutations as being equally likely, and if we want to find the probability that a particular letter, say b, occupies its normal place, we can count the points for which that is true and find that there are six of them. So

$$P(b \text{ is in second place}) = \frac{6}{24}.$$

What if the number of letters is large? An easy way to think about the problem is as follows: b is in its place, and if the set contains n distinct letters, we can arrange the remaining $(n-1)$ letters in $(n-1)!$ ways. Since the entire set can be permuted in $n!$ ways, the probability that b, or any of the other particular letters, occupies its own place is

$$\frac{(n-1)!}{n!} = \frac{1}{n}.$$

This raises the question of other letters also occupying their own position. If we arrange the letters entirely at random, what is the probability that at least one of the letters is in its own place? The problem has been posed in the literature in many different ways, one of which is this: n men enter a restaurant and each checks his hat; the hats become mixed up during the evening and are passed out at the end of the evening in an entirely random way. What is the probability that at least one man gets his own hat? Equivalently, if we assume some natural order for the cards in a deck, what, after thorough shuffling, is the probability that at least one of the cards is in its own position?

One way to solve the problem is to determine the number of *derangements* (where no object occupies its own place) of a set of objects. For the permutations of the preceding set $\{a, b, c, d\}$, we find the derangements are

$$b, a, d, c$$
$$b, c, d, a$$
$$b, d, a, c$$
$$c, a, d, b$$
$$c, d, a, b$$
$$c, d, b, a$$
$$d, a, b, c$$
$$d, c, a, b$$
$$d, c, b, a$$

which is a total of nine derangements in this case. It follows that the probability that at least one object occupies its own place is $1 - \frac{9}{24} = \frac{15}{24} = 0.625$. Surely this is an awkward way to handle larger sets, such as the deck of 52 cards. Surprisingly, we will find that the probability that at least one card occupies its own place after a thorough shuffling of the deck is very close to the probability above for four objects! We'll explain this when we return to this problem later in this section.

For now, consider arranging a set of objects when the objects are not all distinct. If we permute the elements in the set $\{a, a, b, c\}$, we find there are 12 permutations:

$$a, a, b, c$$
$$a, a, c, b$$
$$a, b, a, c$$
$$a, b, c, a$$
$$a, c, a, b$$
$$a, c, b, a$$
$$b, a, a, c$$
$$b, a, c, a$$
$$b, c, a, a$$
$$c, a, a, b$$
$$c, a, b, a$$
$$c, b, a, a$$

So the set of 24 permutations (where the objects were distinct) has been cut in half. We can arrive at this result by starting with the set $\{a, a, b, c\}$. Let R denote the number of distinct permutations. If we tag the a 's with subscripts, say as a_1 and a_2, then each of the permutations in the above list yields 2! permutations with the subscripted a's. Hence, $2! \cdot R = 4!$ so $R = \frac{4!}{2!} = 12$. We could do exactly the same procedure with any set. Consider, for example, the set $\{a, a, b, b, b, c, c, c, c\}$. By subscripting the a's, b's, and c's and again letting R denote the number of distinct permutations, we conclude that

$$2! \cdot 3! \cdot 4! \cdot R = 9!$$

so

$$R = \frac{9!}{2! \cdot 3! \cdot 4!} = 1260.$$

The example is perfectly typical of the general situation: if the set has n_1 objects of one kind, n_2 of another, and so on until we have, say n_k of the kth kind where $\sum_{i=1}^{k} n_i = n$, then there are

$$\frac{n!}{n_1! \cdot n_2! \cdots n_k!}$$

distinct permutations of the n objects.

Example 1.7.1

In how many distinct ways can 10 A's, 5 B's and 2 C's be awarded to a class of 17 students?

Put the students in some order. Then each distinct permutation of the letters leads to a different assignment of the grades. So there are

$$\frac{17!}{10! \cdot 5! \cdot 2!} = 408,408$$

different ways to assign the grades.   ◄

We turn now to *combinations* – the distinguishable sets or samples of objects that can be chosen from a set of n distinct objects, without regard for order. We denote these combinations of r objects chosen from n distinct objects by $\binom{n}{r}$ which we read "n choose r". We have already seen that

$$\binom{n}{2} = \frac{n \cdot (n-1)}{2}$$

in Example 1.5.3. Now suppose we have a set of objects and we want to choose a subset or sample of size 3. To be specific, suppose there are four items: a, b, c, and d. It is easy to write down the four combinations of size 3: a, b, c; a, b, d; a, c, d; and b, c, d. However, if we were dealing with larger set, it might be very difficult to write down a complete list without a procedure in mind. As a suggestion, to create the samples of size 3, we could choose each of the samples of size 2 and then attach a third item. The resulting list is

$$a, b, c$$
$$a, b, d$$
$$b, c, a$$
$$b, c, d$$
$$a, c, b$$
$$a, c, d$$
$$a, d, b$$
$$a, d, c$$
$$b, d, a$$
$$b, d, c$$
$$c, d, a$$
$$c, d, b.$$

Since we have 2 choices for the third item, the resulting list contains $2 \cdot \binom{4}{2}$ items. But each of the *combinations* has occurred three times. Therefore

$$2 \cdot \binom{4}{2} = 3 \cdot \binom{4}{3},$$

so

$$\binom{4}{3} = \frac{2 \cdot \binom{4}{2}}{3} = \frac{2 \cdot 6}{3} = 4.$$

This would appear to be a difficult way to arrive at $\binom{4}{3}$. The reasoning however, can easily be extended and therein lies its advantage. Suppose we have a set of n distinct items and we wish to choose a sample of size r. If we choose all the possible samples of size $r - 1$ and then attach one of the $n - r + 1$ remaining items to each, the resulting list has $(n - r + 1) \cdot \binom{n}{r-1}$ items. But this counts each of the $\binom{n}{r}$ combinations r times. So,

$$(n - r + 1) \cdot \binom{n}{r-1} = r \cdot \binom{n}{r}$$

or   (1.4)

$$\binom{n}{r} = \frac{(n - r + 1) \cdot \binom{n}{r-1}}{r}.$$

This is a *recurrence formula* because it expresses some values of a function, here $\binom{n}{r}$, in terms of other values of the same function. If we have a starting place, we can calculate any value of the function we want. In this case, because $\binom{n}{1} = n$, formula 1.4 shows that

$$\binom{n}{2} = \frac{n - 2 + 1}{2} \cdot \binom{n}{1} = \frac{n \cdot (n - 1)}{2} = \frac{n!}{2! \cdot (n - 2)!}$$

verifying our previous result. We continue to apply formula 1.4 to find

$$\binom{n}{3} = \frac{n-3+1}{3} \cdot \binom{n}{2} = \frac{n \cdot (n-1) \cdot (n-2)}{3 \cdot 2}$$

$$= \frac{n \cdot (n-1) \cdot (n-2) \cdot (n-3)!}{3! \cdot (n-3)!}$$

$$\binom{n}{3} = \frac{n!}{3! \cdot (n-3)!}.$$

It can be concluded by an inductive proof that

$$\binom{n}{r} = \frac{n!}{r! \cdot (n-r)!}, \quad r = 0, 1, \ldots, n$$

using the recurrence formula 1.4.

If we have a set of n distinct objects and r are chosen, then $n - r$ objects must remain unchosen. Each time the chosen set is altered, so is the unchosen set. It follows that

$$\binom{n}{r} = \binom{n}{n-r}.$$

The quantities $\binom{n}{r}$ are often called *binomial coefficients* because they occur in the binomial expansion:

THEOREM 6:   (Binomial Theorem):

$$(a+b)^n = \sum_{r=0}^{n} \binom{n}{r} \cdot a^{n-r} \cdot b^r = \sum_{r=0}^{n} \binom{n}{r} \cdot a^r \cdot b^{n-r} \qquad (1.5)$$

∎

For example,

$$(a+b)^5 = \binom{5}{0}a^5 + \binom{5}{1}a^4b + \binom{5}{2}a^3b^2 + \binom{5}{3}a^2b^3 + \binom{5}{4}ab^4 + \binom{5}{5}b^5$$

$$= a^5 + 5a^4b + 10a^3b^2 + 10a^2b^3 + 5ab^4 + b^5.$$

Many interesting identities are known concerning the binomial coefficients. If $a = 1$ and $b = 1$ are substituted in formula 1.5, the result is

$$(1+1)^n = 2^n = \sum_{r=0}^{n} \binom{n}{r} = \binom{n}{0} + \binom{n}{1} + \binom{n}{2} + \cdots + \binom{n}{n}.$$

Each side of this result may be recognized as the number of possible subsets (including the null set) that can be chosen from a set of n distinct items.

If we differentiate formula 1.5 with respect to a and then let $a = 1$ and $b = 1$, the result is

$$n \cdot 2^n = \sum_{r=0}^{n} \binom{n}{r} \cdot r = 0 \cdot \binom{n}{0} + 1 \cdot \binom{n}{1} + 2 \cdot \binom{n}{2} + \cdots + n \cdot \binom{n}{n}.$$

We show one more fact concerning the binomial coefficients. Suppose we want to choose a committee of size r from a group of n people, one of whom is Sam. Sam is a member of

$$\binom{n-1}{r-1}$$

committees and he is not a member of

$$\binom{n-1}{r}$$

committees; so, because we have exhausted the possibilities,

$$\binom{n}{r} = \binom{n-1}{r-1} + \binom{n-1}{r}.$$

This is often known as *Pascal's Identity* because it occurs in Pascal's triangle of binomial coefficients.

It is also necessary, although this may seem unnatural to the reader, to ascribe some meaning to a symbol such as $\binom{-7}{3}$. Clearly we cannot interpret this as the choice of 3 objects from -7 objects! The following definition, while including our previous interpretation of $\binom{n}{r}$, allows us to extend its meaning as well.

Definition:

$$\binom{n}{r} = \frac{n \cdot (n-1) \cdot (n-2) \cdots (n-r+1)}{r!}$$

provided that r is a non-negative integer and $r \leq n$, $\binom{n}{r} = 0$ otherwise if $r \leq n$; $\binom{n}{r} = 0$ if $r > n$.

Using the above definition,

$$\binom{-7}{3} = \frac{(-7) \cdot (-8) \cdot (-9)}{3!} = -84.$$

We will need facts such as this in subsequent chapters.

With this definition, the binomial theorem can also be used with negative exponents. For example,

$$(a+b)^{-5} = a^{-5} + \binom{-5}{1}a^{-6}b + \binom{-5}{2}a^{-7}b^2 + \binom{-5}{3}a^{-8}b^3 + \cdots$$

or

$$(a+b)^{-5} = a^{-5} - \binom{5}{1}a^{-6}b + \binom{6}{2}a^{-7}b^2 - \binom{7}{3}a^{-8}b^3 + \cdots.$$

We now use some of the results found here in some examples.

Example 1.7.2

A box of manufactured items contains eight items that are good and three that are not usable. What is the probability that a sample of five items contains exactly one unusable item?

Suppose that each of the samples has probability $\frac{1}{\binom{11}{5}}$. There are $\binom{8}{4}$ ways to choose the four good items and $\binom{3}{1}$ ways to choose the unusable item. The multiplication principle gives $\binom{3}{1} \cdot \binom{8}{4}$ ways to choose exactly one unusable item. So the probability we seek is

$$\frac{\binom{3}{1} \cdot \binom{8}{4}}{\binom{11}{5}} = \frac{5}{11}.$$

◄

Finally in this chapter we consider the *general addition law for n events*, having established the addition law for two and for three events. So we seek to prove

THEOREM 7: $P(A_1 \cup A_2 \cup \ldots \cup A_n) = \sum P(A_i) - \sum P(A_i \cap A_j)$
$+ \sum P(A_i \cap A_j \cap A_k) - \cdots + (-1)^{n-1} P(A_1 \cap A_2 \cap \ldots \cap A_n)$ where the sums are over all the *distinct* items in the summand; that is, where $i > j > k > \ldots$

PROOF: We again use the principle of inclusion and exclusion. Consider a point in $A_1 \cup A_2 \cup \ldots \cup A_n$ which is in exactly k of the events A_i. It will be convenient to renumber the A_i's if necessary so that the point is in the first k of these events. We will now show that the right side of Theorem 7 counts this point exactly once, showing the theorem to be correct.

The point is counted on the right side of Theorem 7

$$\binom{k}{1} - \binom{k}{2} + \binom{k}{3} - \cdots \pm \binom{k}{k} \text{ times.}$$

But the binomial expansion of $0 = [1 + (-1)]^k = \sum_{i=0}^{k} \binom{k}{i} \cdot (-1)^i$ shows that

$$\binom{k}{1} - \binom{k}{2} + \binom{k}{3} - \cdots \pm \binom{k}{k} = \binom{k}{0} = 1,$$

establishing the result.   ∎

Example 1.7.3

We return to the matching problem stated earlier in this section: If n integers are randomly arranged in a row, what is the probability that at least one of them occupies its own place? The general addition law can be used to provide the solution.

Let A_i denote the event, "number i is in the ith place." We seek $P(A_1 \cup A_2 \cup \ldots \cup A_n)$.

Here

$$P(A_i) = \frac{(n-1)!}{n!},$$

because, after i is put in its own place, there are $(n-1)!$ ways to arrange the remaining numbers,

$$P(A_i \cap A_j) = \frac{(n-2)!}{n!},$$

because if i and j occupy their own places, we can permute the remaining $n - 2$ objects in $(n-2)!$ ways. Also, in general,

$$P(A_1 \cap A_2 \cap \ldots \cap A_k) = \frac{(n-k)!}{n!}.$$

Now we note that there are $\binom{n}{1}$ choices for an individual number i; there are $\binom{n}{2}$ choices for pairs of numbers i and j; and, in general, there are $\binom{n}{k}$ choices for k of the numbers. So, applying Theorem 7,

$$P(A_1 \cup A_2 \cup \ldots \cup A_n) = \binom{n}{1} \cdot \frac{(n-1)!}{n!} - \binom{n}{2} \cdot \frac{(n-2)!}{n!}$$

$$+ \binom{n}{3} \cdot \frac{(n-3)!}{n!} - \cdots \pm \binom{n}{n} \cdot \frac{(n-n)!}{n!}.$$

This simplifies to

$$P(A_1 \cup A_2 \cup \ldots \cup A_n) = \frac{1}{1!} - \frac{1}{2!} + \frac{1}{3!} - \cdots \pm \frac{1}{n!}.$$

Following is a table of values of this expression.

n	P
1	1.000000
2	0.500000
3	0.666667
4	0.625000
5	0.633333
6	0.631944
7	0.632143
8	0.632118
9	0.632121

To six decimal places, the probability that at least one number is in its natural position remains at 0.632121 for $n \geq 9$. An explanation for this comes from a series expansion for e^x:

$$e^x = 1 + x + \frac{x^2}{2!} + \frac{x^3}{3!} + \frac{x^4}{4!} + \cdots$$

So

$$e^{-1} = 1 - 1 + \frac{(-1)^2}{2!} + \frac{(-1)^3}{3!} + \frac{(-1)^4}{4!} + \cdots$$

or

$$e^{-1} = \frac{1}{2!} - \frac{1}{3!} + \frac{1}{4!} + \cdots.$$

So we see that

$$\frac{1}{1!} - \frac{1}{2!} + \frac{1}{3!} - \cdots \pm \frac{1}{n!}$$

approaches

$$1 - \frac{1}{e} = 0.632120559\ldots$$

This is our first, but certainly not our last, encounter with e in a probability problem. This also explains why we remarked that the probability that at least 1 card in a shuffled deck of 52 cards was in its natural position differed little from that for a deck consisting of only 9 cards. ◀

We turn now to some examples using the results established in this section.

Example 1.7.4

Five red and four blue marbles are arranged in a row. What is the probability that both end marbles are blue?

A basic decision in the solution of the problem concerns the type of sample space to be used. Clearly the problem involves order, but should we consider the marbles to be distinct or not?

Initially consider the marbles to be alike, except for color. There are $\frac{9!}{5!\cdot4!} = 126$ possible orderings of the marbles and we consider each of these to be equally likely. Since the blue marbles are indistinct from each other, and since our only choice here is the arrangement of the seven marbles in the middle, it follows that there are $\frac{7!}{5!\cdot2!} = 21$ arrangements with blue marbles at the ends. Thus, the probability we seek is $\frac{21}{126} = \frac{1}{6}$.

Now if we consider each of the marbles to be distinct, there are 9! possible arrangements. Of these we have $\binom{4}{2} \cdot 2! = 12$ ways to arrange the blue marbles at the ends and 7! ways to arrange the marbles in the middle. This produces a probability of $\frac{7!\cdot12}{9!} = \frac{1}{6}$.

The two methods must produce the same result, but the reader may find one method easier to use than the other. In any event, it is crucial that the sample space be established as a first step in the solution of the problem, and that the events of interest be dealt with consistently for this sample space.

The reader may enjoy showing that, if we have n marbles, r of which are red and b of which are blue, then the probability that both ends are blue in a random arrangement of the marbles is given by the product

$$(1 - \frac{r}{n}) \cdot (1 - \frac{r}{n - 1}).$$

This answer may indicate yet another way to solve the problem, namely this: The probability that the first marble is blue is $(\frac{n-r}{n})$. Given that the first end is blue, the conditional probability that the other end is also blue is $(\frac{n-r-1}{n-1})$. Often probability problems involving counting techniques can be solved in a variety of ways.   ◄

Example 1.7.5

Ten race cars, numbered from 1 to 10, are running around a track. An observer sees three cars go by. If the cars appear in random order, what is the probability that the largest number seen is six?

The choice of the sample space here is natural: consider all the $\binom{10}{3}$ samples of three cars that could be observed. If the largest is to be six, then six must be in the sample, together with two cars chosen from the first five, so the probability of the event "Maximum = 6" is

$$P(\text{Maximum} = 6) = \frac{\binom{1}{1} \cdot \binom{5}{2}}{\binom{10}{3}} = \frac{1}{12}.$$

It is also interesting now to look at the *median*, or the number in the middle when the three observed numbers are arranged in order. What is the probability that the median of the group of three is 6?

For the median to be 6, 6 must be chosen and we must choose exactly one number from the set {1, 2, 3, 4, 5} and exactly one number from {7, 8, 9, 10}. Then

$$P(\text{Median} = 6) = \frac{\binom{1}{1} \cdot \binom{5}{1} \cdot \binom{4}{1}}{\binom{10}{3}} = \frac{1}{6}.$$

This can be generalized to

$$P(\text{Median} = k) = \frac{\binom{1}{1} \cdot \binom{k-1}{1} \cdot \binom{10-k}{1}}{\binom{10}{3}}$$

$$= \frac{(k-1)(10-k)}{120}, \quad k = 2, 3, \ldots, 9.$$

Figure 1.23 shows a graph of $P(\text{Median} = k)$ for $k = 2, 3, \ldots, 9$. It reveals a symmetry in the function around $k = 5.5$.

The problem is easily generalized with a result that may be surprising. Suppose there are 100 cars and we observe a sample of 9 of them. The median of the sample must be at least 5 and can be at most 96. Thus the probability that the median is k is

$$P(\text{Median} = k) = \frac{\binom{1}{1}\binom{k-1}{4}\binom{100-k}{4}}{\binom{100}{9}}, \quad k = 5, 6, \ldots, 96.$$

A graph of this function (an eighth degree polynomial in k) is shown in Figure 1.24.

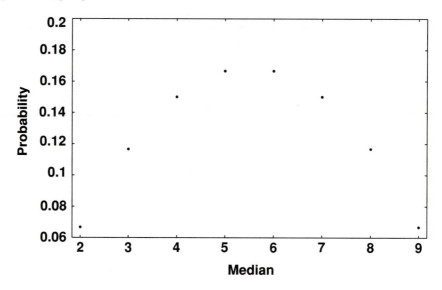

Figure 1.23 $P(\text{Median} = k)$ for a sample of size 3 chosen from 10 cars.

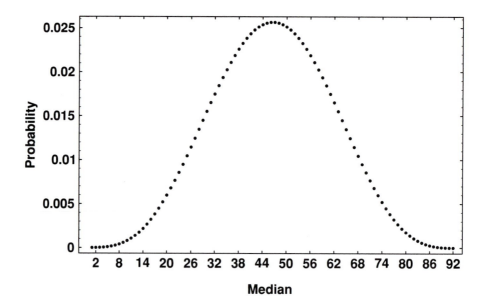

Figure 1.24 $P(\text{Median} = k)$ for a sample of 9 chosen from 100 cars.

The graph here shows a "bell shape" that, as we will see, is very common in probability problems. The curve is very close to what we will call a *normal* curve. Larger values for the number of cars involved will, surprisingly, not change the approximately normal shape of the curve! An approximation for the actual curve involved here can be found when we study the normal curve thoroughly in Chapter 3.   ◄

Example 1.7.6

We can use the result of Example 1.7.3 to count the number of *derangements* of a set of n objects. That is, we want to count the number of permutations in which no object occupies its own place. Example 1.7.3 shows that the number of permutations of n distinct objects in which at least one object occupies its own place is

$$n!(\frac{1}{1!} - \frac{1}{2!} + \frac{1}{3!} - \cdots \pm \frac{1}{n!}).$$

It follows that the number of derangements of n distinct objects is

$$n! - n!(\frac{1}{1!} - \frac{1}{2!} + \frac{1}{3!} - \cdots \pm \frac{1}{n!}) = n!(\frac{1}{2!} - \frac{1}{3!} + \cdots \pm \frac{1}{n!}). \quad (1.6)$$

Using this formula we find that if $n = 2$, there is 1 derangement; if $n = 3$, there are 2 derangements, and if $n = 4$, there are 9 derangements.

Formula (1.6) also suggests that the number of derangements of n distinct objects is approximately $n! \cdot e^{-1}$. (See the series expansion for e^{-1} in Example 1.7.3.) The following table compares the results of formula 1.6 and the approximation.

n	Number of derangements	$n! \cdot e^{-1}$
2	1	0.7358
3	2	2.207
4	9	8.829
5	44	44.146
6	265	264.83
7	1854	1854.11

We see that in every case, the number of derangements is given by $\lfloor n! \cdot e^{-1} + 0.5 \rfloor$ where the symbols indicate the greatest integer function.   ◄

Exercises 1.7

1. The integers $1, 2, 3, \ldots, 9$ are arranged in a row, resulting in a 9-digit integer. What is the probability that

   a. the integer resulting is even?

   b. the integer resulting is divisible by 5?

   c. the digits 6 and 4 are next to each other?

2. License plates in Indiana consist of a number from 1 to 99 (indicating the county of registration), a letter of the alphabet, and finally an integer from 1 to 9999. How many cars may be licensed in Indiana?

3. Prove that at least 2 people in Indianapolis, Indiana, have the same 3 initials.

4. In a small school, 5 centers, 8 guards, and 6 forwards try out for the basketball team.

   a. How many 5-member teams can be formed from these players? (Assume a team has 2 guards, 2 forwards, and 1 center.)

   b. Intercollegiate regulations require that no more than 8 players can be listed for the team roster. How many rosters can be formed consisting of exactly 8 players?

5. A restaurant offers 5 appetizers, 7 main courses and 8 desserts. How many meals can be ordered

   a. assuming all 3 courses are ordered?

   b. assuming not all 3 courses are necessarily ordered?

6. A club of 56 people has 40 men and 16 women. What is the probability that the board of directors, consisting of 8 members, contains no women?

7. In a controlled experiment, 12 patients are to be randomly assigned to each of three different drug regimens. In how many ways can this be done if each drug is to be tested on 4 patients?

 8. In the game Keno, the casino draws 20 balls from a set of balls numbered from 1 to 80. A player must choose 10 numbers in advance of this drawing. What is the probability the player has exactly 5 of the 20 numbers drawn?

9. A lot of 10 refrigerators contains 3 that are defective. The refrigerators are randomly chosen and shipped to customers. What is the probability that by the seventh shipment, none of the defective refrigerators remains?

10. In how many different ways can the letters in the word "repetition" be arranged?

11. In a famous correspondence in the very early history of probability, the Chevalier de Méré wrote to the mathematician Blaise Pascal and asked the following question: "Which is more likely – at least one 6 in 4 rolls of a fair die or at least one sum of 12 in 24 rolls of a pair of dice?"

   a. Show that the two questions above have nearly equal answers. Which is more likely?

   b. A generalization of the Pascal–de Méré problem is: What is the probability that the sum $6n$ occurs at least once in $4 \cdot 6^{n-1}$ rolls of n fair dice? Show that the answer is very nearly $\frac{1}{2}$ for $n \leq 5$.

c. Show that in part b] the probability approaches $1 - e^{-2/3}$ as $n \to \infty$.

12. A box contains 8 red and 5 yellow marbles from which a sample of 3 is drawn.

    a. Find the probability that the sample contains no yellow marbles if

        i. the sampling is done without replacement; and,

        ii. if the sampling is done with replacement.

    b. Now suppose the box contains 24 red and 15 yellow marbles (so that the ratio of reds to yellows is the same as in part **a**). Calculate the answers to part **a**. What do you expect to happen as the number of marbles in the box increases but the ratio of reds to yellows remains the same?

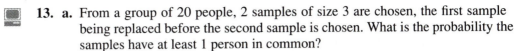

13. **a.** From a group of 20 people, 2 samples of size 3 are chosen, the first sample being replaced before the second sample is chosen. What is the probability the samples have at least 1 person in common?

    b. Show that 2 bridge hands (13 cards selected from a deck of 52), the first being replaced before the second is drawn, are virtually certain to contain at least 1 card in common.

14. A shipment of 20 components will be accepted by a buyer if a random sample of 3 (chosen without replacement) contains no defectives. What is the probability the shipment will be rejected if 2 of the components are defective?

15. A deck of cards is shuffled and the cards turned up one at a time. What is the probability that all the aces will appear before any of the tens?

16. In how many distinguishable ways can 6 A's, 4 B's and 8 C's be assigned as grades to 18 students?

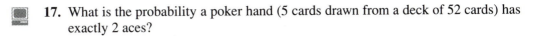

17. What is the probability a poker hand (5 cards drawn from a deck of 52 cards) has exactly 2 aces?

18. In how many ways can 6 students be seated in 10 chairs?

19. Ten children are to be grouped into 2 clubs, say the Lions and the Tigers, with 5 children in each club. Each club is then to elect a president and a secretary. In how many ways can this be done?

20. A small pond contains 50 fish, 10 of which have been tagged. If a catch of 7 fish is made, in how many ways can the catch contain exactly 2 tagged fish?

21. From a fleet of 12 limousines, 6 are to go to hotel I, 4 to hotel II, and the remainder to hotel III. In how many different ways can this be done?

22. The following grid shows a region of city blocks defined by 9 streets running north-south and 8 streets running east-west. Joe will walk from corner A to corner B. At each corner between A and B Joe will choose to walk either north or east.

a. How many possible routes are there?

b. Assuming that each route is equally likely, find the probability that Joe will pass through intersection C.

23. Suppose that N people are arranged in a line. What is the probability that 2 particular people, say A and B, are not next to each other?

24. The Hawaiian language has only 12 letters: the vowels a, e, i, o and u and the consonants h, k, l, m, n, p, and w.

a. How many possible 3-letter Hawaiian "words" are there? (Some of these may be nonsense words.)

b. How many 3-letter "words" have no repeated letter?

c. What is the probability a randomly selected 3-letter "word" begins with a consonant and ends with 2 different vowels?

d. What is the probability that a randomly selected 3-letter "word" contains all vowels?

25. How many partial derivatives of order 4 are there for a function of 4 variables?

26. A set of 15 marbles contains 4 red and 11 green marbles. They are selected, one at a time, without replacement. In how many ways can the last red marble be drawn on the seventh selection?

27. A true-false test has 4 questions. A student is not prepared for the test and so must guess the answer to each question.

    a. What is the probability the student answers at least half of the questions correctly?

    b. Now suppose, in a sudden flash of insight, the student knows the answer to question 2 is false. What is the probability the student answers at least half of the questions correctly?

 28. What is the probability of being dealt a bridge hand (13 cards selected from a deck of 52 cards) that does not contain a heart?

29. Explain why the number of derangements of n distinct objects is given by $\lfloor n! \cdot e^{-1} + 0.5 \rfloor$. Explain why $n! \cdot e^{-1}$ sometimes underestimates and sometimes overestimates the number of derangements. $\lfloor x \rfloor$ denotes the greatest integer in x.

Chapter Review

In dealing with an experiment or situation involving random or chance elements, it is reasonable to begin an analysis of the situation by asking the question, "What *can* happen?" An enumeration of all the possibilities is called a *sample space*. Generally, situations admit of more than one sample space; the appropriate one chosen is usually governed by the probabilities that one wants to compute. Several examples of sample spaces are given in this chapter, each of them *discrete*; that is, either the sample space has a finite number of points or a countably infinite number of points.

    Tossing two dice yields a sample space with a finite number of points; observing births until a girl is born gives a sample space with an infinite (but countable) number of points. In Chapter 3 we will encounter *continuous* sample spaces that are characterized by a non-countably infinite number of points.

    Assessing the *long-range relative frequency*, or *probability*, of any of the points or sets of points (which we refer to as *events*) was the primary goal of this chapter. We used the set symbols $\cup$ for the union of two events and $\cap$ for the intersection of two events. We began with three assumptions, or axioms, concerning sample spaces:

    1. $P(A) \geq 0$ where A is an event;

    2. $P(S) = 1$ where S is the entire sample space; and,

    3. $P(A \text{ or } B) = P(A \cup B) = P(A) + P(B)$ if A and B are disjoint, or mutually exclusive; that is, they have no sample points in common.

    From these assumptions we derived several theorems concerning probability, among them:

    1. $P(A) = \sum_{a_i \in A} P(a_i)$ where the a_i are distinct points in S

    2. $P(A \cup B) = P(A) + P(B) - P(A \cap B)$ (the addition law for two events)

3. $P(\overline{A}) = 1 - P(A)$.

We showed the Law of Total Probability:

THEOREM: (Law of Total Probability):   If the sample space $S = A_1 \cup A_2 \cup \ldots \cup A_n$ where A_i and A_j have no sample points in common if $i \neq j$, then, if B is an event,

$$P(B) = P(A_1) \cdot P(B|A_1) + P(A_2) \cdot P(B|A_2) + \cdots + P(A_n) \cdot P(B|A_n). \quad \blacksquare$$

We then turned our attention to problems of *conditional probability* where we sought the probability of some event, say A, on the condition that some other event, say B, has occurred. We showed that

$$P(A|B) = \frac{P(A \cap B)}{P(B)} = \frac{P(A) \cdot P(B|A)}{P(A) \cdot P(B|A) + P(\overline{A}) \cdot P(B|\overline{A})}.$$

This can be generalized using the Law of Total Probability as

THEOREM: (Bayes' Theorem):   If $S = A_1 \cup A_2 \cup \ldots \cup A_n$ where A_i and A_j have no sample points in common for $i \neq j$, then, if B is an event,

$$P(A_i|B) = \frac{P(A_i)P(B|A_i)}{P(A_1)P(B|A_1) + P(A_2)P(B|A_2) + \cdots + P(A_n)P(B|A_n)}$$

which can be written as

$$P(A_i|B) = \frac{P(A_i)P(B|A_i)}{\sum_{j=1}^{n} P(A_j)P(B|A_j)} \quad \blacksquare$$

Bayes' Theorem has a simple geometric interpretation. The chapter contains many examples of this.

We defined the *independence* of two events A and B as

$$A \text{ and } B \text{ are } independent \text{ if } P(A \cap B) = P(A) \cdot P(B).$$

We then applied the results of this chapter to some specific probability problems, such as the well-known birthday problem and a geometric problem involving the sides and diagonals of a polygonal figure.

Finally we considered some very special counting techniques that are useful, it is to be emphasized, only if the points in the sample space are equally likely. If that is so, then the probability of an event, say A, is

$$P(A) = \frac{\text{Number of points in } A}{\text{Number of points in } S}.$$

If order is important, then all the *permutations* of objects may well comprise the sample space. We showed that there are $n! = n \cdot (n - 1) \cdot (n - 2) \cdots 3 \cdot 2 \cdot 1$ permutations of n distinct objects.

If order is not important, then the sample space may well comprise various *combinations* of items. We showed that there are

$$\binom{n}{r} = \frac{n!}{r!(n - r)!}$$

samples of size r that can be selected from n distinct objects, and we applied this formula to several examples. A large number of identities are known concerning these combinations, or *binomial coefficients*, among them:

1. $\sum_{r=0}^{n} \binom{n}{r} = 2^n$

2. $\binom{n}{r} = \binom{n-1}{r-1} + \binom{n-1}{r}.$

One very important result from this section is the *general addition law*:

THEOREM: $P(A_1 \cup A_2 \cup \ldots \cup A_n) = \sum P(A_i) - \sum P(A_i \cap A_j)$
$$+ \sum P(A_i \cap A_j \cap A_k) - \cdots + (-1)^{n-1} P(A_1 \cap A_2 \cap \ldots \cap A_n)$$
where the summations are over $i > j > k > \ldots$ ∎

A few typical problems from each section of this chapter are listed here for the convenience of the reader who wishes to review the material.

Problems for Review

Exercises 1.1 # 1, 2, 5, 7, 9, 11
Exercises 1.3 # 1, 2, 6, 7, 9, 13
Exercises 1.4 # 1, 2, 3, 6, 10, 15, 16, 18, 19, 21, 24
Exercises 1.5 # 2, 3, 6, 7
Exercises 1.6 # 1, 3
Exercises 1.7 # 1, 6, 8, 10, 12, 13, 16, 17, 20, 23, 28

Supplementary Exercises for Chapter 1

1. A hat contains slips of paper on which each of the integers $1, 2, 3, \ldots, 20$ is written. A sample of size 6 is drawn (without replacement) and the sample values, x_i, put in order so that $x_1 < x_2 < \ldots < x_6$. Find the probability that $x_3 = 12$.

2. Show that $(n - k)\binom{n}{n-k} = (k + 1)\binom{n}{k+1}$.

3. Suppose that events A, B, and C are independent with $P(A) = \frac{1}{3}$, $P(B) = \frac{1}{4}$, and $P(A \cup B \cup C) = \frac{3}{4}$. Find $P(C)$.

4. Events A and B are such that $P(A \cup B) = 0.8$ and $P(A) = 0.2$. For what value of $P(B)$ are

   a. A and B independent?

   b. A and B mutually exclusive?

5. Events A, B, and C in a sample space have $P(A) = 0.2$, $P(B) = 0.4$, and $P(A \cup B \cup C) = 0.9$. Find $P(C)$ if A and B are mutually exclusive, A and C are independent, and B and C are independent.

6. How many distinguishable arrangements are there of the letters in the word **PROBABILITY**?

7. How many people must be in a group so that the probability that at least 2 were born on the same day of the week is at least $\frac{1}{2}$?

8. A and B are special dice. The faces on die A are 2, 2, 5, 5, 5, 5, and the faces on die B are 3, 3, 3, 6, 6, 6. The 2 dice are rolled. What is the probability that the number showing on die B is greater than the number showing on die A?

9. A committee of 5 is chosen from a group of 8 men and 4 women. What is the probability the group contains a majority of women?

10. A college senior finds he needs one more course for graduation and finds only courses in mathematics, chemistry, and computer science available. On the basis of interest he assigns probabilities of 0.1, 0.6, and 0.3 respectively, to the events of choosing each of these. After considering his past performance, his advisor estimates his probabilities of passing these courses as 0.8, 0.7, and 0.6 respectively, regarding the passing of courses as independent events.

    a. What is the probability he passes the course if he chooses a course at random?

    b. Later we find that the student graduated. What is the probability he took chemistry?

11. A number X is chosen at random from the set $\{10, 11, 12, \ldots, 99\}$.

    a. Find the probability that the tens digit in X is less than the units digit.

    b. Find the probability that X is at least 50.

    c. Find the probability that the tens digit in X is the square of the units digit.

12. If the integers 1, 2, 3, and 4 are randomly permuted, what is the probability that 4 is to the left of 2?

13. In a sample space, events A and B are such that $P(A) = P(B)$, $P(\overline{A} \cap \overline{B}) = P(A \cap B) = \frac{1}{6}$. Find

    a. $P(A)$.

    b. $P(\overline{A} \cup \overline{B})$.

    c. P(exactly one of the events A or B).

14. A fair coin is tossed 4 times. Let A be the event "second toss is heads," B be the event "exactly 3 heads," and C be the event "fourth toss is tails if the second toss is heads." Are A, B, and C independent?

15. An instructor has decided to grade each of his students A, B, or C. He wants the probability that a student receives a grade of B or better to be 0.7, and the probability that a student receives at most a grade of B to be 0.8. Is this possible? If so, what proportions of each letter grade must be assigned?

16. How many bridge hands are there containing 3 hearts, 4 clubs, and 6 spades?

17. A day's production of 100 fuses is inspected by a quality control inspector who tests 10 fuses at random, sampling without replacement. If he finds 2 or fewer defective fuses, he accepts the entire lot of 100 fuses. What is the probability the lot is accepted if it contains 20 defective fuses?

18. Suppose that A and B are events for which $P(A) = a$, $P(B) = b$, and $P(A \cap B) = c$. Express each of the following in terms of a, b, and c.

    a. $P(\overline{A} \cup \overline{B})$

    b. $P(\overline{A} \cap B)$

    c. $P(\overline{A} \cup B)$

    d. $P(\overline{A} \cap \overline{B})$

    e. P(exactly one of A or B occurs)

19. An elevator starts with 10 people on the first floor of an 8 story building and stops at each floor.

    a. In how many ways can all the people get off the elevator?

    b. How many ways are there for everyone to get off if no one gets off on some 2 specific floors?

c. In how many ways are there for everyone to get off if at least one person gets off at each floor?

20. A manufacturer of calculators buys integrated circuits from suppliers A, B, and C. Fifty percent of the circuits come from A, 30% from B, and 20% from C. One percent of the circuits supplied by A have been defective in the past, 3% of B's have been defective, and 4% of C's have been defective. A circuit is selected at random and found to be defective. What is the probability it was manufactured by B?

21. Suppose that E and T are independent events with $P(E) = P(T)$ and $P(E \cup T) = \frac{1}{2}$. What is $P(E)$?

22. A quality control inspector draws parts one at a time and without replacement from a set containing 5 defective and 10 good parts. What is the probability the third defective is found on the eighth drawing?

23. If A, B, and C are independent events, show that the events A and $B \cup C$ are independent.

24. Bean seeds from supplier A have an 85% germination rate and those from supplier B have a 75% germination rate. A seed company purchases 40% of their bean seeds from supplier A and the remaining 60% from supplier B and mixes these together. If a seed germinates, what is the probability it came from supplier A?

25. An experiment consists of choosing 2 numbers without replacement from the set $\{1, 2, 3, 4, 5, 6\}$ with the restriction that the second number chosen must be greater than the first.

a. Describe the sample space.
b. What is the probability the second number is even?
c. What is the probability the sum of the 2 numbers is at least 5?

26. What is the probability a poker hand contains exactly one pair?

27. A box contains 6 good and 8 defective light bulbs. The bulbs are drawn out one at a time, without replacement, and tested. What is the probability that the fifth good item is found on the ninth test?

28. An individual tried by a three-judge panel is declared guilty if at least 2 judges cast votes of guilty. Suppose that when the defendant is guilty, each judge will independently vote guilty with probability 0.7, but, if the defendant is, in fact, innocent, each judge will independently vote guilty with probability 0.2. Assume that 70% of the defendants are actually guilty. If a defendant is judged guilty by the panel of judges, what is the probability he is actually innocent?

29. What is the probability a bridge hand is missing cards in at least one suit?

30. Suppose 0.1% of the population is infected with a certain disease. On a medical test for the disease, 98% of those infected give a positive result while 1% of those not infected give a positive result. If a randomly chosen person is tested and gives a positive result, what is the probability the person has the disease?

 31. A committee of 50 politicians is to be chosen from the 100 U.S. senators (2 are from each state). If the selection is done at random, what is the probability that each state will be represented?

32. In a roll of a pair of dice (one red and one green), let A be the event "red die shows 3, 4, or 5," B the event "green die shows a 1 or a 2," and C the event "dice total 7." Show that A, B, and C are independent.

33. A prospector for oil thinks there is an even chance that oil is on his property. He has a test for oil that is 80% reliable: that is, if there is oil, it indicates this with probability 0.80, and if there is no oil, it indicates that with probability 0.80. The test indicates oil on the property. What is the probability there really is oil on the property?

34. Given: A and B are events with $P(A) = 0.3$, $P(B) = 0.7$, and $P(A \cup B) = 0.9$. Find

  a. $P(A \cap B)$.
  b. $P(B|\overline{A})$.

35. Two good transistors become mixed up with 3 defective ones. A person is assigned to sampling the mixture by drawing out 3 items without replacement. However, the instructions are not followed and the first item is replaced, but the second and third items are not replaced.

  a. What is the probability the sample contains exactly 2 items that test as good?
  b. What is the probability the 2 items finally drawn are both good transistors?

36. How many lines are determined by 8 points, no 3 of which are collinear?

37. Show that if A and B are independent then $\overline{A}$ and $\overline{B}$ are independent.

38. How many tosses of a fair coin are needed so that the probability of at least one head is at least 0.99?

39. A lot of 24 tubes contains 13 defective ones. The lot is randomly divided into 2 equal groups, and each group is placed in a box.

  a. What is the probability that one box contains only defective tubes?
  b. Suppose the tubes are divided so that one box contains only defective tubes. A box is chosen at random and one tube is chosen from that box and is found to be defective. What is the probability a second tube chosen from the same box is also defective?

40. A machine is composed of 2 components, A and B, which function (or fail) independently. The machine works only if both components work. It is known that component A is 98% reliable and the machine is 95% reliable. How reliable is component B?

41. Suppose A and B are events. Explain why

$$P(\text{exactly one of events } A, B \text{ occurs}) = P(A) + P(B) - 2P(A \cap B).$$

42. A box contains 8 red, 3 white, and 9 blue balls. Three balls are to be drawn, without replacement. What is the probability that more blues than whites are drawn?

43. A marksman, whose probability of hitting a moving target is 0.6, fires 3 shots. Suppose the shots are independent.

   a. What is the probability the target is hit?
   b. How many shots must be fired to make the probability at least 0.99 that the target will be hit?

44. A box contains 6 green and 11 yellow balls. Three are chosen at random. The first and third balls are yellow. Which method of sampling – with replacement or without replacement – gives the higher probability of this event?

45. A box contains slips of paper numbered from 1 to m. One slip is drawn from the box; if it is 1, it is kept; otherwise, it is returned to the box. A second slip is drawn from the box. What is the probability that the second slip is numbered 2?

46. Three integers are selected at random from the set $\{1, 2, \ldots, 10\}$. What is the probability that the largest of these is 5?

47. A pair of dice is rolled until a 5 or a 7 appears. What is the probability that a 5 occurs first?

48. A door-to-door salesman will say he had a good day when, in fact, he did with probability 1, but the probability is only 0.6 that he will say he had a good day when, in fact, he did not. Only $\frac{1}{4}$ of his selling days are actually good ones. What is the probability that he had a good day if he says he had a good day?

49. An inexperienced employee mistakenly samples n items from a lot of N items, with replacement. What is the probability the sample contains at least one duplicate?

50. A roulette wheel has 38 slots – 18 red, 18 black, and 2 green (the house wins on green). Suppose the spins of the wheel are independent and that the wheel is fair. The wheel is spun twice and we know that at least one spin is green. What is the probability that both spins are green?

51. A "rook" deck of cards consists of 4 suits of cards: red, green, black, and yellow, each suit having 14 cards. In addition, the deck has an uncolored rook card. A hand contains 14 cards.

    a. How many different hands are possible?

    b. How many hands have the rook card?

    c. How many hands contain only 2 colors with equal numbers of cards of each color?

    d. How many hands have at most 3 colors and no rook card?

52. Find the probability that a poker hand contains three of a kind (exactly 3 cards of one face value and 2 cards of different face values).

53. A box contains tags numbered $1, 2, \ldots, n$. Two tags are chosen without replacement. What is the probability they are consecutive integers?

54. In how many different ways can n people be seated around a circular table?

 55. A production lot has 100 units of which 25 are known to be defective. A random sample of 4 units is chosen without replacement. What is the probability that the sample will contain no more than 2 defective units?

 56. A recent issue of a newspaper said that given a 5% probability of an unusual event in a one-year study, one should expect a 35% probability in a seven-year study. This is obviously faulty. What is the correct probability?

57. Independent events A and B have probabilities p_A and p_B respectively. Show that the probability of either 2 successes or 2 failures in 2 trials has probability $\frac{1}{2}$ if and only if at least one of p_A and p_B is $\frac{1}{2}$.

▶ *Chapter 2*

Discrete Random Variables and Probability Distributions

At this point we have considered discrete sample spaces and derived theorems concerning probabilities for any discrete sample space and some of the events within it. Often, however, events are most easily described by performing some *operation* on the sample points. For example, if two dice are tossed, we might consider the *sum* showing on the two dice; but when we find the sum, we have operated on the sample point seen. Other operations, as we will see, are commonly encountered.

We want to consider some properties of the sum; we start with the sample space. In this example a natural sample space shows the result on each die and, if the dice are fair, leads to equally likely sample points. The sample space consists then of the 36 points in S_1 :

$$S_1 = \{(1, 1), (1, 2), \ldots, (1, 6), (2, 1), \ldots, (6, 6)\}.$$

If we consider the *sum* on the two dice, then a sample space

$$S_2 = \{2, 3, 4, 5, 6, 7, 8, 9, 10, 11, 12\}$$

might be considered, but now the sample points are not equally likely. We call the sum in this example a *random variable*.

Definition: A *random variable* is a real-valued function defined on the points of a sample space.

Various functions occur commonly and we will be interested in a variety of them; sums are among the most interesting of these functions, as we will see. We will soon determine the probabilities of various sums, but the determination of these is probably evident now to the reader. We first need, for this problem as well as for others, some ideas and some notation.

2.1 ▶ Random Variables

We have considered only discrete sample spaces to this point; we discuss discrete random variables in this chapter.

First consider another example. It is convenient to let X denote the number of times an examination is attempted until it is passed. X in this case denotes a random variable; we will use capital letters to denote random variables. We show some of the infinite sample space here, indicating the value of X, x, at each point.

Event	*x*
P	1
F P	2
F F P	3
F F F P	4
⋮	⋮

Clearly we see that the event $'X = 3'$ is equivalent to the event $'FFP'$ and so their probabilities must be equal. Therefore,

$$P(X = 3) = P(FFP) = \frac{1}{8}.$$

The terminology "random variable" is curious since we could, in the above example, define a variable, say Y, to be 6 regardless of the outcome of the experiment. Y would carry no information whatsoever, and would be neither random nor variable! There are other curiosities with terminology in probability theory as well, but they have become, alas, standard in the field and so we accept them. What we call here a random variable is in reality a *function* whose domain is the sample space and whose range is the real line. The random variable here, as in all cases, provides a mapping from the sample space to the real line. While being technically incorrect, the phrase "random variable" seems to convey the correct idea. This perhaps becomes a bit more clear when we use functional notation to define a function $f(x)$ to be

$$f(x) = P(X = x),$$

where x denotes a *value* of the random variable X. In the example above we could then write $f(3) = \frac{1}{8}$. The function $f(x)$ is called a *probability distribution function* (abbreviated as *pdf*) for the random variable X.

Since probabilities must be non-negative and since the probabilities must sum to 1, we see that

1] $f(x) \geq 0$

and

2] $\sum_S f(x) = 1$ where S is the sample space.

We turn now to some examples of random variables.

Example 2.1.1

Throw a fair die once and let X denote the result. The random variable X can assume the values 1, 2, 3, 4, 5, 6 and so

$$P(X = x) = \begin{cases} \frac{1}{6} \text{ for } x = 1, 2, 3, 4, 5, 6 \\ 0, \text{ otherwise.} \end{cases}$$

A graph of this function is of course flat; it is shown in Figure 2.1. This is an example of a *discrete uniform probability distribution*.

The use of a computer algebra system for sampling from this distribution is explained in Appendix 1.   ◄

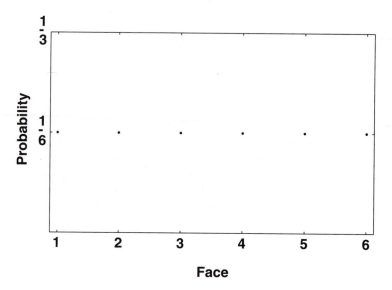

Figure 2.1   Discrete uniform probability distribution.

Example 2.1.2

In the previous example the die is fair, so now we consider an unfair die. In particular, could the die be weighted so that the probability a face appears is proportional to the face?

Suppose that X denotes the face that appears and let $P(X = x) = k \cdot x$ where k denotes the constant of proportionality. The probability distribution function is then

$$P(X = x) = \begin{cases} k & \text{if } x = 1 \\ 2k & \text{if } x = 2 \\ 3k & \text{if } x = 3 \\ 4k & \text{if } x = 4 \\ 5k & \text{if } x = 5 \\ 6k & \text{if } x = 6 \end{cases}.$$

The sum of these probabilities must be 1, so

$$k + 2k + 3k + 4k + 5k + 6k = 1,$$

hence $k = \frac{1}{21}$ and the weighting is possible.

The probability distribution function is

$$P(X = x) = \begin{cases} \dfrac{x}{21} & x = 1, 2, 3, 4, 5, 6 \\ 0, & \text{otherwise.} \end{cases}$$

A procedure for selecting a random sample from this distribution is explained in Appendix 1.   ◀

Example 2.1.3

Now we return to the experiment consisting of throwing two fair dice. We want to investigate the probabilities of the various sums that can occur. Let the random variable X denote the sum that appears. Then, for example,

$$P(X = 5) = P[(1, 4) \text{ or } (2, 3) \text{ or } (3, 2) \text{ or } (4, 1)]$$

$$= \frac{4}{36} = \frac{1}{9}.$$

So we have determined the probability of one sum. Others can be determined in a similar way.

The experiment could be described by giving all the values for the probability distribution function (or pdf), $P(X = x)$, where, as before, x denotes a *value* for the random variable X, as we saw in Example 2.1.2. In this example, it is easy to find that

$$P(X = x) = \begin{cases} \frac{1}{36} & \text{if } x = 2 \text{ or } 12 \\ \frac{2}{36} & \text{if } x = 3 \text{ or } 11 \\ \frac{3}{36} & \text{if } x = 4 \text{ or } 10 \\ \frac{4}{36} & \text{if } x = 5 \text{ or } 9 \\ \frac{5}{36} & \text{if } x = 6 \text{ or } 8 \\ \frac{6}{36} & \text{if } x = 7 \\ 0, & \text{otherwise.} \end{cases}$$

We see that

$$P(X = x) = P(X = 14 - x) = \frac{x - 1}{36} \text{ for } x = 2, 3, 4, 5, 6, 7$$
$$\text{and}$$
$$P(X = x) = 0, \text{ otherwise.}$$

A graph of this function shows a tent-like shape, as in Figure 2.2.   ◀

The sums when two dice are thrown then behave quite differently from the behavior of the individual dice. In fact, we note that if the random variable X_1 denotes the result showing on the first die, and the random variable X_2 denotes the result showing on the second die, then $X = X_1 + X_2$. The random variable X can be expressed as a *sum* of random variables. While X_1 and X_2 are uniform, X most decidedly is *not* uniform.

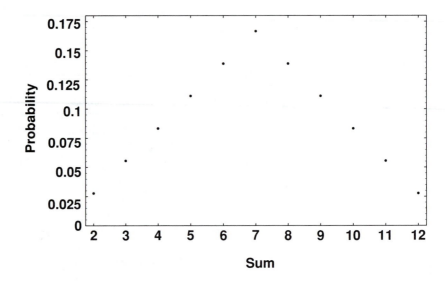

Figure 2.2   Sums on two fair dice.

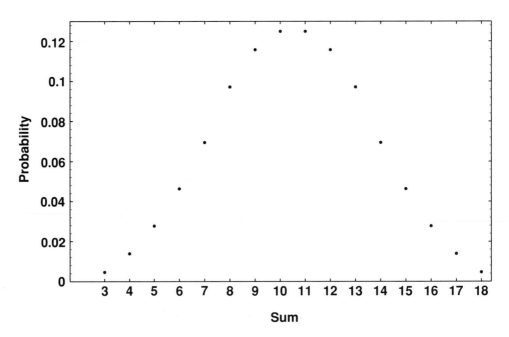

Figure 2.3   Sums on three fair dice.

There is a theoretical reason for this behavior which we will discuss in a later chapter. It is sufficient to note here that this is, in fact, not unusual, but very typical behavior for a sum of random variables.

A natural inquiry at this point is, "What is the probability distribution of the sum on three fair dice?" It is more difficult to work out the distribution here than it was for two dice. While we will show another solution later, we give one approach to the problem at this time. Consider, for example, a sum of 10 on three dice. The sum could have arisen from these *combinations* of results showing on the individual dice (which do not indicate which die showed which face):

$$(2, 2, 6), (3, 3, 4), (2, 4, 4),$$
$$(3, 1, 6), (3, 2, 5), (5, 1, 4).$$

Each of the first three of these combinations could occur in 3 different orders (corresponding to the three different dice), while each of the last three could occur in 6 different orders. This gives a total of 27 possibilities, each of which has probability $\frac{1}{216}$. Therefore $P(X = 10) = \frac{27}{216}$. A similar process could be followed for other values of the sum; the complete probability distribution can be found to be

$$P(X = x) = \begin{cases} \frac{1}{216} & \text{if } x = 3 \text{ or } 18 \\ \frac{3}{216} & \text{if } x = 4 \text{ or } 17 \\ \frac{6}{216} & \text{if } x = 5 \text{ or } 16 \\ \frac{10}{216} & \text{if } x = 6 \text{ or } 15 \\ \frac{15}{216} & \text{if } x = 7 \text{ or } 14 \\ \frac{21}{216} & \text{if } x = 8 \text{ or } 13 \\ \frac{25}{216} & \text{if } x = 9 \text{ or } 12 \\ \frac{27}{216} & \text{if } x = 10 \text{ or } 11 \\ 0, & \text{otherwise.} \end{cases}$$

A computer algebra system may also be used to find the probability distribution for X. Many systems will give all the permutations, each of which may be summed and the relative frequencies recorded. This is shown in Appendix 1. There are other methods that can be used to solve the problem; one of these will be discussed in Chapter 4.

A graph of this function is shown in Figure 2.3. It begins to show what we will call a *normal* probability distribution shape. As the number of dice increases, the "curve" the eye sees smooths out to resemble a normal probability distribution; the distribution for six or more dice is remarkably close to the normal distribution. We will discuss the normal distribution in Chapter 3.

Example 2.1.4

We saw in Example 2.1.2 that a single die could be loaded so that the probability of the occurrence of a face is proportional to the face. Can we load a die so that when the die is thrown twice the probability of a sum is proportional to the sum?

If $P(X = i)$ is denoted by P_i, for $i = 1, 2, 3, 4, 5, 6$, and if k is the constant of proportionality, then $P_1^2 = 2k$, $2P_1 P_2 = 3k$, $2P_1 P_3 + P_2^2 = 4k$, and so on, together with the restriction that $\sum_{i=1}^{6} P_i = 1$, giving a system of 12 equations in 7 unknowns. Unfortunately this set of equations has no solution, so we can't load the die in the manner suggested.   ◄

Example 2.1.5

Let's look now at the sum when two loaded dice are thrown. First let each die be loaded so that the probability a face occurs is proportional to that face, as in Example 2.1.2. The sample space of 36 points can be used to determine the probabilities of the various sums. Figure 2.4 shows these probabilities. We see that the symmetry we noticed in Figures 2.1 and 2.3 is now gone.

Now suppose one die is loaded so that the probability a face appears is proportional to that face, while a second die is loaded so that the probability face i appears is proportional to $7 - i$, $i = 1, 2, \ldots, 6$. The probabilities of various sums are shown in Figure 2.5. Now symmetry around $x = 7$ has returned.

The appearance, once more, of the normal-like shape is striking. Readers with access to a computer algebra system may want to find the probability distribution of the sums on four dice, two loaded in each manner as in this example. The result is remarkably normal.   ◄

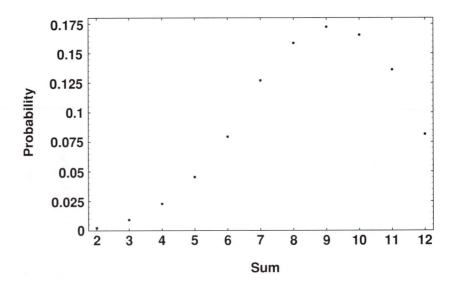

Figure 2.4   Sums on two loaded dice.

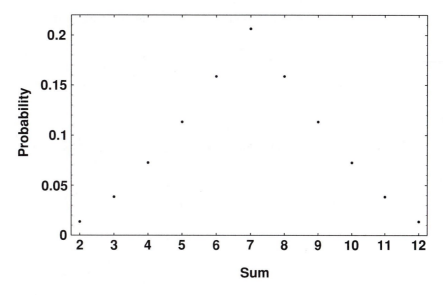

Figure 2.5   Sums on two loaded dice.

Example 2.1.6

Sample spaces in the examples in this chapter so far have been finite. Our final example involves a countably infinite sample space. Consider observing single births until a girl is born. Let the random variable X denote the number of births necessary. Assuming the births to be independent,

$$P(X = x) = \left(\frac{1}{2}\right)^x, \quad x = 1, 2, 3, \ldots.$$

To check that $P(X = x)$ is a probability distribution, note that $P(X = x) \geq 0$ for all x. The sum of all the probabilities is

$$S = \sum_{x=1}^{\infty} P(X = x) = \left(\frac{1}{2}\right) + \left(\frac{1}{2}\right)^2 + \left(\frac{1}{2}\right)^3 + \cdots.$$

To calculate this sum, note that

$$\left(\frac{1}{2}\right) S = \left(\frac{1}{2}\right)^2 + \left(\frac{1}{2}\right)^3 + \left(\frac{1}{2}\right)^4 \cdots.$$

Subtracting the second series from the first gives

$$\left(\frac{1}{2}\right) S = \left(\frac{1}{2}\right)$$

so

$$S = 1.$$

Another way to sum the series is to recognize that it is an infinite geometric series of the form

$$S = a + ar + ar^2 + \cdots$$

and the sum of this series is known to be

$$S = \frac{a}{1 - r}, \text{ if } |r| < 1.$$

In this case a is $\frac{1}{2}$ and r is also $\frac{1}{2}$, so the sum is 1.

Here X is called a *geometric* random variable. A graph of $P(X = x)$ appears in Figure 2.6.

Since $P(X = x + 1) = (\frac{1}{2}) P(X = x)$, the probabilities decline rapidly in size. ◄

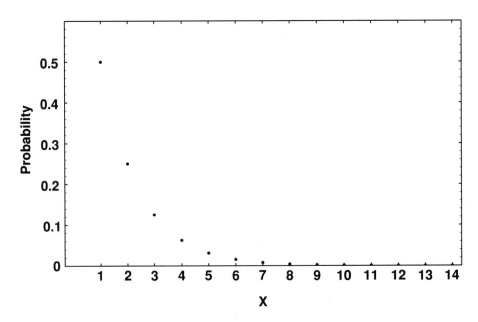

Figure 2.6   Geometric distribution.

2.2 ▶ Distribution Functions

Another function often useful in probability problems is called the *distribution function*. For a discrete random variable we denote this function by $F(x)$ where

$$F(x) = P(X \leq x),$$

so

$$F(x) = \sum_{t \leq x} f(t).$$

$F(x)$ is also known as a *cumulative distribution function (abbreviated cdf)* because it accumulates probabilities. Note the distinction now between $f(x)$, the *probability distribution function (pdf)*, and $F(x)$, the *cumulative distribution function (cdf)*.

In Chapter 1 we used the *reliability* of a component where $R(t) = P(T > t)$, so

$$R(t) = 1 - F(t),$$

establishing a relationship between $R(t)$ and the distribution function.

Example 2.2.1

For the fair die whose probability distribution function is given in Example 2.1.1, we find

$$F(1) = \frac{1}{6}, F(2) = \frac{2}{6}, F(3) = \frac{3}{6}, F(4) = \frac{4}{6}, F(5) = \frac{5}{6}, F(6) = 1.$$

It is also customary to show this function for *any* value of the random variable X. Here, for example, $F(3.4) = P(X \leq 3.4) = \frac{3}{6}$. Since $F(x)$ is defined for any value of X, we draw a continuous graph, unlike the graph of the probability distribution function. We see that in this case,

$$F(x) = \begin{cases} 0, & \text{if } x < 1 \\ \frac{1}{6}, & \text{if } 1 \leq x < 2 \\ \frac{2}{6}, & \text{if } 2 \leq x < 3 \\ \frac{3}{6}, & \text{if } 3 \leq x < 4 \\ \frac{4}{6}, & \text{if } 4 \leq x < 5 \\ \frac{5}{6}, & \text{if } 5 \leq x < 6 \\ 1, & \text{if } 6 \leq x \end{cases}.$$

A graph of this function is shown in Figure 2.7. It is a series of step functions, since, when $f(x)$ is scanned from the right, $F(x)$ can increase only at those points where $f(x)$ is not zero.

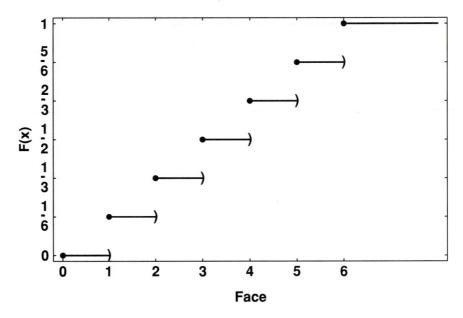

Figure 2.7   Distribution function for one toss of a fair die.

It is clear from the definition of $F(x)$ and from the fact that probabilities are in the interval [0,1] that

$$0 \le F(x) \le 1$$

and that

$$F(a) \ge F(b) \text{ if } a \ge b.$$

It is also true, for *discrete* random variables taking integer values, that

$$P(a \le X \le b) = P(X \le b) - P(X < a) = F(b) - F(a-1).$$

Individual probabilities, say $P(X = a)$, can be found by

$$P(X = a) = P(X \le a) - P(X \le a - 1) = F(a) - F(a-1).$$

These probabilities are the size of the "steps" in the distribution function.   ◄

Exercises 2.2

1. A fair coin is tossed 4 times.

    a. Show a sample space for the experiment and assign probabilities to the sample points.

b. Suppose a count of the total number of heads (X) and the total number of tails (Y) is made after each toss. What is the probability that X always exceeds Y?

c. What is the probability, after 4 tosses, that X is even if we know that $Y \geq 1$?

2. A single expensive electronic part is to be manufactured, but the manufacture of a successful part is not guaranteed. The first attempt costs \$100 and has a 0.7 probability of success. Each attempt thereafter costs \$60 and has a 0.9 probability of success. The outcomes of various attempts are independent, but at most 3 attempts can be made at successful manufacture. The finished part sells for \$500. Find the probability distribution for N, the net profit.

3. An automobile dealer has found that X, the number of cars customers buy each week, follows the probability distribution

$$f(x) = \begin{cases} \dfrac{kx^2}{x!}, & x = 1, 2, 3, 4 \\ 0, & \text{otherwise.} \end{cases}$$

a. Find k.

b. Find the probability the dealer sells at least 2 cars in a week.

c. Find $F(x)$, the cumulative distribution function.

4. Job interviews last one-half hour. The interviewer knows that the probability an applicant is qualified for the job is 0.8. The first person interviewed who is qualified is selected for the job. If the qualifications of any one applicant are independent of the qualifications of any other applicant, what is the probability that 2 hours is sufficient time to select a person for the job?

5. Verify the probability distribution for the sum on 3 fair dice as given in Example 2.1.3.

6. **a.** Since $\left(\frac{1}{2} + \frac{1}{2}\right)^5 = 1$ and since each term in the binomial expansion of $\left(\frac{1}{2} + \frac{1}{2}\right)^5$ is greater than 0, it follows that the individual terms in the binomial expansion are probabilities. Suggest an experiment and a sample space for which these terms represent probabilities of the sample points.

b. Answer part **a** for $(p + q)^n$, $q = 1 - p$, $0 \leq p \leq 1$.

7. Two loaded dice are tossed. Each die is loaded so the probability that a face, i, appears, is proportional to $7 - i$. Find the probability distribution for the sum that appears. Draw a graph of the probability distribution function.

8. Suppose that X is a random variable giving the number of tosses necessary for a fair coin to turn up heads. Find the probability that X is even.

9. The random variable Y has the probability distribution

$$g(y) = \frac{1}{4} \text{ if } y = 2, 3, 4, \text{ or } 5.$$

Find $G(y)$, the distribution function for Y.

10. Find the distribution function for the geometric distribution $f(x) = \left(\frac{1}{2}\right)^x$, $x = 1, 2, 3 \ldots$.

11. A random variable, X, has the distribution function

$$F(x) = \begin{cases} 0, & x < -1 \\ \frac{1}{3}, & -1 \leq x < 0 \\ \frac{5}{6}, & 0 \leq x < 2 \\ 1, & x \geq 2 \end{cases}.$$

Find the probability distribution function, $f(x)$.

12. A random variable X is defined on the integers $0, 1, 2, 3, \ldots$, and has distribution function $F(x)$. Find expressions, in terms of $F(x)$, for

   a. $P(a < X < b)$
   b. $P(a \leq X < b)$
   c. $P(a < X \leq b)$
   d. $P(a \leq X \leq b)$

13. If $f(x) = \frac{1}{n}$, $x = 1, 2, 3, \ldots, n$ (so that each value of X has the same probability) then X is called a *discrete uniform* random variable. Find the distribution function for this random variable.

2.3 ▶ Expected Values of Discrete Random Variables

2.3.1   Expected Value of a Discrete Random Variable

Random variables are easily distinguished by their probability distribution functions. They are also often characterized or described by measures that summarize these distributions. Usually, "average" values, or measures of centrality, and some measure of their dispersion, or variability, are found as values characteristic of the distribution.

    We begin with the definition of an average value for a discrete random variable, X, denoted by $E(X)$, or μ_x, which we will call the *expectation*, or *expected value*, or *mean*, or *mean value* (all of these terms are in common usage) of X.

Definition:

$$E(X) = \mu_x = \sum_x x \cdot P(X = x),$$

provided the sum converges, where the summation occurs over all the discrete values of the random variable, X. Note that each value of the random variable X is weighted by its probability in the sum.

The provision that the sum be convergent cautions us that the sum may, indeed, be infinite. There are random variables, otherwise seemingly well-behaved, that have no mean value.

This definition is, in reality, a simple extension of what the reader would recognize as an average value. Consider an example.

Example 2.3.1.1

A student has examination grades of 82, 91, 79, and 96 in a course in probability. We would no doubt calculate the average grade as

$$\frac{82 + 91 + 79 + 96}{4} = 87.$$

This could also be calculated as

$$82 \cdot \frac{1}{4} + 91 \cdot \frac{1}{4} + 79 \cdot \frac{1}{4} + 96 \cdot \frac{1}{4} = 87,$$

where the examination scores have now been equally weighted. Should the instructor decide to weight the fourth examination three times as much as any one of the other examinations, this simply changes the weights and the average examination grade is then

$$82 \cdot \frac{1}{6} + 91 \cdot \frac{1}{6} + 79 \cdot \frac{1}{6} + 96 \cdot \frac{3}{6} = 90.$$

So the idea of adding scores multiplied by their probabilities is not a new one. This is exactly what we do when we calculate $E(X)$.   ◄

Example 2.3.1.2

If a fair die is thrown once, as in Example 2.1.1, the average result is

$$\mu_x = 1 \cdot \frac{1}{6} + 2 \cdot \frac{1}{6} + 3 \cdot \frac{1}{6} + 4 \cdot \frac{1}{6} + 5 \cdot \frac{1}{6} + 6 \cdot \frac{1}{6} = \frac{7}{2}.$$

So we recognize $\frac{7}{2}$, or 3.5, as the average result, although 3.5 is not a possible value for the face showing on the die. What is the meaning of this? The interpretation is as follows: If we threw a fair die a large number of times, we would expect each of the faces from 1 to 6 to occur about $\frac{1}{6}$ of the time, so the average result would be given by μ_x. We could, of course, expect some deviation from this result in actual practice, the size of the deviation decreasing as the number of tosses of the die increases. Later we will see that a deviation of more than about 0.11 in the average is highly unlikely in 1000 tosses of the die; that is, the average is almost certain to fall in the interval from 3.39 to 3.61. If the deviation is more than 0.11 we would no doubt conclude that the die is an unfair one.   ◀

Example 2.3.1.3

What is the average result on the loaded die where $P(X = i) = \frac{i}{21}$, for $i = 1, 2, 3, 4, 5, 6$? Here

$$E(X) = 1 \cdot \frac{1}{21} + 2 \cdot \frac{2}{21} + 3 \cdot \frac{3}{21} + 4 \cdot \frac{4}{21} + 5 \cdot \frac{5}{21} + 6 \cdot \frac{6}{21} = \frac{13}{3}.$$ ◀

Example 2.3.1.4

In Example 2.1.3 we determined the probability distribution for X, the sum showing on two fair dice. Next we find

$$E(X) = 2 \cdot \frac{1}{36} + 3 \cdot \frac{2}{36} + 4 \cdot \frac{3}{36} + \cdots + 12 \cdot \frac{1}{36} = 7.$$

Now let X_1 denote the face showing on the first die and let X_2 denote the face showing on the second die. We found in Example 2.3.1.2 that $E(X_i) = \frac{7}{2}$, for $i = 1, 2$. We note here that

$$E(X) = E(X_1) + E(X_2),$$

so that the expectation of the sum is the sum of the expectations of the sum's components; this is, in fact, generally true and so is no coincidence. We will discuss this further in Chapter 5.   ◀

Example 2.3.1.5

Sometimes the calculation of an expected value will involve an infinite series. Suppose we toss a coin, loaded to come up heads with probability p, until a head occurs. Since the tosses are independent, and since the event, "First head on toss x," is equivalent to $x - 1$ tails followed by a head, it follows that

$$P(X = x) = q^{x-1}p, \; x = 1, 2, 3, \ldots, \text{ where } q = 1 - p.$$

We check first that $\sum_{x} P(X = x) = 1$. Here

$$\sum_{x} P(X = x) = p + q \cdot p + q^2 \cdot p + q^3 \cdot p + \cdots$$
$$= p \cdot (1 + q + q^2 + q^3 + \cdots)$$
$$= p \cdot \frac{1}{1 - q} = 1.$$

Then

$$E(X) = \sum_{x=1}^{\infty} x \cdot q^{x-1} p = p + 2 \cdot q \cdot p + 3 \cdot q^2 \cdot p + 4 \cdot q^3 \cdot p + \cdots.$$

To simplify this, notice that

$$q \cdot E(X) = q \cdot p + 2 \cdot q^2 \cdot p + 3 \cdot q^3 \cdot p + 4 \cdot q^4 \cdot p + \cdots.$$

By subtracting $q \cdot E(X)$ from $E(X)$ we find that

$$E(X) - q \cdot E(X) = p + q \cdot p + q^2 \cdot p + q^3 \cdot p + q^4 \cdot p + \cdots$$

where the right side is $\sum_{x} P(X = x) = 1$. So

$$(1 - q) \cdot E(X) = 1$$

hence

$$E(X) = \frac{1}{p}.$$

The reader is cautioned that the "trick" for summing the series is valid only because the series is absolutely convergent. $E(X)$ could also be found by integrating, with respect to q, the series for $E(X)$ term by term.

With a fair coin, then, since $p = \frac{1}{2}$, an average of two tosses is necessary to find the first occurrence of a head. Since $P(X = x)$ involves a geometric series, X here, as in Example 2.1.6, is often called a *geometric* random variable.

Mean values generally show a central value for the random variable. Now we turn to a discussion of the dispersion, or variability, of the random variable. ◀

2.3.2 Variance of a Random Variable

Figure 2.8 shows two random variables with the same mean value, $\mu = 3$. If we didn't know μ and wanted to estimate μ by selecting an observation from one of these

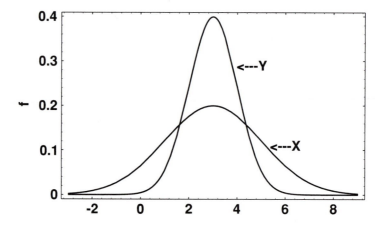

Figure 2.8   Two random variables with mean value 3.

probability distributions, we would no doubt choose Y since the values of Y are less disperse and generally closer to μ than those for X.

There are many ways to measure the fact that Y is less disperse than X. We could look at the *range* (the largest possible value minus the smallest possible value). Another possibility is to calculate the deviation of each value of X from μ and then calculate the average value of these deviations from the mean, $E(X - \mu)$. This, however, is 0 for any random variable and hence carries absolutely no information whatsoever regarding X. Here is a demonstration that this is so.

$$E(X - \mu) = \sum_x (x - \mu) \cdot P(X = x)$$
$$= \sum_x x \cdot P(X = x) - \mu \cdot \sum_x P(X = x)$$
$$= \mu - \mu = 0.$$

So the positive deviations from the mean exactly compensate for the negative deviations.

One way to avoid this is to consider the *mean deviation*, $E \mid X - \mu \mid$, but this is not commonly done. Yet another way to prevent the positive deviations from compensating for the negative deviations is to square each value of $X - \mu$ and then sum the result. This is the usual solution; we call the result the *variance*, denoted by σ^2, which we define as

Definition: $$\sigma^2 = Var(X) = E(X - \mu)^2$$

so   (2.1)

$$\sigma^2 = \sum_x (x - \mu)^2 \cdot P(X = x),$$

provided the sum converges, and where the summation is over all the possible values of X.

The quantity σ^2 is then a weighted average of the squared deviations of the values of X from its mean value. The variance may appear to be much more complex than the range or mean deviation. This is true, but the variance also has remarkable properties that we cannot describe now and that do not hold for the range or for the mean deviation. We will consider some of the properties of the variance subsequently.

Example 2.3.2.1

Consider the random variable X with probability distribution function

$$f(x) = \begin{cases} \frac{1}{2} & \text{if } x = 1 \\ \frac{1}{3} & \text{if } x = 2 \\ \frac{1}{6} & \text{if } x = 3. \end{cases}$$

Here

$$E(X) = \mu = 1 \cdot \frac{1}{2} + 2 \cdot \frac{1}{3} + 3 \cdot \frac{1}{6} = \frac{5}{3},$$

so

$$E(X - \mu)^2 = \sigma^2 = \left(1 - \frac{5}{3}\right)^2 \cdot \frac{1}{2} + \left(2 - \frac{5}{3}\right)^2 \cdot \frac{1}{3} + \left(3 - \frac{5}{3}\right)^2 \cdot \frac{1}{6} = \frac{5}{9}. \quad \blacktriangleleft$$

Before turning to some more examples, we show another formula for σ^2. This formula is often very useful.

Expand Formula 2.1 as follows:

$$\sigma^2 = \sum_x (x - \mu)^2 \cdot P(X = x)$$

$$= \sum_x (x - \mu)^2 \cdot f(x)$$

$$= \sum_x (x^2 - 2\mu x + \mu^2) \cdot f(x)$$

$$= \sum_x x^2 \cdot f(x) - 2\mu \sum_x x \cdot f(x) + \mu^2$$

because $\sum_x f(x) = 1$. Now $\sum_x x \cdot f(x) = \mu$, so

$$\sigma^2 = \sum_x x^2 \cdot f(x) - \mu^2$$

so (2.2)

$$\sigma^2 = E(X^2) - \mu^2 = E(X^2) - [E(X)]^2.$$

Formula 2.2 is often easier to use for computational purposes than formula 2.1. σ is called the *standard deviation of X*.

Example 2.3.2.2

Refer again to throwing a single die, as in Examples 2.1.1 and 2.3.1.2. We calculate

$$E(X^2) = 1^2 \cdot \frac{1}{6} + 2^2 \cdot \frac{1}{6} + 3^2 \cdot \frac{1}{6} + 4^2 \cdot \frac{1}{6} + 5^2 \cdot \frac{1}{6} + 6^2 \cdot \frac{1}{6} = \frac{91}{6}$$

so that

$$\sigma^2 = \frac{91}{6} - (\frac{7}{2})^2 = \frac{35}{12}.$$ ◀

Example 2.3.2.3

What is the variance of the geometric random variable whose probability distribution function is $P(X = x) = q^{x-1} \cdot p$, $x = 1, 2, 3, \ldots$?

Starting with $\sigma^2 = E(X^2) - \mu^2$, since we know that $\mu = \frac{1}{p}$, we only need to compute $E(X^2)$.

$$E(X^2) = \sum_{x=1}^{\infty} x^2 q^{x-1} p = p(1^2 + 2^2 q + 3^2 q^2 + \cdots),$$

from which no easily seen pattern emerges.

Another thought is to consider $E[X(X - 1)]$. If we write

$$E[X(X - 1)] = \sum_{x=1}^{\infty} (x^2 - x) \cdot P(X = x)$$

we see that

$$E[X(X - 1)] = \sum_{x=1}^{\infty} x^2 \cdot P(X = x) - \sum_{x=1}^{\infty} x \cdot P(X = x)$$

or

$$E(X^2 - X) = E(X^2) - E(X).$$

So if we know $E[X(X - 1)]$ we can find $E(X^2)$ and hence calculate σ^2. In this example, a trick will help as it did in determining $E(X)$.

$$E[X(X - 1)] = 1 \cdot 0 \cdot p + 2 \cdot 1 \cdot q \cdot p + 3 \cdot 2 \cdot q^2 \cdot p + 4 \cdot 3 \cdot q^3 \cdot p + \cdots$$

so, multiplying through by q, we have

$$q \cdot E[X(X-1)] = 2 \cdot 1 \cdot q^2 \cdot p + 3 \cdot 2 \cdot q^3 \cdot p + 4 \cdot 3 \cdot q^4 \cdot p + \cdots.$$

Subtract the second series from the first and, because $p = 1 - q$, it follows that

$$p \cdot E[X(X-1)] = 2 \cdot q \cdot p + 4 \cdot q^2 \cdot p + 6 \cdot q^3 \cdot p + \cdots$$
$$= 2q(1p + 2qp + 3q^2p + \cdots)$$

so

$$p \cdot E[X(X-1)] = 2q \cdot E(X) = \frac{2q}{p}.$$

So

$$E[X(X-1)] = \frac{2q}{p^2},$$
$$\text{and}$$
$$E(X^2) = \frac{2q}{p^2} + \frac{1}{p},$$
$$\text{giving}$$
$$\sigma^2 = \frac{2q}{p^2} + \frac{1}{p} - \frac{1}{p^2} = \frac{q}{p^2}. \qquad \blacktriangleleft$$

The value of the variance is quite difficult to interpret at this point, but, as we proceed, we will find more and more uses for the variance. Patience is requested of the reader now, with the promise that these calculations are in fact useful and meaningful. We pause to consider the question, "Does σ measure variability?" We can show a general result, albeit a very crude one, in the following inequality.

2.3.3 Tchebycheff's Inequality

THEOREM: Suppose the random variable X has mean μ and standard deviation σ. Choose a positive quantity, k. Then

$$P(|X - \mu| \le k \cdot \sigma) \ge 1 - \frac{1}{k^2}. \qquad \blacksquare$$

Tchebycheff's Inequality gives a lower bound on the probability that a value of X is within $k \cdot \sigma$ units of the mean, μ.

Before offering a proof, we consider some special cases. If $k = 2$, the inequality is

$$P(|X - \mu| \le 2 \cdot \sigma) \ge 1 - \frac{1}{2^2} = \frac{3}{4},$$

so $\frac{3}{4}$ of *any* probability distribution lies within two standard deviations – that is, 2σ – units of the mean while, if $k = 3$, the inequality states that

$$P(|X - \mu| \le 3 \cdot \sigma) \ge 1 - \frac{1}{3^2} = \frac{8}{9},$$

showing that $\frac{8}{9}$ of *any* probability distribution lies within 3σ units of the mean. We will see later that if the specific distribution is known, these inequalities can be sharpened considerably. Now we show a proof.

PROOF: Let $P(X = x) = f(x)$. Consider two sets of points,

$$A = \{x \mid |x - \mu| \ge k \cdot \sigma\}$$

and

$$B = \{x \mid |x - \mu| < k \cdot \sigma\}.$$

We could then write the variance as

$$\sigma^2 = \sum_{x \in A} (x - \mu)^2 \cdot f(x) + \sum_{x \in B} (x - \mu)^2 \cdot f(x).$$

Now for every point x in A, replace $|x - \mu|$ by $k \cdot \sigma$, and in B, replace $|x - \mu|$ by 0. The crudity of the result is now evident! So

$$\sigma^2 \ge \sum_{x \in A} (k \cdot \sigma)^2 f(x) + \sum_{x \in B} 0^2 \cdot f(x).$$

Since

$$\sum_{x \in A} f(x) = P(A)$$

$$= P(|X - \mu| \ge k \cdot \sigma), \sigma^2 \ge k^2 \cdot \sigma^2 \cdot P(|X - \mu| \ge k \cdot \sigma),$$

from which we conclude that

$$P(|X - \mu| \ge k \cdot \sigma) \le \frac{1}{k^2}$$

or

$$P(|X - \mu| \le k \cdot \sigma) \ge 1 - \frac{1}{k^2}. \qquad \blacksquare$$

While the theorem is far from precise, it does verify that as we move farther away from the mean, in terms of standard deviations, the more of the probability distribution we cover; hence σ is indeed a measure of variability.

Exercises 2.3

1. A small manufacturing firm sells 1 machine per month with probability 0.3; it sells 2 machines per month with probability 0.1; it never sells more than 2 machines per month. If X represents the number of machines sold per month,

   a. find the mean and variance of X.
   b. If the monthly profit is $2X^2 + 3X + 1$ (in thousands of dollars), find the expected monthly profit.

2. Bolts are packaged in boxes so that the mean number of bolts per box is 100 with standard deviation 3. Use Tchebycheff's Inequality to find a bound on the probability that the box has between 95 and 105 bolts.

3. Graduates of a distinguished undergraduate mathematics program received graduate school fellowships as follows: 20% received $10,000, 10% received $12,000, 30% received $14,000, 30% received $13,000, 5% received $15,000, and 5% received $17,000.
   Find the mean and the variance of the value of a graduate fellowship.

4. A fair coin is tossed 4 times; let X denote the number of heads that occur. Find the mean and variance of X.

5. A batch of 15 electric motors contains 3 defective ones. An inspector chooses 3 (without replacement). Find the mean and variance of X, the number of defective motors in the sample.

6. A coin, loaded to show heads with probability $\frac{2}{3}$, is tossed until a head appears or until 5 tosses have been made. Let X denote the number of tosses made. Find the mean and variance of X.

7. Suppose X is a discrete uniform random variable so that $f(x) = \frac{1}{n}$, $x = 1, 2, 3, \ldots, n$. Find the mean and variance of X.

8. In exercise 5, suppose the batch of motors is accepted if no more than one defective motor is in the sample. If each motor costs $100 to manufacture, how much should the manufacturer charge for each motor in order to make the expected profit for the batch be $200?

9. A physicist makes several independent measurements of the specific gravity of a substance. The limitations of his equipment are such that the standard deviation of each measurement is σ units. Suppose μ is the true specific gravity of the

substance. Approximate the probability that a particular one of the measurements is within $5\sigma/4$ units of μ.

10. A manufacturer ships parts in lots of 1000 and makes a profit of $50 per lot sold. The purchaser, however, subjects the product to a sampling inspection plan as follows: 10 parts are selected at random. If none of these parts is defective, the lot is purchased; if one part is defective, the manufacturer returns $10 to the buyer; if 2 or more parts are found to be defective, the entire lot is returned at a net loss of $25 to the manufacturer. What is the manufacturer's expected profit if 10% of the parts are defective? (Assume that the sampling is done with replacement.)

11. In a lot of 6 batteries, one is worn out. A technician tests the batteries one at a time until the worn-out battery is found. Tested batteries are put aside, but after every third test the tester takes a break and another worker, unaware of the test, returns one of the tested batteries to the set of batteries not yet tested.

a. Find the probability distribution for X, the number of tests required to identify the worn-out battery.

b. Assume the first test of each set of 3 tests costs $5 and that each of the next 2 tests in each set of three tests costs $2. Find the increase in the expected cost of locating the worn-out battery due to the unaware worker.

12. A carnival game consists of hitting a lever with a sledge hammer to propel a weight upward toward a bell. Because the hammer is quite heavy, the chance of ringing the bell declines with the number of attempts; in particular, the probability of ringing the bell on the ith attempt is $\left(\frac{3}{4}\right)^i$. For a fee, the carnival sells you the privilege of swinging the hammer until the bell rings or until you have made 3 attempts, whichever occurs first.

a. Find the probability distribution of X, the number of hits taken.

b. The prize for ringing the bell on the ith try is $ $(4 - i), i = 1, 2, 3$. How much should the carnival charge for playing the game if it wants an expected profit of $1 per customer?

13. Suppose X is a random variable defined on the points $x = 0, 1, 2, 3,\ldots$ Calculate

$$\sum_{x=0}^{\infty} P(X > x).$$

There are many very important specific discrete probability distribution functions that arise in practical applications. Having established some general properties, we now turn to discussions of several of the most important of these distributions.

Occasionally random variables in apparently different situations actually arise from common assumptions and hence lead to the same probability distribution function. We now investigate some of these special circumstances and the probability distribution functions that result.

2.4 ▶ Binomial Distribution

Among all discrete probability distribution functions, the most commonly occurring one, arising in a great variety of applications, is called the *binomial* probability distribution function.

Consider an experiment where, on each trial of the experiment, one of only two outcomes occurs, which we describe as *success*, (S) or *failure*, (F). For example, a manufactured part is either good or does not meet specifications; a student's examination score is passing or it is not; a team wins a basketball game or it does not – these are some examples, and the reader can no doubt think of many more. One of these outcomes can be associated with success and the other with failure; it does not matter which is which.

In addition to the restriction that there be two and only two outcomes on each trial of the experiment, suppose further that the trials are independent, and that the probabilities of success or failure at each trial remain constant from trial to trial and do not change with subsequent performances of the experiment.

The individual trials of such an experiment are often called *Bernoulli trials*.

Consider, as a specific example, five independent trials with probability $\frac{2}{3}$ of success at any trial. Then, if interest centers on the occurrence of exactly three successes, we note that exactly three successes can occur in ten different ways:

$$SSSFF, SSFSF, SFSSF, FSSSF, SFSFS,$$
$$SSFFS, FSSFS, SFFSS, FSFSS, FFSSS.$$

There are $\binom{5}{3} = 10$ of these mutually exclusive orders. Each has probability $\left(\frac{2}{3}\right)^3 \cdot \left(\frac{1}{3}\right)^2$, so

$$P \text{ (exactly 3 } S's \text{ in 5 trials)} = \binom{5}{3} \cdot \left(\frac{2}{3}\right)^3 \cdot \left(\frac{1}{3}\right)^2 = \frac{80}{243}.$$

Now return to the general situation. Let the probabilities be $P(S) = p$ and $P(F) = q = 1 - p$, and let the random variable X denote the number of successes in n trials of the experiment. Any specific sequence of exactly x successes and $n - x$ failures has probability $p^x \cdot q^{n-x}$. The successes in such a sequence can occur at $\binom{n}{x}$ positions so, since the sequences are mutually exclusive,

$$P(X = x) = \binom{n}{x} p^x \cdot q^{n-x}, \quad x = 0, 1, 2, \ldots, n, \tag{2.3}$$

giving the probability distribution function for a binomial random variable.

Although the binomial random variable occurs in many different situations, a perfect model for any binomial situation is that of observing the number of heads when a coin loaded so that the probability of a head is p and that of a tail is $q = 1 - p$ is tossed n times.

Now does Formula 2.3 define a probability distribution? Since $P(X = x) \geq 0$ and

$$\sum_{x=0}^{n} P(X = x) = \sum_{x=0}^{n} \binom{n}{x} p^x \cdot q^{n-x} = (q + p)^n = 1$$

by the binomial theorem, we conclude that Formula 2.3 defines a probability distribution.

It is interesting to note that individual terms in the binomial expansion of $(q + p)^n$, if $p + q = 1$, represent binomial probabilities.

Example 2.4.1

A student has no knowledge whatsoever of the material to be tested on a true-false examination, and so the student flips a fair coin in order to determine the response to each question. What is the probability that the student scores at least 60% on a ten-item examination?

Here the binomial variable, X, the number of correct responses, has $n = 10$, and $p = q = \frac{1}{2}$. We need

$$P(X \geq 6) = \sum_{x=6}^{10} \binom{10}{x} \left(\frac{1}{2}\right)^x \left(\frac{1}{2}\right)^{10-x}$$

Now we find that $P(X \geq 6) = \frac{193}{512} = 0.376953$.

These calculations can easily be done with a pocket computer. If we want to investigate the probability that at least 60% of the questions are answered correctly as the number of items on the examination increases, then use of a computer algebra system is recommended for aiding in the calculation. Many computer algebra systems contain the binomial probability distribution as a defined probability distribution; for other systems, the probability distribution function may be entered directly. The following results can be found where n is the number of trials and P is the probability of at least 60% correct:

n	10	40	80	100
P	0.376953	0.134094	0.0464559	0.028444

Clearly, guessing is not a sensible strategy on a test with a large number of items. ◄

Example 2.4.2

Graphs of $P(X = x) = \binom{n}{x} p^x \cdot q^{n-x}$ for $x = 0, 1, 2, \ldots, n$ are interesting. We show in Figure 2.9 the graphs of $P(X = x)$ for $n = 10$ and also for

$n = 100$ with $p = \frac{1}{2}$ in each case. We see that each curve is bell-shaped or normal-like, and the distributions are symmetric about $x = 5$ and $x = 50$, respectively.

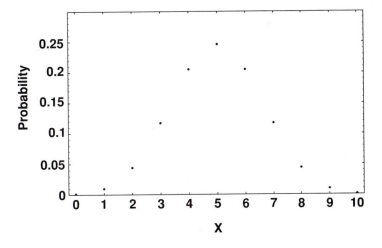

Binomial distribution, $n = 10$, $p = 1/2$.

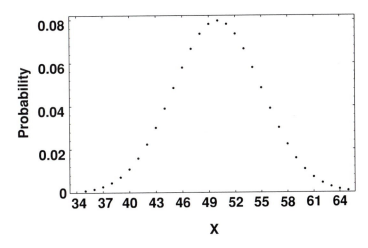

Figure 2.9   Binomial distribution, $n = 100$, $p = 1/2$.

Again we find the bell-shaped or normal appearance here, but the reader may wonder if the appearance is still normal for $p \neq \frac{1}{2}$. Figure 2.10 shows a graph of $P(X = x)$ for $n = 50$ and $p = \frac{3}{4}$. This curve indicates that the bell shape survives even though $p \neq \frac{1}{2}$. The maximum point on the curve has shifted to the right, however.

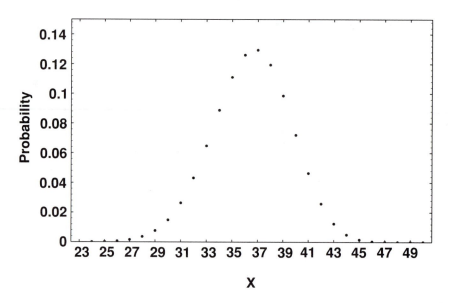

Figure 2.10   Binomial distribution, $n = 50$, $p = \frac{3}{4}$.

We will discuss the reason for the normal appearance of the binomial distribution in the next chapter. Appendix 1 contains a procedure for selecting a sample from a binomial distribution and for simulating an experiment consisting of flipping a loaded coin.   ◄

2.5 ► A Recursion

If a computer algebra system is not available, calculating values of $P(X = x) = \binom{n}{x}p^x \cdot q^{n-x}$ can certainly become difficult, especially for large values of n and small values of p. In any event, $\binom{n}{x}$ becomes large while $p^x \cdot q^{n-x}$ becomes small. By calculating the ratio of successive terms we find an interesting result, which will aid in making these calculations (and which has other interesting consequences as well).

$$\frac{P(X = x)}{P(X = x - 1)} = \frac{\binom{n}{x}p^x \cdot q^{n-x}}{\binom{n}{x-1}p^{x-1} \cdot q^{n-x+1}}, x = 1, 2, \ldots, n$$

This can be simplified to

$$\frac{P(X = x)}{P(X = x - 1)} = \frac{n - x + 1}{x} \cdot \frac{p}{q},$$

so

$$P(X = x) = \frac{n - x + 1}{x} \cdot \frac{p}{q} \cdot P(X = x - 1) , \, x = 1, 2, \ldots, n. \qquad (2.4)$$

Formula 2.4 is another example of a *recursion* since it expresses one value of a function, here $P(X = x)$, in terms of another value of the function, here $P(X = x - 1)$. Given a starting point and the recursion, any value of the function can be computed. In this case, since n failures has probability q^n, $P(X = 0) = q^n$ is a natural starting value. We then find that

$$P(X = 0) = q^n,$$

so

$$P(X = 1) = n \cdot \frac{p}{q} \cdot P(X = 0) = n \cdot \frac{p}{q} \cdot q^n = \binom{n}{1} \cdot p \cdot q^{n-1},$$

so

$$P(X = 2) = \frac{(n - 1)}{2} \cdot \frac{p}{q} \cdot P(X = 1) = \frac{(n - 1)}{2} \cdot \frac{p}{q} \cdot n \cdot p \cdot q^{n-1}$$

$$= \binom{n}{2} \cdot p^2 \cdot q^{n-2}$$

and so on, giving the expected result that $P(X = x) = \binom{n}{x} p^x \cdot q^{n-x}, x = 0, 1, \ldots, n$. So we can recover the probability distribution function from the recursion.

Recursions can be easily programmed, and recursions such as Formula 2.4 are also of some interest for theoretical purposes. For example, consider locating the maximum, or most frequently occurring value, of $P(X = x)$.

If we require that $P(X = x) \geq P(X = x - 1)$ then, from Formula 2.4,

$$\frac{n - x + 1}{x} \cdot \frac{p}{q} \geq 1.$$

This reduces to $x \leq p \cdot (n + 1)$, so we can conclude that the value of X with the maximum probability is $X = \lfloor p \cdot (n + 1) \rfloor$ where $\lfloor x \rfloor$ denotes the largest integer in x.

2.5.1   The Mean and Variance of the Binomial

The recursion 2.4 can be used to determine the mean and variance of a binomial random variable.

Consider first $\mu = \sum_{x=0}^{n} x \cdot P(X = x)$. Recursion 2.4 is

$$P(X = x) = \frac{n - x + 1}{x} \cdot \frac{p}{q} \cdot P(X = x - 1), x = 1, 2, \ldots, n.$$

Multiplying through by x and summing from 1 to n gives

$$\sum_{x=1}^{n} x \cdot P(X = x) = \sum_{x=1}^{n} [n - (x-1)] \cdot \frac{p}{q} \cdot P(X = x - 1),$$

so

$$\mu = \frac{p}{q} \cdot n \cdot [1 - P(X = n)] - \frac{p}{q} \cdot \sum_{x=1}^{n} (x-1) \cdot P(X = x - 1),$$

or

$$\mu = \frac{p}{q} \cdot n \cdot (1 - p^n) - \frac{p}{q} \cdot [\mu - n \cdot P(X = n)],$$

which reduces to

$$\mu = n \cdot p.$$

This result makes a good deal of intuitive sense: if we toss a coin, loaded to come up heads with probability $\frac{3}{4}$, 1000 times, we expect $1000 \cdot \frac{3}{4} = 750$ heads. So in n trials of a binomial experiment with p as the probability of success, we expect $n \cdot p$ successes.

The variance can also be found using Formula 2.4. We first calculate $E(X^2)$.

$$E(X^2) = \sum_{x=1}^{n} x^2 \cdot P(X = x) = \sum_{x=1}^{n} x \cdot [n - (x-1)] \cdot \frac{p}{q} \cdot P(X = x - 1)$$

$$= n \cdot \frac{p}{q} \cdot \sum_{x=1}^{n} [(x-1) + 1] \cdot P(X = x - 1) - \frac{p}{q} \cdot \sum_{x=1}^{n} x \cdot (x-1) \cdot P(X = x - 1)$$

Then, because

$$\sum_{x=1}^{n} (x - 1) \cdot P(X = x - 1) = \mu - n \cdot P(X = n)$$

and since

$$\sum_{x=1}^{n} x \cdot (x-1) \cdot P(X = x - 1) = \sum_{x=1}^{n} [(x-1)^2 + (x-1)] \cdot P(X = x - 1),$$

it follows that

$$E(X^2) = p \cdot (n-1) \cdot (np - np^n) + n \cdot p \cdot (1 - p^n) + n^2 \cdot p^{n+1},$$

and this reduces to $E(X^2) = np^2(n-1) + np$. Therefore,

$$\sigma^2 = E(X^2) - [E(X)]^2 = np^2(n-1) + np - (np)^2 = npq.$$

Example 2.5.1.1

We apply the above results to a binomial experiment in which $p = q = \frac{1}{2}$ and $n = 100$ trials. Here $E(X) = \mu = n \cdot p = 50$ and $\sigma^2 = npq = 25$. Tchebycheff's Inequality with $k = 3$ then gives

$$P[n \cdot p - k \cdot \sqrt{n \cdot p \cdot q} \le X \le n \cdot p + k \cdot \sqrt{n \cdot p \cdot q}] \ge 1 - \frac{1}{k^2}$$

so

$$P[50 - 3 \cdot 5 \le X \le 50 + 3 \cdot 5] \ge \frac{8}{9}$$

or

$$P[35 \le X \le 65] \ge \frac{8}{9}.$$

But we find exactly that

$$\sum_{x=35}^{65} \binom{100}{x} \cdot \left(\frac{1}{2}\right)^{100} = 0.99821,$$

verifying Tchebycheff's Inequality in this case.   ◄

Exercises 2.5

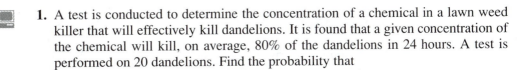

1. A test is conducted to determine the concentration of a chemical in a lawn weed killer that will effectively kill dandelions. It is found that a given concentration of the chemical will kill, on average, 80% of the dandelions in 24 hours. A test is performed on 20 dandelions. Find the probability that

   a. exactly 14 are killed in 24 hours.

   b. at least 10 are killed in 24 hours.

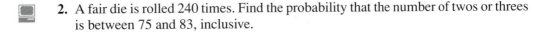

2. A fair die is rolled 240 times. Find the probability that the number of twos or threes is between 75 and 83, inclusive.

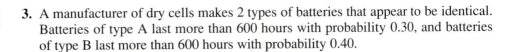

3. A manufacturer of dry cells makes 2 types of batteries that appear to be identical. Batteries of type A last more than 600 hours with probability 0.30, and batteries of type B last more than 600 hours with probability 0.40.

   a. What is the probability that 5 out of 10 of the type A batteries last more than 600 hours?

   b. Of 50 type B batteries, how many are expected to last at least 600 hours?

   c. What is the probability that 3 type A batteries have more batteries lasting 600 hours than 2 type B batteries?

4. X and Y play the following game: X tosses 2 fair coins and Y tosses 3. The player throwing the greater number of heads wins. In case of a tie, the throws are repeated until a winner is determined.

  a. What is the probability that X wins on the first play?
  b. What is the probability that X wins the game?

5. In a political race it is known that 40% of the voters favor candidate C. In a random sample of 100 voters, what is the probability that

  a. between 30 and 45 voters favor C?
  b. exactly 36 voters favor C?

6. A gambling game is played as follows. A player, who pays $4 to play the game, tosses a fair coin 5 times. The player wins as many dollars as heads are tossed.

  a. Find the probability distribution for N, the player's net winnings.
  b. Find the mean and variance of the player's net winnings.

7. A red die is fair, and a green die is loaded so that the probability it comes up 6 is $\frac{1}{10}$.

  a. What is the probability of rolling exactly 3 sixes in 3 rolls with the red die?
  b. What is the probability of at least 30 sixes in 100 rolls of the red die?
  c. The green die is thrown 5 times and the red die is thrown 4 times. Find the probability that a total of 3 sixes occurs.

8. What is the probability of one head twice in 3 tosses of 4 fair coins?

9. A commuter's drive to work includes 7 stoplights. Assume the probability that a light is red when the commuter reaches it is 0.20, and that the lights are far enough apart to operate independently.

  a. If X is the number of red lights the commuter stops for, find the probability distribution function for X.
  b. Find $P(X \geq 5)$.
  c. Find $P(X \geq 5 \mid X \geq 3)$.

10. The probability of being able to log on a computer system from a remote terminal during a busy period is 0.7. Suppose that 10 independent attempts are made and that X denotes the number of successful attempts.

  a. Write an expression for the probability distribution function, $f(x)$.
  b. Find $P(X \geq 5)$.
  c. Now suppose that Y represents the number of attempts up to and including the first successful attempt. Write an expression for the probability distribution function, $g(y)$.

11. An experimental rocket is launched 5 times. The probability of a successful launch is 0.9. Let X denote the number of successful launches. A study has shown that the net cost of the experiment, in thousands of dollars, is $2 - 3X^2$. Find the expected net cost of the experiment.

12. Twenty percent of the IC chips made in a plant are defective. Assume that a binomial model is appropriate.

    a. Find the probability that, at most, 13 defective chips occur in a sample of 100.

    b. Find the probability that 2 samples, each of size 100, will have a total of exactly 26 defective chips.

13. A coin, loaded to come up heads with probability $\frac{2}{3}$, is tossed 5 times. If the number of heads is odd, the player is paid \$5. If the number of heads is 2 or 4 the player wins nothing; if no heads occur, the player tosses the coin 5 more times and wins, in dollars, the number of heads thrown. If the game costs \$3 to play, find the probability distribution of N, the player's net winnings.

14. **a.** Show that the probability of being dealt a full house (3 cards of one value and 2 of another value) in poker is about 0.0014.

    b. Find the probability that in 1000 hands of poker you will be dealt at least 2 full houses.

15. An airline knows that the probability a person holding a reservation on a certain flight will not appear is 10%. The plane holds 90 people.

    a. If 95 reservations have been sold, find the probability that the airline will be able to accommodate everyone appearing for the flight.

    b. How many reservations should be sold so that the airline can accommodate everyone who appears for the flight 99% of the time?

16. The probability that an individual seed of a certain type will germinate is 0.9. A nurseryman sells flats of this type of plant and wants to "guarantee" (with probability 0.99) that at least 100 plants in the flat will germinate. How many plants should he put in each flat?

17. A coin with $P(H) = \frac{1}{2}$ is flipped 4 times and then a coin with $P(H) = \frac{2}{3}$ is tossed twice. What is the probability that a total of 5 heads occurs?

18. **a.** Each of 2 persons tosses 3 fair coins. What is the probability each gets the same number of heads?

    b. In part **a**, what is the probability that $X_1 + X_2$ is odd, where X_1 is the number of heads the first person tosses and X_2 is the number of heads the second person tosses?

    c. Repeat part **a** if each person tosses n fair coins. Simplify the result as much as possible.

19. Find the probability that more than 520 heads occur in 1000 tosses of a fair coin.

20. How many times must a fair coin be tossed if the probability of obtaining at least 40 heads is at least 0.95?

21. Samples of 100 are selected each hour from an assembly line that produces items, 20% of which are defective.

    a. What is the probability that at most 15 defectives are found in an hour?
    b. What is the probability that a total of 47 defectives is found in the first 2 hours?

22. A small engineering college would like to have an entering class of 360 students. Past data indicates that 85% of those accepted actually enroll in the class. How many students should be accepted if the probability the class will be at least 360 is to be approximately 0.95?

23. A fair coin is tossed repeatedly. What is the probability that the number of heads tossed reaches 6 before the number of tails tossed reaches 4?

24. Evaluate the sums

$$\sum_{x=0}^{n} x \cdot \binom{n}{x} p^x \cdot q^{n-x} \text{ and } \sum_{x=0}^{n} x \cdot (x-1) \cdot \binom{n}{x} p^x \cdot q^{n-x}$$

directly and use these to verify the formulas for μ and σ^2 for the binomial distribution.

$$\left[\text{Note that } \sum_{x=0}^{n} \binom{n}{x} \cdot p^x \cdot (1-p)^{n-x} = [p + (1-p)]^n = 1. \right]$$

25. In exercise 4, show that the game is fair if X wins if he tosses at least as many heads as Y.

2.6 ► Some Statistical Considerations

We pause here and in the next two sections to show some statistical applications of the probability theory we have developed so far. From time to time in this book we will show some applications of probability theory to statistics and the statistical analysis of data as well as to other applied situations; this is our first consideration of statistical problems.

From the previous section, we know what can happen when n observations are taken from a binomial distribution with known parameter p. Generally, however, p is unknown. We might, for example, be interested in the proportion of unacceptable items arising from a production line. Normally, this proportion would not be known. So we suppose now that p is unknown. How can we *estimate* the unknown p? We certainly would observe the binomial process that the production line represents; the result of this would be a number of good items from the process, say X, and we would surely use X in some way to estimate p. How precisely can we use X to estimate p?

It would appear natural to estimate p by the proportion of good items in the sample, $\frac{X}{n}$. Since X is a random variable, so is $\frac{X}{n}$. We can calculate the expected value of this random variable as follows:

$$E\left[\frac{X}{n}\right] = \sum_{x=0}^{n} \frac{x}{n} \cdot P(X = x) = \frac{1}{n} \cdot \sum_{x=0}^{n} x \cdot P(X = x),$$

so

$$E\left[\frac{X}{n}\right] = \frac{1}{n} \cdot n \cdot p = p.$$

This indicates that, on average, our estimate for p gives the true value, p. We say that our estimator, $\frac{X}{n}$, is an *unbiased estimator* for p.

This gives us a way of estimating p by a single value. This single value is dependent upon the sample, and if we choose another sample, we are likely to find another value of X, and hence arrive at another estimate of p. Could we also find a "likely" *range* for the value of p?

To answer this, consider a related question. If we have a binomial situation with probability p and sample size n, what is a likely range for the observed values of the random variable, X? The answer of course depends upon the meaning of the word likely. Suppose that a likely range for the values of a random variable is a range in which the values of the variable occur with probability 0.95.

With the considerable aid of our computer algebra system, we can evaluate a number of different binomial distributions. We vary n, the number of observations, and p, the probability of success. In each case we find the proportion of the values of X that lie within two standard deviations of the mean; that is, the proportion of the values of X that lie in the interval $\mu \pm 2\sigma = n \cdot p \pm 2\sqrt{n \cdot p \cdot (1 - p)}$. We select the constant two because we need to find a range that includes a large portion – 95% – of the values of X, and two appears to be a reasonable multiplier for the standard deviation. Table 2.6.1 shows the results of these calculations. Here P represents the probability that an observed value of the random variable X lies in the interval $\mu \pm 2\sigma = n \cdot p \pm 2\sqrt{n \cdot p \cdot (1 - p)}$. The values of n and p have been chosen so that the end-points of the intervals are integers.

We are led to believe from the table, regardless of the value of p, that at least 95% of the values of the variable X lie in the interval $\mu \pm 2\sigma$. (Later we will show, for large values of n, regardless of the value of p, that the probability is approximately 0.9545, a result supported by our calculations.) So we have

$$P(\mu - 2\sigma \leq X \leq \mu + 2\sigma) \geq 0.95. \tag{2.5}$$

Solving the inequalities for μ, we have

$$P(X - 2\sigma \leq \mu \leq X + 2\sigma) \geq 0.95. \tag{2.6}$$

TABLE 2.6.1

n	p	σ	μ ± 2σ	P
36	$\frac{1}{2}$	3	12,24	0.971183
64	$\frac{1}{2}$	4	24,40	0.967234
100	$\frac{1}{2}$	5	40,60	0.964800
144	$\frac{1}{2}$	6	60,84	0.963148
196	$\frac{1}{2}$	7	84,112	0.961530
18	$\frac{1}{3}$	2	2,10	0.978800
72	$\frac{1}{3}$	4	16,32	0.967288
162	$\frac{1}{3}$	6	42,66	0.963177
288	$\frac{1}{3}$	8	80,112	0.961066
48	$\frac{1}{4}$	3	6,18	0.971345
192	$\frac{1}{4}$	6	36,60	0.963214
432	$\frac{1}{4}$	9	90,126	0.960373
10000	$\frac{1}{2}$	50	4900,5100	0.954494
11250	$\frac{1}{3}$	50	3650,3850	0.954497
13872	$\frac{1}{4}$	51	3366,3570	0.954499

Replacing μ and σ by $n \cdot p$ and $\sqrt{n \cdot p \cdot q}$, respectively, Formula 2.6 becomes

$$P(X - 2\sqrt{n \cdot p \cdot q} \le n \cdot p \le X + 2\sqrt{n \cdot p \cdot q}) \ge 0.95. \qquad (2.7)$$

The inequalities in Formula 2.7 can now be solved for p. The result is

$$P\left(\frac{nX + 2n - 2\sqrt{n^2X + n^2 - nX^2}}{n^2 + 4n} \le p \le \frac{nX + 2n + 2\sqrt{n^2X + n^2 - nX^2}}{n^2 + 4n}\right)$$
$$\ge 0.95. \qquad (2.8)$$

Our thinking here is as follows: If we find an interval that contains at least 95% of the values of X and if p is unknown, then those same values of X will produce an interval in which p, in some sense, is likely to lie. The end points produced by Formula 2.8 comprise what we call a *95% confidence interval for p*.

While Formula 2.5 gives a likely range of values of X if p is known, Formula 2.8 gives a likely range of values of p if X is known. So we have a response to a variant of our first question: If X successes are observed in n binomial trials, what is a likely value for p?

We note that Formula 2.5 is a legitimate probability statement since X is a random variable and 95% of its values lie in the stated interval. Formula 2.8, however, is *not* a probability statement! Why not? The reason is that p is an unknown constant. It either lies in the stated interval or it doesn't. Then what does the 95% mean?

Here's a way of looking at this. Consider samples of fixed size, say $n = 100$. If we find 25 successes in these 100 trials (so $X = 25$), then Formula 2.8 gives the interval $0.174152 \leq p \leq 0.345079$. However, the next time we perform the experiment, we are most likely to find another value of X, and hence another confidence interval. For example, if $X = 30$, the confidence interval is $0.217492 \leq p \leq 0.397893$. From Formula 2.8 we see that these confidence intervals are centered about the value $\frac{X+2}{n+4}$ and have width

$$\frac{4\sqrt{n^2 X + n^2 - nX^2}}{n^2 + 4n},$$

so both the center and width of the intervals change as X changes for a fixed sample size n. This gives us a proper interpretation of Formula 2.8: 95% of these intervals will contain the unknown, and fixed, value p.

As another example, 15 observations were taken from a binomial distribution with $n = 100$ and gave the following values for X: 40, 44, 29, 43, 43, 42, 39, 40, 43, 42, 36, 44, 35, 39, and 42. Formula 2.8 was then used to compute a confidence interval for p for each of these values of X.

Figure 2.11 shows these confidence intervals. As expected, they vary both in position and width. The actual value of p used to generate the X values was 0.40. As it happens here, $p = 0.40$ is contained in 14 of the 15 confidence intervals, but in larger samples we would expect that 0.40 would be contained in about 95% of the confidence intervals produced.

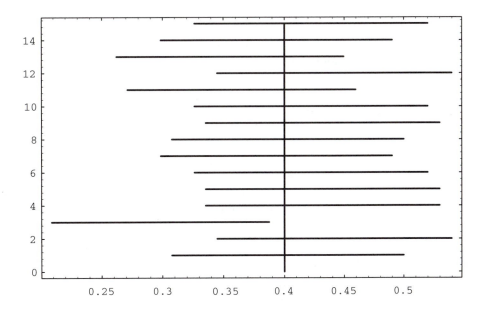

Figure 2.11   Some confidence intervals.

Exercises 2.6

1. If a sample of size 30 is chosen from a binomial distribution with $p = \frac{1}{2}$, and if X denotes the number of successes obtained, find an interval in which 95% of the values of X will lie.

2. Use your computer algebra system to verify the results in Table 2.6.1 for

   a. $p = \frac{1}{2}$, $n = 36$.
   b. $p = \frac{1}{3}$, $n = 18$.
   c. $p = \frac{1}{4}$, $n = 48$.

3. Use your computer algebra system to verify the result in Table 2.6.1 for

   a. $p = \frac{1}{2}$, $n = 10000$.
   b. $p = \frac{1}{3}$, $n = 11250$.
   c. $p = \frac{1}{4}$, $n = 13872$.

4. A survey of 300 college students found that 50 are thinking about changing their majors. Find a 95% confidence interval for the true proportion of college students thinking about changing their majors.

5. A random sample of 1250 voters was asked whether or not they voted in favor of a school bond issue, and 325 replied that they favored the issue. Find a 95% confidence interval for the true proportion of voters who favor the school bond issue.

6. Find 90% confidence intervals by constructing a table similar to Table 2.6.1. One should find that $P(\mu - 1.645\sigma \leq X \leq \mu + 1.645\sigma) = 0.90$.

7. A newspaper survey of 125 of its subscribers found that 40% of the respondents knew someone who was killed or injured by a drunk driver. Find a 90% confidence interval for the true proportion of people in the population who know someone who was killed or injured by a drunk driver.

8. As a project in a probability course, a student discovered that among a random sample of 80 families, 25% did not have checking accounts. Use this information to construct a 90% confidence interval for the true proportion of families in the population who do not have checking accounts.

9. A study showed that $\frac{1}{8}$ of American workers worked in management or in administration, while $\frac{1}{27}$ of Japanese workers worked in management or administration. The study was based on 496 American workers and 810 Japanese workers.

   Is it possible that the same proportion of American and Japanese workers are in management or administration and that the apparent differences found by the study are simply due to the variation inherent in sampling? [Hint: Compare 90% confidence intervals.]

10. n values of X, the number of successes in a binomial process, are used to compute n 95% confidence intervals for the unknown parameter p. Find the probability that p lies in exactly k of the n confidence intervals.

2.7 ▶ Hypothesis Testing: Binomial Random Variables

In the previous section we considered confidence intervals for binomial random variables. The problem of estimating a parameter, in this case the value of p by means of an interval, is part of *statistics* or *statistical inference*. Statistical inference, in simplest terms, is concerned with drawing inferences from data that have been gathered by a sampling process. Statistical inference is comprised of the theory of *estimation* and that of *hypothesis testing*. In the preceding section, we considered the construction of a confidence interval, which is part of the theory of estimation. The remaining portion of the theory of drawing inferences from samples is called *hypothesis testing*. We begin with a somewhat artificial example in order to fix ideas and define some vocabulary before proceeding to other applications.

Example 2.7.1

The manufacturing process of a sensitive component has been producing items of which 20% must be reworked before they can be used. A recent sample of 20 items shows 6 items that must be reworked. Has the manufacturing process changed so that 30% of the items must be reworked?

Assume that the production process is binomial, with p, which is of course unknown to us, denoting the probability an item must be reworked. We begin with a *hypothesis* or conjecture about the binomial process, that the process has not in fact changed and that the proportion of items that must be reworked is 20%. We denote this by H_o and call it the *null hypothesis*. As a result of a *test* – in this case the result of a sample of the items – this hypothesis will be *accepted* (that is we will believe that H_o is true) or it will be *rejected* (that is, we will believe that H_o is not true). In the latter case, when the null hypothesis is rejected, we agree to accept an *alternative hypothesis*, H_a. Here the hypotheses are chosen as follows:

$$H_o : p = 0.20$$
$$H_a : p = 0.30.$$

How are sample results (in this case, 6 items that must be reworked) to be interpreted? Does this information lead to the acceptance or the rejection of H_o? We must decide what sample results lead to the acceptance of H_o

and what sample results lead to its rejection (and hence the acceptance of H_a).

The sampling is, of course, subject to variability, and our conclusions cannot be reached without running the risk of error. There are two risks: that we will reject H_o even though it is, in reality, true; or that we will accept H_o even though it is, in reality, false. The following table may help in seeing the four possibilities that exist whenever a hypothesis is tested:

	Reality H_o True	H_o False
H_o Rejected	Type I error (α)	Correct decision
H_o Accepted	Correct decision	Type II error (β)

We never will know reality, but the table does indicate the consequences of the decision process. It is customary to denote the two types of errors by

$$\alpha = \text{probability of a Type I error}$$
$$= P[H_o \text{ is rejected when it is true}]$$

and

$$\beta = \text{probability of a Type II error}$$
$$= P[H_o \text{ is accepted when it is false}].$$

Both α and β are conditional probabilities, and each is highly dependent on the set of sample values that lead to the rejection of the hypothesis. This set of values is called the *critical region*.

What should the critical region be? We are free to choose any critical region we want; it would appear sensible in this case to conclude that the percentage of product to be reworked has increased when the number of items to be reworked in the sample is large. Therefore, we *arbitrarily* take as a critical region $\{x \mid x \geq 9\}$, where X is the random variable denoting the number of items in the sample that must be reworked.

What are the consequences of this choice for the critical region? We can calculate α, the size of the Type I error.

$$\alpha = P\,[X \geq 9 \text{ if } H_o \text{ is true}]$$

$$= P[X \geq 9 \text{ if } p = 0.2]$$

$$= \sum_{x=9}^{20} \binom{20}{x}(0.2)^x(0.8)^{20-x}$$

$$= 0.00998179 \approx 0.01$$

So about 1% of the time this critical region will reject a true hypothesis. This means that the manufacturing process is such, that if $p = 0.20$, about 1% of the time it will behave as if $p = 0.30$ with this critical region. α is called the *size* or the *significance level* of the test.

What is β?

$$\beta = P[\text{accept } H_o \text{ if it is false}]$$

$$= P[X < 9 \text{ if } H_o \text{ is false}]$$

$$= P[X < 9 \text{ if } p = 0.30]$$

$$= \sum_{x=0}^{8} \binom{20}{x}(0.30)^x(0.70)^{20-x}$$

$$= 0.886669$$

These calculations are shown in Appendix 1.

So, with this critical region, about 89% of the time a process producing 30% items to be reworked behaves as if it were producing only 20% of such items. This might appear to be a very high risk. Can it be reduced? One way to reduce β would be to change the critical region to, say, $\{x \mid x \geq 8\}$. We now find that

$$\beta = \sum_{x=0}^{7} \binom{20}{x}(0.30)^x(0.70)^{20-x}$$

$$= 0.772272,$$

but then

$$\alpha = \sum_{x=8}^{20} \binom{20}{x}(0.20)^x(0.80)^{20-x}$$

$$= 0.032147$$

So the cost in decreasing β comes at the cost of an increase in α. We will see later than one way to decrease both errors is to increase the sample size.

What are the consequences of other choices for the critical region? We could choose $x = 0$ for the critical region so that the hypothesis is rejected only if $x = 0$. Then

$$\alpha = P[X = 0 \text{ if } p = 0.20]$$

$$= (.8)^{20}$$

$$= 0.0115292,$$

producing a Type I error of about the same size as it was before. But then

$$\beta = \sum_{x=1}^{20} \binom{20}{x}(0.30)^x(0.70)^{20-x}$$

$$= 0.999202.$$

These two critical regions then have roughly equal Type I errors, but β is larger for the second choice of critical region.

We will choose one more critical region whose Type I error is about 0.01: the critical region $X = 9, 10,$ or 11. Then

$$\alpha = P[X = 9, 10, \text{ or } 11 \text{ if } p = 0.20]$$

$$\alpha = \sum_{x=9}^{11} \binom{20}{x}(0.20)^x(0.80)^{20-x}$$

$$= 0.00998,$$

again roughly 0.01. β, however, is now

$$\beta = 1 - \sum_{x=9}^{11} \binom{20}{x}(0.30)^x(0.70)^{20-x}$$

$$= 0.891807.$$

The preceding four cases illustrate that there are several choices for critical regions that give the same size for the Type I error; we will call the critical region *best*, if for a given Type I error, it *minimizes* the Type II error. In this case the best critical region for a test with $\alpha \approx 0.01$ is $\{x \mid x \geq 9\}$. Best critical regions can often, but not always, be constructed.

So, to return to the original problem where the sample yielded six items for reworking, we conclude that the process has not changed because it is not in the critical region $\{x \mid x \geq 9\}$ for $\alpha \approx 0.01$.

Finally, we note that the size of the Type II error, β, is a function of the alternative, $p = 0.30$, in this example. If the alternative hypothesis were $H_a : p > .20$, then β could be calculated for any *particular* alternative in H_a. That is, if $p > 0.20$, then

$$\beta = \sum_{x=0}^{8} \binom{20}{x}p^x(1-p)^{20-x}, \text{ a function of } p.$$

As p increases, β decreases quite rapidly, reflecting the fact that it is increasingly unlikely that the hypothesis will be accepted if it is false. A graph of β as a function of p is shown in Figure 2.12.

It is customary to graph $1 - \beta = P$ (a false H_o is rejected). This is called the *power function* for the test.

The hypothesis $H_o : p = 0.20$ is called a *simple hypothesis* because it completely specifies the probability distribution of the variable under consideration. The hypothesis $H_a : p > 0.20$ is composed of an infinity of simple hypotheses. It is called a *composite hypothesis*.   ◄

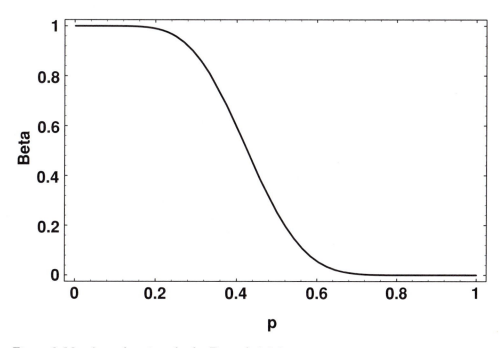

Figure 2.12 β as a function of p for Example 2.7.1.

Example 2.7.2

In the previous example, the critical region was specified and then values for α and β were found. It is common, however, for experimenters to specify α and β before the experiment is done; often the sample size necessary to achieve these probabilities can be found, at least approximately. One of the consequences of the binomial model in the preceding example is that a change in the critical region by a single unit produces large changes in α and β. Suppose, in the preceding example, that it is desired to have, approximately, $\alpha = 0.05$ and $\beta = 0.10$. If we assume that the best critical region is of the form $\{x \mid x \geq k\}$, then

$$\alpha = \sum_{x=k}^{n} \binom{n}{x}(0.20)^x (0.80)^{n-x} = 0.05$$

and

$$\beta = \sum_{x=0}^{k-1} \binom{n}{x}(0.30)^x (0.70)^{n-x} = 0.10.$$

These equations are difficult to solve without the aid of extensive binomial tables or a computer algebra system. We find that

$$\alpha = \sum_{x=40}^{156} \binom{156}{x}(0.20)^x (0.80)^{156-x} = 0.05145$$

and

$$\beta = \sum_{x=0}^{39} \binom{156}{x}(0.30)^x (0.70)^{156-x} = 0.09962,$$

so $n \approx 156$ and $k \approx 40$. These values are probably close enough for all practical purposes. Other solutions are possible, of course, depending upon the closeness with which we want to solve the equations for α and β. It may well be that we cannot carry out an experiment with this large sample size; such a restriction would obviously have implications for the sizes of α and β that can be entertained.   ◄

Exercises 2.7

1. It is thought that 80% of VCR owners do not know how to program their VCR for taping a TV program. To test this hypothesis, a sample of 20 VCR owners is chosen and the proportion, p, who cannot program a VCR is recorded. The hypotheses are

$$H_o : p = 0.20$$
$$H_a : p < 0.20.$$

a. Find α if the critical region is $X < 14$ where X is the number in the sample who cannot program a VCR.

b. Find β for the alternative $H_a : p = 0.70$.

c. Graph β as a function of p, $0 \leq p \leq 0.80$.

2. A researcher speculates that 20% of the people in a very large group under study are left-handed, a proportion much larger than the 10% of people in the population who are left-handed. A sample is chosen to test

$$H_o : p = 0.10$$
$$H_a : p = 0.20.$$

The critical region is $X \geq k$, where X is the number of left-handed people in the sample. It is desired to have $\alpha = 0.07$ and $\beta = 0.13$, approximately. How large a sample should be chosen?

3. In exercise 2, show that β is larger for the critical region $X \leq c$ where c is chosen so that the test has size α.

4. A drug is thought to cure $\frac{2}{3}$ of the patients with a disease; without the drug, $\frac{1}{3}$ of the patients recover. The hypothesis

$$H_o : p = \tfrac{1}{3} \text{ is tested against } H_a : p = \tfrac{2}{3}$$

on the basis of a sample of 12 patients. H_o is rejected if X, the number of patients in the sample who recover, is greater than 5. Find α and β for this test.

5. In exercise 4, find the sample size for which $\alpha = 0.05$ and $\beta = 0.13$, approximately.

6. A recent survey showed that 46% of Americans feel that they are "being left behind by technology." To test this hypothesis, a sample of 36 Americans showed that 18 of them agreed that they were being left behind by technology. Do the data support the hypothesis $H_o : p = 0.46$ against the alternative $H_a : p > 0.46$? (Use $\alpha = 0.05$.)

7. A publisher thinks that 57% of the magazines on newsstands are unsold. To test this hypothesis, a sample of 1000 magazines put on the newsstand resulted in 495 unsold magazines. Do these data support $H_o : p = 0.57$ or the alternative $H_a : p < 0.57$ if $\alpha = 0.05$? *Use equation 2.8*

8. A survey indicates that 41% of the people interviewed think that holders of Ph.D. degrees have attended medical school. In a sample of 88 people, 50 agreed that Ph.D.'s attended medical school. Is this evidence, using $\alpha = 0.05$, that the percentage of people thinking that Ph.D.'s are M.D.'s is greater than 41%?

9. In a survey of questions concerning health issues, 59% of the respondents thought that at some time in their lives they would develop cancer. If a sample of 200 people showed that 89 agreed that they would develop cancer at some time, is this evidence to support the hypothesis that the percentage thinking they will develop cancer is less than 59%? (Use $\alpha = 0.05$.)

10. Among Americans earning more than $50,000 per year, $\frac{2}{3}$ agree that Americans are "materialistic." If 70 out of 100 people interviewed agree that Americans are materialistic, is this evidence that the true proportion thinking Americans are materialistic is greater than $\frac{2}{3}$? (Use $\alpha = 0.05$.)

∎ 2.8 ▶ Distribution of a Sample Proportion

Before considering some important probability distributions in addition to the binomial distribution, we consider here a common problem: A sample survey of n individuals indicates that the proportion P_s of the respondents favor a certain candidate in an election. P_s is clearly a random variable because our sampling will not always produce exactly the same proportion of voters favoring the candidate if the sampling is repeated. P_s is called a *sample proportion*. What is its probability distribution? How can we expect P_s to vary from sample to sample? If we observe a value of P_s – say 51% of the voters favor a candidate – what does this tell us about the true proportion, p, of all voters who favor the candidate? We consider these questions now.

Let us suppose that in reality a proportion p of the voters favor a candidate. Let's also assume that the sample is taken so that the responses can be assumed to be independent among the people interviewed. The *number* of voters favoring the candidate, say X, is a binomial random variable because a voter either favors the candidate or the opponent. The sample proportion favoring the candidate, P_s, is also a random variable. If we take a random sample of size n, then

$$P_s = \frac{X}{n}.$$

So our random variable P_s is related to a binomial random variable. We considered confidence intervals for binomial random variables in section 2.6. We now extend that theory somewhat.

We calculate the mean and variance of the variable P_s. We let the sample proportion be $p_s = \frac{x}{n}$. Clearly,

$$P(P_s = p_s) = P(\frac{X}{n} = p_s) = P(X = n \cdot p_s) = P(X = x),$$

so

$$E(P_s) = \sum_{p_s} p_s \cdot P(P_s = p_s)$$

$$= \sum_{x=0}^{n} \frac{x}{n} \cdot P(\frac{X}{n} = p_s)$$

$$= \frac{1}{n} \sum_{x=0}^{n} x \cdot P(X = x).$$

Therefore,

$$E(P_s) = \frac{1}{n} \cdot E(X) = \frac{n \cdot p}{n} = p.$$

So, as might be expected, the average value of the variable P_s is the true proportion, p. This is precisely the same result we saw in section 2.6.

The variance of P_s can be calculated using the variance of a binomial random variable as follows:

$$Var(P_s) = Var(\frac{X}{n}) = \sum_{P_s}(P_s - p)^2 \cdot P(P_s = p_s)$$

$$= \sum_{x=0}^{n}(\frac{X}{n} - p)^2 \cdot P(\frac{X}{n} = p_s)$$

$$= \frac{1}{n^2} \sum_{x=0}^{n}(X - n \cdot p)^2 \cdot P(X = n \cdot p_s)$$

$$= \frac{1}{n^2} \sum_{x=0}^{n}(x - n \cdot p)^2 \cdot P(X = x)$$

showing that

$$Var(P_s) = \frac{1}{n^2} \cdot Var(X)$$

or that

$$Var(P_s) = \frac{1}{n^2} \cdot n \cdot p \cdot q = \frac{p \cdot q}{n}.$$

The above considerations also show a more general result: If random variables X and Y are related by $Y = k \cdot X$, where k is a constant, then $E(Y) = k \cdot E(X)$, and $Var(Y) = k^2 \cdot Var(X)$.

Using the facts we derived in section 2.6 regarding binomial confidence intervals, we can say that

$$P\left(P_s - 2 \cdot \sqrt{\frac{p \cdot q}{n}} \le p \le P_s + 2 \cdot \sqrt{\frac{p \cdot q}{n}}\right) \ge 0.95,$$

giving a 95% confidence interval for the true population proportion, p. But, as occurred in the binomial situation, the standard deviation is a function of the unknown p, so we must solve for p. There are two ways to do this. One method is to solve the quadratic equations which arise exactly. However, if $0.3 \le p \le 0.7$, then a good approximation to $p \cdot q$ is $\frac{1}{4}$. This approximation is far from exact, but often yields acceptable results when p is in the indicated range.

Example 2.8.1

A sample survey of 400 voters showed that 51% of the voters favored a certain candidate. Find a 95% confidence interval for p, the true proportion of voters in the population favoring the candidate.

We have that

$$P\left(0.51 - 2 \cdot \sqrt{\frac{p \cdot q}{400}} \le p \le 0.51 + 2 \cdot \sqrt{\frac{p \cdot q}{400}}\right) \ge 0.95.$$

If we solve the inequality $p \le p_s + 2 \cdot \sqrt{\frac{p \cdot q}{400}}$ for p, noting that $q = 1 - p$, we find that

$$\frac{np_s + 2 - 2\sqrt{1 + np_s - np_s^2}}{n + 4} \le p \le \frac{np_s + 2 + 2\sqrt{1 + np_s - np_s^2}}{n + 4}.$$

This result is equivalent to Formula 2.8 in section 2.6.
Substituting $n = 400$ and $p_s = 0.51$ gives

$$P(0.46016 \le p \le 0.55964) \ge 0.95.$$

Using the approximation $p \cdot q = \frac{1}{4}$ gives $P(0.46 \le p \le 0.56) \ge 0.95$.
The difference in the confidence intervals is very small, but this is because the observed proportion, 0.51, is close to $\frac{1}{2}$. The two confidence intervals will deviate more markedly as the difference between p_s and $\frac{1}{2}$ increases. The candidate certainly cannot feel confident of winning the election on the basis of the sample, but we can make this observation only because we have created a confidence interval for p. In the popular press, half the width of the confidence interval is referred to as the *sampling error*. So a survey may be reported with a sampling error of 3%, meaning that a 95% confidence interval for p is $p_s \pm 0.03$.

If the sampling error is given, then the sample size can be inferred. If the sampling error is stated as 3%, then

$$2 \cdot \sqrt{\frac{p \cdot q}{n}} = 0.03.$$

Of course, the difficulty is that p is unknown. Note that $p \cdot q \approx \frac{1}{4}$ if $0.3 \le p \le 0.7$. Using this approximation here we conclude that $\sqrt{\frac{1}{n}} \approx 0.03$, so that $n \approx 1111$.

The approximation $p \cdot q = \frac{1}{4}$ is usually used only if p is in the interval $0.3 \le p \le 0.7$; otherwise p is replaced by the sample proportion, p_s, in determining sample size.

We presumed earlier that the sample of voters is a simple random one, and we further presumed that the people sampled will actually vote and that they have been candid with the interviewer concerning their voting preference. Samplers commonly call these presumptions into question and have a variety of ways of dealing with them. In addition, such samples are rarely simple random ones; all we can say here is that these variations in the sampling design have some effect on the sampling error. ◄

Exercises 2.8

1. A survey of 300 paperback novels showed that 47% could be classified as romance novels. Find an approximate 95% confidence interval for p, the true proportion of romance paperback novels.

2. Records indicate that $\frac{1}{8}$ of American children receive welfare payments. If this survey was based on 250 records, find an approximate 95% confidence interval for the true proportion of children who receive welfare payments.

3. A random sample of 300 voters showed that 48% favored a candidate. Does an approximate 95% confidence interval indicate that it is possible for the candidate to win the election?

4. A survey of 423 workers found that $\frac{1}{9}$ were union members. Find an approximate 95% confidence interval for the true proportion of union workers.

5. The sampling error of a survey in a magazine was stated to be 5%. What was the sample size for the survey?

6. A student conducted a project for a statistics course and found that $\frac{2}{3}$ of the respondents in interviews of 120 people did not know that the Bill of Rights is the first ten amendments to the Constitution. Find an approximate 90% confidence interval for the true proportion of people who do not know that the Bill of Rights is the first ten amendments to the Constitution.

7. A magazine devoted to health issues discovered that $\frac{3}{5}$ of the time a visit to a physician resulted in a prescription. The survey was based on 130 telephone interviews. Use this data to construct an approximate 90% confidence interval for the true proportion of patients given a prescription as a result of a visit to their physician.

8. According to a recent study, 81% of college students say that they favor drug testing in the workplace. The study was conducted among 400 college students. Find an approximate 90% confidence interval for the true proportion of college students who favor drug testing in the workplace.

9. Interviews of 150 patients recently tested for HIV indicate that among those whose tests indicate the presence of the virus, $\frac{1}{2}$ did not know they had the virus prior to testing. Find an approximate 95% confidence interval for the proportion of people in the population whose tests indicate they have HIV and who did not know this.

 10. A California automobile dealer knows that $\frac{1}{10}$ of California residents own convertibles. Is the dealer likely (with probability 0.95) to sell at least 200 convertibles in the next 1000 sales?

2.9 ▶ Geometric and Negative Binomial Distributions

We considered *geometric* random variables in Examples 2.3.1.5 and 2.3.2.3 where the random variable of interest was the waiting time for the occurrence of a binomial event. A perfect model for the geometric random variable is tossing a coin, loaded so that the probability of coming up heads is p, until a head appears. If X denotes the number of tosses necessary and if $q = 1 - p$, we have seen that

$$P(X = x) = q^{x-1} \cdot p, \ x = 1, 2, 3, \ldots$$

and that

$$E(X) = \frac{1}{p} \text{ and } Var(X) = \frac{q}{p^2}.$$

Now suppose we wait until the *second* head appears when the loaded coin is tossed. Let X denote the number of trials necessary for this event to occur. We want $P(X = x)$, the probability distribution for X. Since the last trial must be a head, the first $x - 1$ trials must contain exactly one head and $x - 2$ tails; and because the trials are independent, and the single head can occur in any of $x - 1$ places, it follows that

$$P(\text{first } x - 1 \text{ trials have exactly 1 head and } x - 2 \text{ tails}) = \binom{x-1}{1} \cdot q^{x-2} \cdot p.$$

So, because the last trial must be a head,

$$P(X = x) = \binom{x-1}{1} \cdot q^{x-2} \cdot p \cdot p, \quad x = 2, 3, 4, \ldots . \qquad (2.9)$$

Since Formula 2.9 exhausts the possibilities, it must be that $\sum\limits_{x=2}^{\infty} P(X = x) = 1$. One way to verify this is to notice that

$$\sum_{x=2}^{\infty} \binom{x-1}{1} \cdot q^{x-2} \cdot p^2 = p^2 \sum_{x=2}^{\infty} \binom{x-1}{1} \cdot q^{x-2} = p^2 \cdot (1-q)^{-2} = p^2 \cdot p^{-2} = 1$$

by the binomial theorem with a negative exponent. This series will arise again in our work. We have established the probability distribution for the waiting time for the second head.

What is the average waiting time for the second head? We might reason as follows: We flip the coin until the first head appears; the average number of flips is $\frac{1}{p}$. But then the situation is exactly the same as it was for the first flip of the coin; the fact that we flipped the coin and waited for the first head has absolutely no influence on subsequent tosses of the coin. We must wait an average of $\frac{1}{p}$ flips again until the second head appears. So the average waiting time for the second head to appear is $\frac{1}{p} + \frac{1}{p} = \frac{2}{p}$. It follows that if we were to wait for the rth head to appear, the average total waiting time would be $\frac{r}{p}$. We will give a more formal derivation of this result later.

What is the probability distribution function for the rth head to appear? Let X denote the number of tosses until the rth head appears. Since, again, the last toss must be a head and the first $x - 1$ tosses must contain exactly $r - 1$ heads,

$$P(X = x) = \binom{x-1}{r-1} \cdot p^{r-1} \cdot q^{x-r} \cdot p, \, x = r, r+1, r+2, \ldots . \qquad (2.10)$$

Since $P(X = x) \geq 0$, we must check the sum of the probabilities to see that we have a probability distribution function. But,

$$\sum_{x=r}^{\infty} \binom{x-1}{r-1} \cdot p^{r-1} \cdot q^{x-r} \cdot p = p^r \sum_{x=r}^{\infty} \binom{x-1}{r-1} q^{x-r} = p^r (1-q)^{-r} = 1,$$

so $P(X = x)$ is a probability distribution.

If $r = 1$ in Formula 2.10, we find that $P(X = x)$ reduces to the geometric probability distribution function. The result in Formula 2.10 is called the *negative binomial distribution* because of the occurrence of the binomial expansion with a negative exponent.

We now calculate the mean and the variance of this negative binomial random variable. We reasoned that the mean is $\frac{r}{p}$ and we now give another derivation of this.

By the definition of expected value,

$$E(X) = \sum_{x=r}^{\infty} x \cdot \binom{x-1}{r-1} \cdot p^r \cdot q^{x-r}$$

$$= \sum_{x=r}^{\infty} r \cdot \frac{x!}{r! \cdot (x-r)!} \cdot p^r \cdot q^{x-r}$$

$$= r \cdot p^r \cdot \sum_{x=r}^{\infty} \binom{x}{r} \cdot q^{x-r}$$

$$= r \cdot p^r \cdot [1 + \binom{r+1}{1} \cdot q + \binom{r+2}{2} \cdot q^2 + \cdots$$

$$= r \cdot p^{r\cdot} \cdot (1-q)^{-(r+1)}$$

$$= \frac{r \cdot p^r}{p^{r+1}} = \frac{r}{p}.$$

Now we seek the variance of this negative binomial random variable. Since $E(X^2)$ is difficult to find directly, we resort to the fact that

$$Var(X) = E[X(X+1)] - E(X) - [E(X)]^2.$$

Now,

$$E[X(X+1)] = \sum_{x=r}^{\infty} x(x+1) \cdot \binom{x-1}{r-1} \cdot p^r \cdot q^{x-r}$$

$$= r(r+1) \cdot p^r \sum_{x=r}^{\infty} \binom{x+1}{r+1} \cdot q^{x-r}$$

$$= r(r+1) \cdot p^r \cdot (1-q)^{-(r+2)}$$

$$= \frac{r(r+1)}{p^2}.$$

Since $E(X) = \frac{r}{p}$, it follows that

$$Var(X) = \frac{r(r+1)}{p^2} - \frac{r}{p} - \left(\frac{r}{p}\right)^2 = \frac{r \cdot q}{p^2}.$$

It is also useful to view the preceding random variable X as a sum of other random variables. Let X_1 denote the number of trials up to and including the first success, X_2 the number of trials after the first success until the second success, and so on. It follows that

$$X = X_1 + X_2 + \cdots + X_r.$$

Each of the X_i's has mean $\frac{1}{p}$ and variance $\frac{q}{p^2}$. We see that

$$E(X) = \frac{r}{p} = E(\sum_{i=1}^{r} X_i) = \sum_{i=1}^{r} E(X_i)$$

and in this case

$$Var(X) = \frac{r \cdot q}{p^2} = Var(\sum_{i=1}^{r} X_i) = \sum_{i=1}^{r} Var(X_i).$$

verifying results previously obtained.

The fact that the expectation of a sum is the sum of the expectations is generally true; the fact that the variance of a sum is the sum of the variances requires independence of the summands. We will discuss these facts in a more thorough manner in Chapter 5.

In Figure 2.13 we show a graph of the negative binomial distribution with $r = 5$ and $p = \frac{1}{2}$. It shows that the probabilities increase to a maximum and then decline to become asymptotic to the x axis as Formula 2.10 would lead us to suspect.

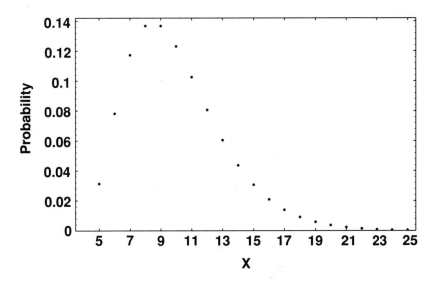

Figure 2.13   A negative binomial distribution.

It is also interesting to consider the *total* number of failures that precede the last success. If Y denotes the number of failures preceding the rth success, then

$$P(Y = y) = \binom{y + r - 1}{y} \cdot p^r \cdot q^y, \, y = 0, 1, 2, \ldots,$$

which is also a negative binomial distribution. Here

$$E(Y) = E(X - r) = E(X) - r = \frac{r}{p} - r = \frac{r \cdot q}{p},$$

and $Var(Y) = \dfrac{r \cdot q}{p^2}$.

We now consider three fairly complex examples involving the negative binomial distribution. Each involves special techniques.

Example 2.9.1 (All Heads)

I have some fair coins. I toss them once, together, and set aside any that come up heads. I continue to toss the coins remaining, on each toss removing those that come up heads, until all of the coins have come up heads. On average, how many (group) tosses will I have to make?

The problem is probably a bit hard at this point, so let's analyze the situation with only two fair coins.

Since the waiting time for a head with either coin is a geometric variable, we are interested in the maximum value of two geometric variables. Let Y be the random variable denoting the number of group tosses that must be made. We seek $P(Y = y)$. The last head can occur at the yth toss in two mutually exclusive ways:

1. Both coins come up tails for $y - 1$ tosses and then both come up heads on the yth toss,

or

2. Exactly one of the coins comes up heads on one of the first $y - 1$ tosses, followed by a head on the remaining coin on the yth toss.

The first of these possibilities has probability $(\frac{1}{4})^{y-1} \cdot (\frac{1}{4})$. To calculate the second, suppose first that there are $j - 1$ tosses where both coins show tails. Then one of the coins comes up heads on the jth toss. Finally the single remaining coin is tossed giving $y - j - 1$ tails followed by a head on the yth toss. This sequence of events has probability

$$\left(\frac{1}{4}\right)^{j-1} \cdot \left\{\binom{2}{1} \cdot \frac{1}{2} \cdot \frac{1}{2}\right\} \cdot \left(\frac{1}{2}\right)^{y-j-1} \cdot \frac{1}{2}.$$

To find the probability for the second possibility we must sum the above expression over all possible values of j. Thus the second possibility has probability

$$\sum_{j=1}^{y-1} \binom{2}{1} \cdot (\tfrac{1}{4})^j \cdot (\tfrac{1}{2})^{y-j}.$$

So, putting these results together,

$$P(Y = y) = \left(\frac{1}{4}\right)^y + \sum_{j=1}^{y-1} \binom{2}{1} \cdot \left(\frac{1}{4}\right)^j \cdot \left(\frac{1}{2}\right)^{y-j}, \quad y = 1, 2, 3, \ldots$$

$$= \left(\frac{1}{4}\right)^y + 2 \cdot \left(\frac{1}{2}\right)^y \sum_{j=1}^{y-1} \left(\frac{1}{2}\right)^{2j} \cdot \left(\frac{1}{2}\right)^{-j}$$

$$= \left(\frac{1}{4}\right)^y + 2 \cdot \left(\frac{1}{2}\right)^y \sum_{j=1}^{y-1} \left(\frac{1}{2}\right)^{j}$$

$$= \left(\frac{1}{4}\right)^y + 2 \cdot \left(\frac{1}{2}\right)^y \left[1 - \left(\frac{1}{2}\right)^{y-1}\right].$$

This reduces to

$$P(Y = y) = \frac{2^{y+1} - 3}{4^y}, \quad y = 1, 2, 3, \ldots .$$

A computer algebra system shows that the mean, and also the variance, of this distribution is $\frac{8}{3}$. ◀

Example 2.9.2

A fair coin is tossed repeatedly and a running count of the number of heads and tails is made. What is the probability that the heads count reaches five before the tails count reaches three?

Clearly the last toss must result in the fifth head which can be preceded by exactly zero, or one or two tails. Each of these probabilities is a negative binomial probability.

Let X denote the total number of tosses necessary and j denote the number of tails. Then, by the negative binomial distribution,

$$P(5 \text{ heads before } 3 \text{ tails}) = \sum_{j=0}^{2} \binom{4+j}{4} \cdot \left(\frac{1}{2}\right)^{5+j}$$

$$= \left(\frac{1}{2}\right)^5 + \binom{5}{4} \cdot \left(\frac{1}{2}\right)^6 + \binom{6}{4} \cdot \left(\frac{1}{2}\right)^7$$

$$= \frac{29}{128}.$$

It may be easier to see the structure of the answer if the coin is loaded. Let p denote the probability of a head. Then, reasoning as above,

$$P(5 \text{ heads before } 3 \text{ tails}) = \sum_{j=0}^{2} \binom{4+j}{4} p^5 q^j$$

and

$$P(\text{heads count reaches } h \text{ before the tails count reaches } t)$$

$$= \sum_{j=0}^{t-1} \binom{h-1+j}{h-1} q^j p^h = p^h \sum_{j=0}^{t-1} \binom{h-1+j}{h-1} q^j. \quad \blacktriangleleft$$

Example 2.9.3 (Candy Jars)

A professor has two jars of candy on his desk. When a student enters his office he or she is invited to choose a jar at random and then select a piece of candy. At some time one of the jars will be found empty. At that time, on average, how many pieces of candy are in the remaining jar?

The problem appears in the literature as *Banach's matchbook problem* after the famous Polish mathematician. It is an instance of Example 2.9.2.

We specialize the problem to two jars, each jar initially containing n pieces of candy, and we further suppose that each jar is selected with probability $\frac{1}{2}$.

Consider either of the jars and call it, for convenience, the first jar. Suppose we empty it and then, at some subsequent selection, choose it again and find that it is empty. Suppose further that the remaining jar at that point has X pieces of candy in it. Thus the first $n + (n - x)$ selections involve choosing the first jar exactly n times, and the last choice must be the first jar. Since the jars are symmetric and it makes no difference which we designate as the first jar,

$$P(X = x) = 2 \cdot \binom{2n - x}{n} \cdot \left(\frac{1}{2}\right)^{2n-x+1}, \quad x = 0, 1, 2, \ldots, n. \quad (2.11)$$

A graph of this probability distribution function, for $n = 15$, is shown in Figure 2.14. It shows that the most probable value for X is $x = 0$ or $x = 1$, and that the probabilities decrease steadily as x increases.

From the arguments used to establish Formula 2.11, it follows that

$$\sum_{x=0}^{n} 2 \cdot \binom{2n - x}{n} \cdot \left(\frac{1}{2}\right)^{2n-x+1} = 1.$$

A direct analytic proof of this is challenging. Finding the mean and variance is similarly difficult, so we show a way to find these using a recursion. (This method was also used to establish the mean and variance of the binomial distribution and is generally applicable to other discrete distributions.)   ◄

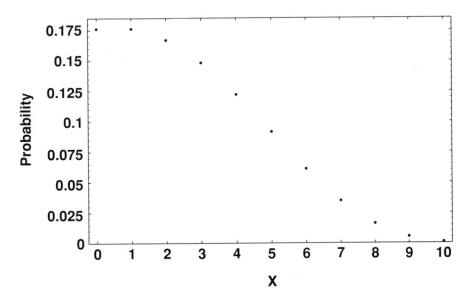

Figure 2.14   The candy jars problem for n = 15.

A Recursion

It is easy to use Formula 2.11 to show that

$$\frac{P(X = x)}{P(X = x - 1)} = 2 \cdot \frac{n - x + 1}{2n - x + 1}, \quad x = 1, 2, \ldots, n. \tag{2.12}$$

This can also be written as

$$\frac{P(X = x)}{P(X = x - 1)} = 1 - \frac{x - 1}{2n - (x - 1)}, \quad x = 1, 2, \ldots, n,$$

showing that the probabilities decrease as x increases, and that the most probable value is $x = 0$ or $x = 1$.

Now we seek the mean and the variance. Rearranging and summing Formula 2.12 from 1 to n (the region of validity for the recursion), we have

$$\sum_{x=1}^{n} (2n - x + 1) \cdot P(X = x) = 2 \cdot \sum_{x=1}^{n} (n - x + 1) \cdot P(X = x - 1).$$

This in turn can be written as

$$(2n + 1) \cdot [1 - P(X = 0)] - E(X) = 2n \cdot [1 - P(X = n)]$$
$$-2 \cdot [E(X) - n \cdot P(X = n)].$$

Simplifying and rearranging gives

$$E(X) = (2n + 1) \cdot \binom{2n}{n} \cdot \left(\frac{1}{2}\right)^{2n} - 1.$$

$E(X)$ is approximately a linear function of n as Figure 2.15 shows.
To find the variance of X, we first find $E(X^2)$. It follows from recursion 2.12 that

$$\sum_{x=1}^{n} x \cdot (2n - x + 1) \cdot P(X = x) = 2 \sum_{x=1}^{n} x \cdot (n - x + 1) \cdot P(X = x - 1).$$

The left side reduces to $(2n + 1) \cdot E(X) - E(X^2)$, while the right side can be written as

$$2n \cdot \sum_{x=1}^{n} (x - 1) \cdot P(X = x - 1) + 2n \cdot \sum_{x=1}^{n} P(X = x - 1)$$
$$- 2 \sum_{x=1}^{n} x \cdot (x - 1) \cdot P(X = x - 1),$$

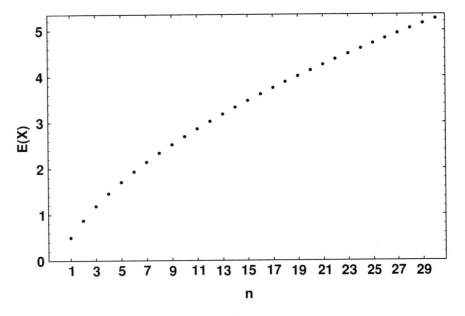

Figure 2.15 $E(X)$ for the candy jars problem.

which becomes

$$2n \cdot [E(X) - n \cdot P(X = n)] + 2n \cdot [1 - P(X = n)] - 2E[X(X+1)]$$
$$+2n(n+1)P(X = n).$$

It then follows that

$$Var(X) = 2(n+1) - (2n+1) \cdot \binom{2n}{n} \cdot \left(\frac{1}{2}\right)^{2n} - [(2n+1) \cdot \binom{2n}{n} \cdot \left(\frac{1}{2}\right)^{2n}]^2.$$

This is an increasing function of n. A graph is shown in Figure 2.16.

Exercises 2.9

1. A coin, loaded to come up heads $\frac{2}{3}$ of the time, is thrown until a head appears. What is the probability that an odd number of tosses is necessary?

2. The coin in exercise 1 is tossed until the fifth head appears. What is the probability this will occur in, at most, 9 tosses?

3. The probability of a successful rocket launching is 0.8, with the process following the binomial assumptions.

   a. Find the probability that the first successful launch occurs at the fourth attempt.

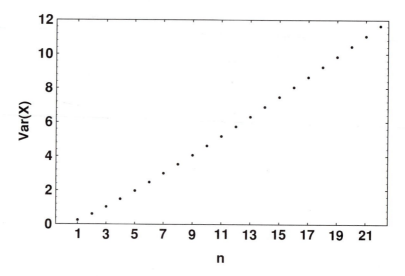

Figure 2.16   Variance in the candy jars problem.

    b. Suppose that attempts are made until 3 successful launchings have occurred. What is the probability that exactly six attempts will be necessary?

4. A box of manufactured parts contains 4 good and 3 defective parts. They are drawn out one at a time, without replacement. Let X denote the number of the drawing on which the first defective part occurs.

    a. Find the probability distribution for X.

    b. Find $E(X)$.

5. The probability that a player wins a game at a single trial is $\frac{1}{3}$. Assume the trials follow the binomial assumptions. If the player plays until he wins, find the probability that the number of trials is divisible by 4.

6. The probability a new driver will pass a driving test is 0.8.

    a. One student takes the test until she passes it. What is the probability it will take at least 2 attempts to pass the test?

    b. Now suppose 3 students take the driving test until each has passed it. What is the probability that exactly 1 of the 3 will take at least two attempts before passing the test? (Assume independence.)

7. To become an actuary, one must pass a series of 9 examinations. Suppose that 60% of those taking each examination pass it and that passing the examinations are independent of one another. What is the probability a person passes the ninth examination, and thus has passed all the examinations, on the 15th attempt?

8. A quality control inspector on a production line samples items until a defective item is found.

   a. If the probability an item is defective is 0.08, what is the probability that at least 10 items must be inspected?

   b. Suppose now that the 16th item inspected is the first defective item found. If p is the probability an item is defective, what is the value of p that makes the probability that the 16th item inspected is the first defective item found most likely?

9. A fair coin is tossed. What is the probability the fourth head is preceded by, at most, 2 tails?

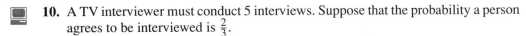

10. A TV interviewer must conduct 5 interviews. Suppose that the probability a person agrees to be interviewed is $\frac{2}{3}$.

   a. What is the probability that the interviewer will ask 9 people in all to be interviewed?

   b. How many people can the interviewer expect to ask to be interviewed?

11. In August, the probability that a thunderstorm will occur on any particular day is 0.1. What is the probability that the first thunderstorm in August will occur on August 12?

12. In a manufacturing process, the probability that a produced item is good is 0.97. Assuming the items produced are independent, what is the probability that exactly 5 defective items precede the 100th good item?

13. A box contains 6 good and 4 defective items. Items are drawn out one at a time, without replacement.

   a. Find the probability that the third defective item occurs on the fifth draw.

   b. On what drawing is it most likely for the third defective to occur?

14. A coin, loaded to come up heads with probability $\frac{3}{4}$, is tossed until a head appears or until it has been tossed 5 times. Find the probability that the experiment will end in an odd-numbered toss, given that the experiment takes more than one toss.

15. Suppose you are allowed to flip a fair coin until the first head appears. Let X denote the total number of flips required.

   a. Suppose you win $\$ 2^X$ if $X \le 19$ and $\$2^{20}$ if $X \ge 20$ for playing the game. A game is *fair* if the amount paid to play the game equals the expected winnings. How much should you pay to play this game if it is fair?

   b. Suppose now that you win $\$2^X$ regardless of the number of flips. Can the game be made fair?

16. Use the recursion 2.12 to find the most likely number of pieces of candy remaining when one of the candy jars is found empty.

17. X is a negative binomial random variable with p as the probability of success at any trial. Suppose the rth success occurs at trial t. Find the value of p that makes this event most likely.

2.10 ▶ The Hypergeometric Random Variable; Acceptance Sampling

2.10.1   Acceptance Sampling

Products produced from industrial processes are often subjected to sampling inspection before they are delivered to the customer. This sampling is done to insure a level of quality and uniformity in delivered manufactured products. Commonly, unacceptable products (products that do not meet the manufacturer's specifications) become mixed up with acceptable product due to changes in the manufacturing process and random events in that process. Modern techniques of statistical process control have greatly improved the quality of manufactured products and, while it is best to produce only flawless products, often the quality of a product can be determined only through sampling. However, determining whether a product is acceptable or unacceptable may destroy the product. Because of the time and money involved in inspecting the product in its entirety even if destruction of the product is not involved, *sampling plans,* which inspect only a sample of the product, are often employed. It has also been found that sampling is often more accurate than 100% inspection since the inspection of each and every item demands constant attention. Boredom or lack of care often sets in, which is not the case when smaller samples are randomly chosen at random times. As we will see, probability theory renders 100% inspection unnecessary even when it is possible, so total inspection of a manufactured product has become rare.

Due to the emphasis on quality in manufacturing and statistical process control, probability theory has become extremely important in industry.

The chance that a sample has a given composition can be determined from probability theory. As an example, suppose we have a lot (a number of produced items) containing eight acceptable, or good, items as well as four unacceptable items. A sample of three items is drawn. What is the probability the sample contains exactly one unacceptable item?

The sampling is done without replacement (since one would not want to inspect the same item repeatedly!), and since the order in which the items are drawn is of no importance, there are $\binom{12}{3} = 220$ samples comprising the sample space. If the sampling plan, that is, the manner in which the sampled items are drawn – is appropriate, we consider each of these samples to be equally likely. Now we must count the number of samples containing exactly one unacceptable item (and so exactly two acceptable items). There are $\binom{4}{1} \cdot \binom{8}{2} = 112$ such samples. So the probability that the sample contains exactly one unacceptable item is

$$\frac{\binom{4}{1} \cdot \binom{8}{2}}{\binom{12}{3}} = \frac{112}{220} = 0.509.$$

The probability that the sample contains no defective items is

$$\frac{\binom{4}{0} \cdot \binom{8}{3}}{\binom{12}{3}},$$

so the probability that the sampling plan will detect *at least* one unacceptable item is

$$1 - \frac{\binom{8}{3}}{\binom{12}{3}} = 0.745.$$

Our sampling plan is then likely to detect at least one of the unacceptable items in the lot, but it is not certain to do so.

Let's suppose that we carry out the inspection plan described and decide to sell the entire lot only if no unacceptable items are found in the sample. The probability that this lot survives this sampling plan and is sold is 0.255. So about 26% of the time, lots with $\frac{4}{12} = 33\frac{1}{3}\%$ unacceptable items will be sold.

Usually the sampling plan will determine some unacceptable items, which are not sent to the customer. One of two courses of action is generally pursued at this point. Either the unacceptable items in the sample are replaced with good items, or the entire lot is inspected and any unacceptable items in the lot are replaced by good items. Either of these plans will improve the quality of the product sold, the second being the better if it can be carried out. In case the testing is destructive, only the first plan can be executed. Let's compare the plans in this case, assuming that either can be carried out.

We start by replacing only the unacceptable items in the sample. The sample contains no unacceptable items with probability

$$\frac{\binom{8}{3}}{\binom{12}{3}} = \frac{14}{55},$$

so the outgoing lot will contain $\frac{4}{12}$ or $\frac{1}{3}$ unacceptable items with this probability.

The sample contains exactly one unacceptable item with probability

$$\frac{\binom{8}{2} \cdot \binom{4}{1}}{\binom{12}{3}} = \frac{28}{55},$$

producing $\frac{3}{12}$ or $\frac{1}{4}$ unacceptable items in the outgoing lot.

The sample contains exactly two unacceptable items with probability

$$\frac{\binom{8}{1} \cdot \binom{4}{2}}{\binom{12}{3}} = \frac{12}{55},$$

producing $\frac{2}{12}$ or $\frac{1}{6}$ unacceptable items in the outgoing lot.

Finally, the sample contains exactly three unacceptable items with probability

$$\frac{\binom{4}{3}}{\binom{12}{3}} = \frac{1}{55},$$

resulting in $\frac{1}{12}$ unacceptable items in the outgoing lot.

The result of this plan is that, on average, the percentage of unacceptable items the lot will contain is

$$\frac{14}{55} \cdot \frac{1}{3} + \frac{28}{55} \cdot \frac{1}{4} + \frac{12}{55} \cdot \frac{1}{6} + \frac{1}{55} \cdot \frac{1}{12} = 25\%.$$

This is considerably less than the $33\frac{1}{3}\%$ unacceptable items in the lot. Sampling cannot improve the quality of the product manufactured, but it can, and does, improve the quality of the product sold. In fact dramatic gains can be made by this process, which we will call *acceptance sampling*.

Even greater gains can be attained if, when the sample contains at least one unacceptable item, the entire lot is inspected and any unacceptable items in the lot are replaced by good items. In that circumstance, either the lot sold is 100% good (with probability 0.745) or the lot contains $\frac{4}{12} = 33\frac{1}{3}\%$ unacceptable items. Then the average percentage of unacceptable items sold is

$$0\% \cdot 0.745 + 33\frac{1}{3}\% \cdot 0.255 = 8.5\%.$$

This is a dramatic gain, and, as we shall see, is often possible if acceptance sampling is employed.

The average percentage of unacceptable product sold is called the *average outgoing quality*. The average outgoing quality, if only unacceptable items in the sample are replaced before the lot is sold, is 25%.

Lots are rarely so small as in our example, so we must investigate the behavior of this sampling plan when the lots are large. Before doing that we define the relevant random variable and determine some of its properties.

2.10.2   The Hypergeometric Random Variable

We generalize the situation in Section 2.10.1 to a lot of N items, D of which are unacceptable. Let X denote the number of unacceptable items in the randomly chosen sample of n items. Then

$$P(X = x) = \frac{\binom{D}{x} \cdot \binom{N-D}{n-x}}{\binom{N}{n}}, x = 0, 1, 2, \ldots, Min\{n, D\}. \qquad (2.13)$$

We assume that $Min\{n, D\} = n$ in what follows. The argument is similar if $Min\{n, D\} = D$.

If X has the probability distribution given by Formula 2.13, then X is called a *hypergeometric* random variable.

Since

$$\sum_{x=0}^{n} \binom{D}{x} \cdot \binom{N-D}{n-x}$$

represents all the mutually exclusive ways in which x unacceptable items and $n - x$ acceptable items can be chosen from a group of N items, this sum must be $\binom{N}{n}$, showing that the sum of the probabilities in Formula 2.13 must be 1.

We will use a recursion to find the mean and variance. Let $G = N - D$. Then, from Formula 2.13,

$$\frac{P(X = x)}{P(X = x - 1)} = \frac{(D - x + 1)(n - x + 1)}{x(G - n + x)}, \ x = 1, 2, \ldots, n. \qquad (2.14)$$

So,

$$(G - n) \sum_{x=1}^{n} x P(X = x) + \sum_{x=1}^{n} x^2 P(X = x)$$
$$= \sum_{x=1}^{n} (D - x + 1)(n - x + 1) P(X = x - 1).$$

After expanding and simplifying the sums involved, we find that

$$E(X) = n \cdot \frac{D}{N}.$$

This result is analogous to the mean of the binomial, np, but here D/N is the probability that only the first item drawn is unacceptable. It is surprising that the non-replacement does not affect the mean value. The drawings for the hypergeometric are clearly dependent, a fact that will affect the variance.

To find $E(X^2)$, multiply Formula 2.14 through by x, giving

$$(G - n) \sum_{x=1}^{n} x^2 P(X = x) + \sum_{x=1}^{n} x^3 P(X = x)$$

$$= \sum_{x=1}^{n} x \cdot (D - x + 1)(n - x + 1)P(X = x - 1).$$

These quantities can be expanded and simplified using the result for $E(X)$. We find that

$$E(X^2) = \frac{nD}{N(N - 1)} \cdot (nD - n - D + N),$$

from which it follows that

$$\text{Var}(X) = n \cdot \frac{D}{N} \cdot \frac{N - D}{N} \cdot \frac{N - n}{N - 1}.$$

This result is analogous to the variance, $n \cdot p \cdot q$, of the binomial but involves a factor, $\frac{N-n}{N-1}$, often called a *finite population correction factor*, due to the fact that the drawings are not independent.

The correction factor, however, approaches 1 as $N \to \infty$ and so the variance of the hypergeometric approaches that of the binomial. This result, together with the mean value, suggest that the hypergeometric distribution can be approximated by the binomial distribution as the population size, N, increases. This is due to the fact that as N increases, the non-replacement of the items drawn has less and less effect on the probabilities involved. We now show that is indeed the case.

We begin with

$$P(X = x) = \frac{\binom{D}{x} \cdot \binom{N - D}{n - x}}{\binom{N}{n}},$$

which can be written as

$$P(X = x) = \frac{D(D-1)(D-2)\cdots(D-x+1)}{x!} \cdot$$

$$\frac{(N-D)(N-D-1)\cdots(N-D-n+x+1)}{(n-x)!} \cdot$$

$$\frac{n!}{N(N-1)(N-2)\cdots(N-n+1)} \cdot$$

This in turn can be rearranged as

$$P(X = x) = \binom{n}{x} \cdot \frac{D}{N} \cdot \frac{D-1}{N-1} \cdots \frac{D-x+1}{N-x+1} \cdot$$

$$\frac{N-D}{N-x} \cdot \frac{N-D-1}{N-x-1} \cdots \frac{N-D-n+x+1}{N-n+1} \cdot$$

Approximating each of the factors

$$\frac{D}{N}, \frac{D-1}{N-1}, \ldots, \frac{D-x+1}{N-x+1} \text{ by } \frac{D}{N},$$

and each of the factors

$$\frac{N-D}{N-x}, \frac{N-D-1}{N-x-1}, \ldots, \frac{N-D-n+x+1}{N-n+1} \text{ by } \frac{N-D}{N},$$

we see that

$$P(X = x) \approx \binom{n}{x} \cdot \left(\frac{D}{N}\right)^x \cdot \left(\frac{N-D}{N}\right)^{n-x},$$

which is the binomial distribution.

2.10.3   Some Specific Hypergeometric Distributions

It is useful at this point to look at some specific hypergeometric distributions. Our initial example, in section 2.10.1, had $N = 12$, $n = 3$, and $D = 4$. A graph of the probability distribution is shown in Figure 2.17.

As the population size increases, we expect the hypergeometric distribution to appear more binomial, or normal-like. Figure 2.18 shows that this is the case. Here $N = 1000$, $D = 400$, and $n = 30$.

While Figure 2.17 on the next page shows no particular features, Figure 2.18 shows the now-familiar normal appearance.

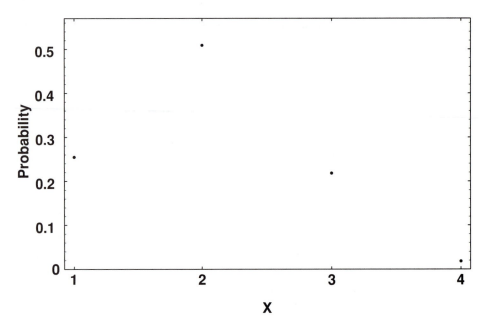

Figure 2.17   Hypergeometric distribution with $N = 12$, $n = 3$, and $D = 4$.

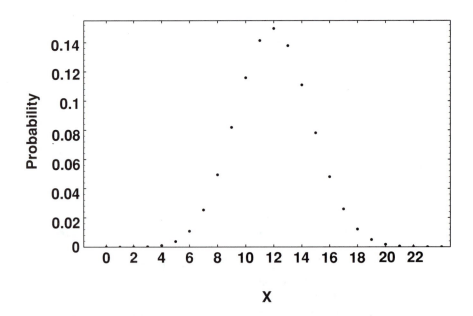

Figure 2.18   Hypergeometric distribution with $N = 1000$, $n = 30$, and $D = 400$.

Exercises 2.10

1. A lot of 50 fuses is known to contain 7 defectives. A random sample of size 10 is drawn without replacement. What is the probability the sample contains at least 1 defective fuse?

2. A collection of 30 gems, all of which are identical in appearance and are supposed to be genuine diamonds, actually contains 8 worthless stones. The genuine diamonds are valued at $1200 each. Two gems are selected.

    a. Let X denote the total actual value of the gems selected. Find the probability distribution function for X.

    b. Find $E(X)$.

3. a. A box contains 3 red and 5 blue marbles. The marbles are drawn out one at a time and without replacement, until all of the red marbles have been selected. Let X denote the number of drawings necessary. Find the probability distribution function for X.

    b. Find the mean and variance for X.

4. a. A box contains 3 red and 5 blue marbles. The marbles are drawn out one at a time and without replacement, until all the marbles left in the box are of the same color. Let X denote the number of drawings necessary. Find the probability distribution function for X.

    b. Find the mean and variance for X.

5. A lot of 400 automobile tires contains 10 with blemishes which cannot be sold at full price. A sampling inspection plan chooses 5 tires at random and accepts the lot only if the sample contains no tires with blemishes.

    a. Find the probability that the lot is accepted.

    b. Suppose any tires with blemishes in the sample are replaced by good tires if the lot is rejected. Find the average outgoing quality of the lot.

6. A sample of size 4 is chosen from a lot of 25 items of which D are defective. Draw the curve showing the probability that the lot is accepted as a function of D if the lot is accepted only when the sample contains no defective items.

7. A lot of 250 items that contains 15 defective items is subject to an acceptance sampling plan that calls for a sample of size 6 to be drawn. The lot is accepted if the sample contains, at most, one defective item.

    a. Find the probability that the lot is accepted.

    b. Suppose any defective items in the sample are replaced by good items. Find the average outgoing quality.

8. In exercise 5, suppose now that the entire lot is inspected and any blemished tires are replaced by good tires if the lot is rejected by the sample. Find the average outgoing quality.

9. In exercise 7, if any defective items in the lot are replaced by good items when the sample rejects the entire lot, find the average outgoing quality.

10. Exercises 3 and 4 can be generalized. Suppose a box has a red and b blue marbles and that X is the number of drawings necessary to draw out all of the red marbles.

   a. Show that
   $$P(X = x) = \frac{\binom{x-1}{a-1}}{\binom{a+b}{a}}, \quad x = a, a + 1, \ldots, a + b.$$

   b. Using the result in part **a**, show that a recursion can be simplified to
   $$\frac{P(X = x)}{P(X = x - 1)} = \frac{x-1}{x-a}, \quad x = a + 1, a + 2, \ldots, a + b.$$

   c. Show that the recursion in part **b** leads to
   $$\sum_{x=a+1}^{a+b} x \cdot (x - a) \cdot P(X = x) = \sum_{x=a+1}^{a+b} x \cdot (x - 1) \cdot P(X = x - 1).$$
   From this conclude that
   $$E(X) = a \cdot \frac{a + b + 1}{a + 1}.$$

   d. Show that
   $$\text{Var}(X) = \frac{a \cdot b \cdot (a + b + 1)}{(a + 1)^2 \cdot (a + 2)}.$$

11. (Exercise 10 continued.) Now suppose X represents the number of drawings until all the marbles remaining in the box are of the same color. Show that
   $$P(X = x) = \frac{\binom{x-1}{a-1} + \binom{x-1}{b-1}}{\binom{a+b}{a}}, \quad x = \text{Min}[a, b], \ldots, a + b - 1.$$
   and that
   $$E(X) = \frac{a \cdot b}{a + 1} + \frac{a \cdot b}{b + 1}.$$

12. A box contains 3 red and 5 blue marbles. The marbles are drawn out one at a time without replacement until a red marble is drawn. Let X denote the total number of drawings necessary.

a. Find the probability distribution function for X.

b. Find the mean and the variance of X.

13. Exercise 12 is generalized here. Suppose a box contains a red and b blue marbles and that X denotes the total number of drawings made without replacement until a red marble is drawn.

   a. Show that

   $$P(X = x) = \frac{\binom{a+b-x}{a-1}}{\binom{a+b}{a}}, x = 1, 2, \ldots, b+1.$$

   b. Using the result in part **a**, show that a recursion can be simplified to

   $$\frac{P(X = x)}{P(X = x-1)} = \frac{b-x+2}{a+b-x+1}, \ x = 2, 3, \ldots, b+1.$$

   c. Use the recursion in part **b** to show that

   $$E(X) = \frac{a+b+1}{a+1}$$

   and

   $$\text{Var}(X) = \frac{a \cdot b \cdot (a+b+1)}{(a+1)^2 \cdot (a+2)}.$$

   d. Show that the mean and variance in part **c** approach the mean and variance of the geometric random variable as both a and b become large.

2.11 ▶ Acceptance Sampling (Continued)

We considered an acceptance sampling plan in section 2.10.1, and we saw that some gains can be made with respect to the average quality delivered when the unacceptable items either in the sample or in the entire lot are replaced with good items. We can now discuss some specific results, dealing with lots that are usually large. We first consider the effect of the size of the sample on the process.

Example 2.11.1

A lot of 200 items is inspected by drawing a sample of size n without replacement; the lot is accepted only if all the items in the sample are good.

Suppose the lot contains 2%, or four, unacceptable items. The probability that the lot is accepted by this sampling plan is

$$\frac{\dbinom{196}{n}}{\dbinom{200}{n}}.$$

This is a steadily decreasing function of n, as we would expect. We find that if $n = 5$, the probability that the lot is accepted is 0.903, while if $n = 30$, this probability is 0.519. A graph of this function is shown in Figure 2.19.

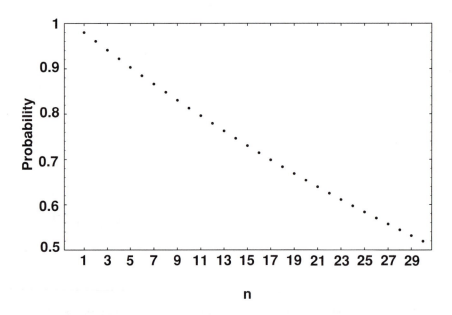

Figure 2.19   Effect of sample size, n, on a sampling plan.

Not surprisingly, large samples yield more accurate results than small samples.   ◄

Example 2.11.2

Now we consider the effect of the quality of the lot on the probability of acceptance. Suppose $p\%$ of a lot of 1000 items are unacceptable. The sampling plan is this: Select a sample of 100 and accept the lot if the sample contains, at most, 4 unacceptable items. The probability that the lot is accepted is then

$$\sum_{x=0}^{4} \frac{\binom{1000p}{x} \cdot \binom{1000 - 1000p}{100 - x}}{\binom{1000}{100}}.$$

This is a decreasing function of the percentage of unacceptable items in the lot. These values are easily calculated. If, for example, the lot contains 10 unacceptable items, then the probability that the lot is accepted is 0.9985.

A graph of this probability as a function of p is shown in Figure 2.20.

The curve in Figure 2.20 is called the *operating characteristic* (or OC) curve for the sampling plan. Sampling plans are often compared by comparing the rapidity with which the OC curves for different plans decrease.

In this case the sample size is small relative to the population size, so we would expect that the non-replacement of the sample items will have little effect on the probability that the lot is accepted. A binomial model approximates the probability that the lot is accepted if in fact it contains 10 unacceptable items as 0.9966 (we found the exact probability above to be 0.9985). ◀

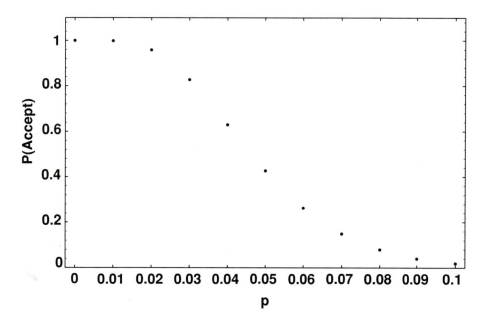

Figure 2.20   Effect of quality in the lot on the probability of acceptance.

2.11.3 Producer's and Consumer's Risks

Acceptance sampling involves two types of risk. The producer would like to guard against a "good" lot being rejected, although this cannot be guaranteed. The consumer, on the other hand, wants to guard against a "poor" lot being accepted by the sampling plan, although, again, this cannot be guaranteed.

The words "good" and "poor" of course must be decided in the context of the practical situation. Often when these are defined and the probability of the risks set, a sampling plan can be devised (specifically, a sample size can be determined) that, at least approximately, meets the risks set.

Consider Example 2.11.1 again. Here the lot size is 200, but suppose that D of these items are unacceptable. Again we draw a sample of size n and accept the lot when the sample contains no unacceptable items. So,

$$P(\text{lot is accepted}) = \frac{\binom{200 - D}{n}}{\binom{200}{n}}.$$

Figure 2.21 shows this probability as a function of the sample size, n, where D has been varied from 0 to 25.

Since the curves are monotonically decreasing, it is often possible to select a curve (thus determining a sample size) that passes through two given points. If the producer would like lots with exactly 1 unacceptable item rejected with probability 0.10 (so

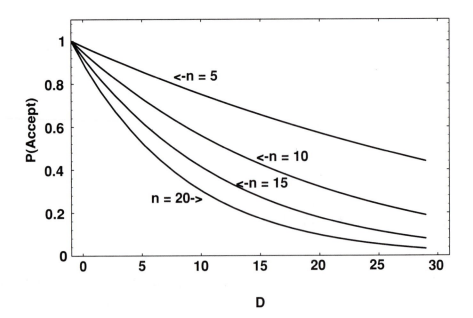

Figure 2.21   Some operating characteristic curves.

such lots are accepted with probability 0.90), and if the consumer would like lots with 24 unacceptable items rejected with probability 0.95 (so such lots are accepted with probability 0.05), we find a sample size of 22 will approximate these restrictions. To check this, note that

$$\frac{\binom{199}{22}}{\binom{200}{22}} = 0.89 \quad \text{and} \quad \frac{\binom{176}{22}}{\binom{200}{22}} = 0.05.$$

A computer algebra system is of great help here in finding an approximate solution to the problem.

2.11.4   Average Outgoing Quality

We saw that considerable improvement in the quality of the product sold can be made if any items in the sample are replaced by good items. This is the most sensible strategy we can follow if the sampling is destructive; in that case we have little choice but to take a sample because destroyed products must be replaced by others. Recall that the *average outgoing quality* (AOQ) is the percentage of unacceptable items sold to the buyer. We want to consider the behavior of the average outgoing quality in this section.

Example 2.11.4.1

Suppose a lot of 100 items contains 4 unacceptable items. A sample of 5 items is drawn and any unacceptable item in the sample is replaced by a good item. On average, what proportion of unacceptable items is sold using this sampling plan?

Let X denote the number of unacceptable items in the sample. Then X is a hypergeometric random variable and

$$P(X = x) = \frac{\binom{4}{x} \cdot \binom{96}{5-x}}{\binom{100}{5}}, \quad x = 0, 1, 2, 3, 4.$$

So the average outgoing quality is

$$AOQ = \sum_{x=0}^{4} \frac{(4-x)}{100} \cdot \frac{\binom{4}{x} \cdot \binom{96}{5-x}}{\binom{100}{5}}$$

since $4 - X$ unacceptable items will be sold. But

$$AOQ = \frac{4}{100} - \frac{1}{100} \sum_{x=0}^{4} x \cdot \frac{\binom{4}{x} \cdot \binom{96}{5-x}}{\binom{100}{5}},$$

where the summation is the mean value of a hypergeometric random variable with $N = 100$, $n = 5$, and $D = 4$. It follows that

$$AOQ = \frac{4}{100} - \frac{1}{100} \cdot 5 \cdot \frac{4}{100} = 0.038.$$

This is less than the population percentage of unacceptable items, 0.04, but not greatly less. The sampling has improved the quality of the product sold, but not by much. The effect the sampling plan has will increase as the percentage of unacceptable items in the lot increases.

Another possible plan is to replace each unacceptable item in the lot with a good item if the sample contains any unacceptable items. Now we either deliver all the unacceptable items in the lot, or none of them. It follows that

$$AOQ = \frac{4}{100} \cdot P \text{ (sample contains no unacceptable items)}$$

$$= \frac{4}{100} \cdot \frac{\binom{96}{5}}{\binom{100}{5}} = 0.032475.$$

So the gain is greater if we happen to inspect the entire lot.

These conclusions are probably not surprising. But more lurks behind the scenes here! Let's consider a general example.   ◄

Example 2.11.4.2

From a lot of N items that contains D unacceptable items we draw a sample of size n. If the sample contains any unacceptable items, we inspect the entire lot, replacing each unacceptable item with an acceptable, or good, item. The resulting lot then contains either D or 0 unacceptable items. The average outgoing quality is then

$$AOQ = \frac{D}{N} \cdot P(\text{sample contains no unacceptable items})$$

$$= \frac{D}{N} \cdot \frac{\binom{N-D}{n}}{\binom{N}{n}}. \qquad (2.15)$$

What happens to this product as D increases? Since $\frac{D}{N}$ increases as D increases and

$$\frac{\binom{N-D}{n}}{\binom{N}{n}}$$

is a decreasing function of D, it follows in this case that the product in Formula 2.15 attains a maximum value. This is true, regardless of the size of D! So, no matter what D is, there is a limit for the percentage of unacceptable product sold. This is called the *average outgoing quality limit*. We illustrate this phenomenon in the next example. ◄

Example 2.11.4.3

Consider again the situation when $N = 1000$ and $n = 100$. Then

$$AOQ = \frac{D}{1000} \cdot \frac{\binom{1000-D}{100}}{\binom{1000}{100}}.$$

A graph of the AOQ is shown in Figure 2.22, illustrating that the maximum value of the average outgoing quality is about 0.35%. ◄

2.11.5 Double Sampling

Occasionally lots are accepted if the sample contains, say, at most, c unacceptable items and are subject to total inspection if a sample has d or more unacceptable items, where $d > c$. Often if the number of unacceptable items falls between c and d, another sample is taken. We illustrate this procedure with a concrete example.

A lot of 500 items contains 40 unacceptable items. A sample of 50 is taken and the lot accepted if the sample contains no more than 3 unacceptable items. If the sample contains 4 or 5 unacceptable items, an additional sample of 30 is taken; the lot is accepted only if this additional sample contains no unacceptable items. Otherwise, the lot is rejected. We want the probability that the lot is accepted.

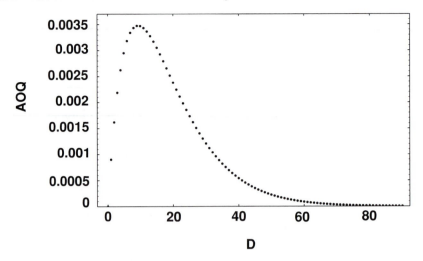

Figure 2.22   AOQ as a function of the number of unacceptable items in the lot.

Let X denote the number of unacceptable items in the first sample. Then the probability that the lot is accepted on the basis of the first sample is

$$P(X \le 3) = \sum_{x=0}^{3} \frac{\binom{40}{x} \cdot \binom{460}{50-x}}{\binom{500}{50}}.$$

Now if the first sample has 4 or 5 unacceptable items, then the second sample is taken. The lot now contains 450 items of which $40 - X$ are unacceptable while $450 - (40 - X) = 410 + X$ are good items. The second sample of size 30 must contain only good items. So the probability that the lot is accepted on the basis of the second sample is

$$\sum_{x=4}^{5} \frac{\binom{40}{x} \cdot \binom{460}{50-x}}{\binom{500}{50}} \cdot \frac{\binom{410+x}{30}}{\binom{450}{30}}.$$

The probability that the lot is accepted is then

$$\sum_{x=0}^{3} \frac{\binom{40}{x} \cdot \binom{460}{50-x}}{\binom{500}{50}} + \sum_{x=4}^{5} \frac{\binom{40}{x}\binom{460}{50-x}}{\binom{500}{50}} \cdot \frac{\binom{410+x}{30}}{\binom{450}{30}} = 0.445334 \, ,$$

and, assuming that none of the unacceptable items found in the samples is sold, the average outgoing quality is

$$\sum_{x=0}^{3} \frac{(40-x)\cdot\binom{40}{x}\cdot\binom{460}{50-x}}{500\cdot\binom{500}{50}} +$$

$$\sum_{x=4}^{5} \frac{(40-x)\cdot\binom{40}{x}\cdot\binom{460}{50-x}}{500\cdot\binom{500}{50}}\cdot\frac{\binom{410+x}{30}}{\binom{450}{30}}$$

$$= 0.0334579.$$

Here the sampling plan has reduced the percentage of unacceptable item sold from $\frac{40}{500} = 0.08$ to 0.033 on average, so the plan is quite effective.

With the aid of a computer we can easily vary the number of unacceptable items in the lot and observe the effect this has on both the probability that the lot is accepted and the average outgoing quality. These graphs are shown in Figures 2.23 and 2.24.

The maximum value for the average outgoing quality is 0.0386071 and occurs when the lot contains 29 unacceptable items.

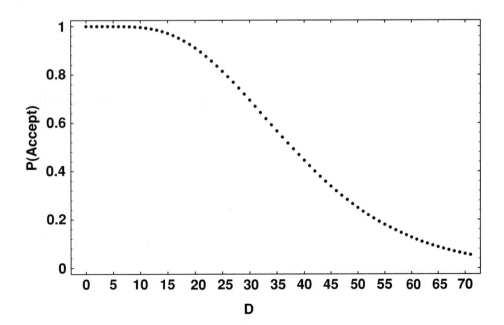

Figure 2.23 Probability lot is accepted in a double sampling plan.

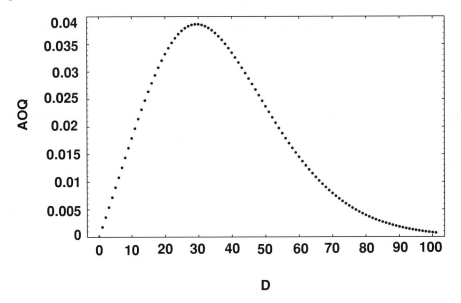

Figure 2.24   Average outgoing quality for a double sampling plan.

Exercises 2.11

1. A day's production of 25 television sets from a small company has 4 which have defects and cannot be sold. The company inspects its product by selecting 2 sets; if at most 1 of these has defects, the lot is shipped. What is the probability the lot is shipped?

2. A shipment of 1500 washers contains 400 defective and 1100 non-defective items. Two hundred washers are chosen at random, without replacement.

    a. Find the probability that exactly 90 defective items are found.
    b. Approximate the probability in part **a** by using the binomial distribution.

3. A lot of 100 fuses is inspected by a quality control engineer who tests 10 fuses selected at random. If 2 or fewer defective fuses are discovered, the entire lot is accepted. Find the probability that the lot is accepted if it actually contains 20 defective fuses.

4. A lot of 25 items contains 4 defective items. A sample of size 2 is chosen; the lot is accepted if the sample shows no defective items.

    a. Find the probability that the lot is shipped.
    b. If any defective items in the sample are replaced by good items before the lot is shipped, find the average outgoing quality.
    c. Now suppose the lot contains D defective items and that the entire lot is rectified if the sample shows any defective items. Plot the operating characteristic curve.

5. A bakery has a batch of 100 cookies, 5 of which are burned. A sample of 3 cookies is chosen and the batch put out for sale if none of the cookies in the sample is burned.

   a. What is the probability that the batch of cookies is put out for sale?
   b. Find the average outgoing quality if any burned cookies in the sample are replaced by good cookies.
   c. Assuming that the batch contains B burned cookies and that the entire batch is rectified if any of the cookies in the sample is burned, show the operating characteristic curve.

6. In exercise 4, suppose that the number of defective items is unknown and that a rejected lot is subject to 100% inspection and any defective item in the population is replaced by a good item. Estimate the average outgoing quality limit from a graph of the average outgoing quality.

7. In exercise 5, suppose that the entire batch of cookies is inspected if the sample should reject the batch. Estimate the average outgoing quality limit from a graph.

8. In inspecting a lot of 500 items, it is desired to accept the lot if the lot contains 1 defective item with probability 0.95, and it is desired to accept the lot if the lot contains 20 defective items with probability 0.05. Suppose the lot is accepted only if the sample contains no defective items. What sample size is necessary?

9. A producer inspects a lot of 400 items and wants the probability that the lot is accepted if the lot contains 1% defectives to be 0.90; the consumer wants the probability a lot containing 5% defective items is accepted to be 0.60. Suppose the lot is accepted only if the sample contains no defective items. Find the sample size so that the sampling plan meets these risks.

10. A double sampling plan is carried out from a lot of 500 items. A sample of 10 is selected and the lot is accepted if this sample contains no unacceptable items; if this sample contains 3 or more unacceptable items, the lot is rejected. If the sample contains 1 or 2 unacceptable items, a second sample of 20 is drawn; the lot is then accepted if the total number of unacceptable items in the 2 samples combined is at most 3. Suppose that at any stage an unacceptable item is replaced by a good item.

   a. Find the probability that a lot containing 15 unacceptable items is accepted.
   b. Graph the probability in part **a** as a function of D, the number of unacceptable items in the lot.
   c. Find the AOQL for this double sampling plan if unacceptable lots are rectified.
   d. Approximate the probability that the lot is accepted using the binomial distribution.

11. A lot of 400 items containing 3 defective items is subject to the following double sampling plan: The lot is accepted if a first sample of 5 contains no defectives; the lot is rejected if this sample contains 2 or more defectives; if the first sample contains

1 defective, a second sample of 7 is drawn; the lot is accepted if the second sample contains no more than 1 defective and, otherwise, the lot is rejected. Suppose that at any stage an unacceptable item is replaced by a good item.

a. What is the probability that the lot is accepted?

b. What is the average outgoing quality if all the defective items in unacceptable lots are replaced by good items ?

c. Show the operating characteristic curve for the sampling plan.

12. A day's production of 200 compact discs is inspected as follows. If an initial sample of 15 shows, at most, 2 defective discs, the lot is accepted and is subject to no more sampling. However, if the first sample shows 3 or more defective discs, then a second sample of 20 discs is chosen and the lot is accepted if the total number of defectives in the two samples is no more than 4.

a. Find the probability that the lot is accepted if, in fact, it contains 10 defective discs.

b. Find the average outgoing quality.

c. Plot the operating characteristic curve.

13. A random sample of 100 items is chosen from a lot of 4500 items that is 2% defective. If the sample contains no more than 4 defective items, the lot is accepted; otherwise, the remainder of the lot is inspected and defective items are replaced by good items.

a. What is the average number of items inspected?

b. Graph the average number of items inspected as a function of the percentage defective in the lot.

2.12 ▶ The Hypergeometric Random Variable; Further Examples

Example 2.12.1 (A Lottery)

Lottery games have become popular in many states. In Indiana, the game is played as follows: A player chooses five different numbers from the integers $1, 2, \ldots, 45$. Another integer is then chosen from the same set; this choice, called a *powerball*, may match one of the first five integers chosen. Lottery officials then choose five integers and the powerball.

The number of integers the player correctly chooses from among the first five is a hypergeometric random variable which we call X. Then

$$P(X = x) = \frac{\binom{5}{x} \cdot \binom{40}{5-x}}{\binom{45}{5}}, \quad x = 0, 1, \ldots, 5.$$

Let Y denote the number of correct powerball choices made. Then

$$P(Y = y) = \frac{\binom{1}{y} \cdot \binom{44}{1-y}}{\binom{45}{1}}, \quad y = 0, 1.$$

Since the choices are independent, it follows that

$$P(X = x \text{ and } Y = y) = \frac{\binom{5}{x} \cdot \binom{40}{5-x}}{\binom{45}{5}} \cdot \frac{\binom{1}{y} \cdot \binom{44}{1-y}}{\binom{45}{1}},$$

$$x = 0, 1, \ldots, 5; \quad y = 0, 1.$$

Here is a table of values of X and Y giving the probabilities with which the possible values occur, along with the payoffs to the player. The jackpot varies from week to week.

X	Y	Probability	Payoff	X	Y	Probability	Payoff
5	1	$\dfrac{1}{54,979,155}$	Jackpot	2	1	$\dfrac{19,760}{10,995,831}$	$5
5	0	$\dfrac{4}{4,998,105}$	$100,000	2	0	$\dfrac{79,040}{999,621}$	$0
4	1	$\dfrac{40}{10,995,831}$	$5,000	1	1	$\dfrac{91,390}{10,995,831}$	$2
4	0	$\dfrac{160}{999,621}$	$100	1	0	$\dfrac{366,560}{999,621}$	$0
3	1	$\dfrac{520}{3,665,277}$	$100	0	1	$\dfrac{73,112}{6,108,795}$	$1
3	0	$\dfrac{2080}{333,207}$	$5	0	0	$\dfrac{292,448}{555,345}$	$0

The probability that the player selects at least one of the five integers correctly is

$$1 - \frac{\binom{5}{0} \cdot \binom{40}{5}}{\binom{45}{5}} = \frac{62,639}{135,751} = 0.461426 \, .$$

This explains why the lottery offers no payoff for players choosing one or two integers correctly from the first five.

The expected payoff is found by multiplying the probabilities by the payoffs and adding the results. If the jackpot is J, this is

$$\frac{140,804}{714,015} + \frac{J}{54,979,155} = 0.1972 + \frac{J}{54,979,155} \, .$$

So the value of a $1 ticket is roughly 20 cents plus the jackpot divided by 55,000,000. The jackpot must reach $16,647,670 before the expected value of a ticket is 50 cents.   ◄

Example 2.12.2 (A Card Game)

A bridge hand consists of 13 cards chosen without replacement from a deck of 52 cards. What is the most likely distribution of suits in the bridge hand?

In the hypergeometric random variable, the sampling is done from a population containing two kinds of items; here, we generalize the distribution somewhat to sample from a population containing four kinds of items, namely the suits in the deck.

It would appear, since the suits all occur with equal frequency in the deck, that the most likely distribution of suits might be four of one suit and three each of the remaining three suits. This has probability

$$P(4, 3, 3, 3) = \frac{\binom{4}{3} \cdot \binom{13}{3}^3 \cdot \binom{13}{4}}{\binom{52}{13}}$$

because we first choose three suits; then we choose three cards from each of those suits; finally we choose four cards from the remaining suit.

However, the distribution of four cards from each of two suits, three cards from one suit, and two cards from the remaining suit has probability

$$P(4, 4, 3, 2) = \frac{\binom{4}{2} \cdot \binom{13}{4}^2 \cdot \binom{2}{1} \cdot \binom{13}{3} \cdot \binom{13}{2}}{\binom{52}{13}}.$$

We find that

$$\frac{P(4, 4, 3, 2)}{P(4, 3, 3, 3)} = \frac{45}{22},$$

so the distribution of four cards from each of two suits, three cards from another suit, and the remaining two cards from the fourth suit is more than twice as likely as the more uniform distribution of suits. Any other combination of suits is less likely than the combination found here.   ◄

2.13 ▶ The Poisson Random Variable

In section 2.4 we considered the binomial random variable whose probability distribution is

$$P(X = x) = \binom{n}{x} \cdot p^x \cdot q^{n-x}, \; x = 0, 1, 2, \ldots, n$$

where $q = 1 - p$.

Events that have small probability – rare events – are of particular interest, and we turn our attention to them.

We calculated the recursion from the binomial probability distribution function:

$$P(X = x) = \frac{p}{q} \cdot \frac{n - x + 1}{x} \cdot P(X = x - 1), \; x = 1, 2, \ldots, n. \qquad (2.16)$$

Notice that the recursion could also be written as

$$P(X = x) = \frac{np - p(x - 1)}{(1 - p) \cdot x} \cdot P(X = x - 1).$$

In this form of the recursion we fix x and let $n \to \infty$ and $p \to 0$ while keeping np, which we denote by λ, fixed. These presumptions allow us to concentrate on events that are rare. We see that, under these conditions, in the limit we have

$$P(X = x) = \frac{\lambda}{x} \cdot P(X = x - 1), x = 1, 2, 3, \ldots \qquad (2.17)$$

Our task now is to determine the function $P(X = x)$ that satisfies the recursion 2.17. Applying the recursion repeatedly, we find that

$$P(X = 1) = \frac{\lambda}{1} \cdot P(X = 0),$$

$$P(X = 2) = \frac{\lambda}{2} \cdot P(X = 1) = \frac{\lambda^2}{1 \cdot 2} \cdot P(X = 0),$$

$$P(X = 3) = \frac{\lambda}{3} \cdot P(X = 2) = \frac{\lambda^3}{1 \cdot 2 \cdot 3} \cdot P(X = 0)$$

and so on. Since $\sum_{n=0}^{\infty} P(X = x) = 1$, it follows that

$$P(X = 0) \cdot \{1 + \lambda + \frac{\lambda^2}{2!} + \frac{\lambda^3}{3!} + \cdots\} = 1$$
or
$$P(X = 0) \cdot e^\lambda = 1$$
so
$$P(X = 0) = e^{-\lambda}$$
and
$$P(X = 1) = \lambda \cdot e^{-\lambda}$$
and
$$P(X = 2) = \frac{\lambda^2 \cdot e^{-\lambda}}{2!} \; .$$

We conjecture then that

$$P(X = x) = \frac{e^{-\lambda} \cdot \lambda^x}{x!}, \quad x = 0, 1, 2, \ldots \tag{2.18}$$

Formula 2.18 defines the *Poisson* probability distribution with parameter λ. The reader should check that Formula 2.18 satisfies Formula 2.17. It is also easy to check that Formula 2.18 is a probability distribution. Obviously, $P(X = x) \geq 0$ and

$$\sum_{x=0}^{\infty} P(X = x) = \sum_{x=0}^{\infty} \frac{e^{-\lambda} \cdot \lambda^x}{x!} = e^{-\lambda} \sum_{x=0}^{\infty} \frac{\lambda^x}{x!}$$
$$= e^{-\lambda}(1 + \lambda + \frac{\lambda^2}{2!} + \frac{\lambda^3}{3!} + \cdots) = e^{-\lambda} \cdot e^\lambda = 1.$$

So Formula 2.18 is a probability distribution function.

2.13.1   Mean and Variance of the Poisson

It should be no surprise that the mean of the Poisson distribution is np since we found the Poisson by taking the limit of the binomial with mean np and keeping np fixed. The calculation is as follows:

$$\mu = \sum_{x=0}^{\infty} x \cdot P(X = x) = \sum_{x=0}^{\infty} x \cdot \frac{e^{-\lambda} \cdot \lambda^x}{x!}$$

$$= e^{-\lambda} \cdot \lambda \cdot \sum_{x=1}^{\infty} \frac{\lambda^{x-1}}{(x-1)!}$$

$$= e^{-\lambda} \cdot \lambda \cdot e^{\lambda} = \lambda.$$

For the variance, it is easiest to calculate $E[X \cdot (X - 1)]$ and then make use of the fact that

$$\text{Var}(X) = E[X \cdot (X - 1)] + E(X) - [E(X)]^2.$$

Here

$$E[X \cdot (X - 1)] = \sum_{x=0}^{\infty} x \cdot (x - 1) \cdot \frac{e^{-\lambda} \cdot \lambda^x}{x!}$$

$$= e^{-\lambda} \cdot \lambda^2 \cdot e^{\lambda} = \lambda^2,$$

from which it follows that

$$\text{Var}(X) = \lambda.$$

That shouldn't be much of a surprise. After all, the variance of the binomial is

$$n \cdot p \cdot q = n \cdot p \cdot (1 - p) = \lambda \cdot (1 - p).$$

To find the Poisson, we let λ stay fixed and $p \to 0$. So $\lambda \cdot (1 - p) \to \lambda$.

2.13.2   Some Comparisons

The Poisson distribution was derived here as an approximation to the binomial distribution. It is interesting to compare some binomial distributions and their Poisson approximations in order to measure, to some extent, how close the approximation is. We use a computer algebra system to make the calculations and the graphs.

First consider a binomial variable with $n = 20$ and $p = 0.03$. The value of n is not particularly large here, nor is p particularly small.

We show $P(X = x)$ using both the binomial distribution and the Poisson approximation.

X	Binomial	Poisson
0	0.548812	0.543794
1	0.329287	0.336358
2	0.0987861	0.0988297
3	0.0197572	0.0183395
4	0.00296358	0.00241061
5	0.00035563	0.000238576

So the values are very close. The graphs in Figure 2.25 reveal the same observation:

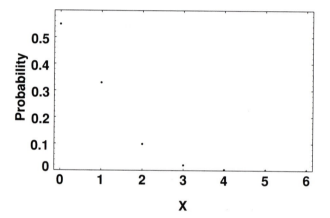

Poisson distribution with parameter 0.60.

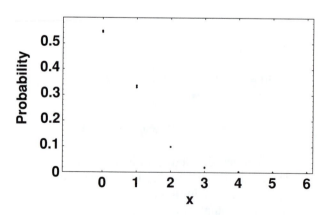

Figure 2.25   Comparison of the binomial and the Poisson distributions.

As n increases, the approximation is generally very good. Figure 2.26 shows a comparison between a binomial with $n = 100$ and $p = 0.03$ and a Poisson distribution with $\lambda = 3$.

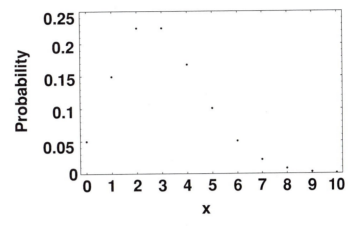

Binomial distribution with $n = 100$, $p = 0.03$.

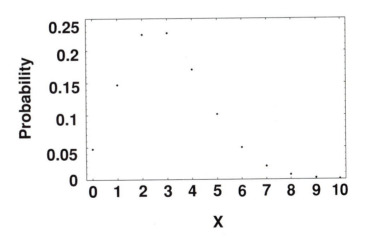

Poisson distribution with parameter 3.

Figure 2.26 Comparison of the binomial and Poisson distributions with $np = 3$.

The curves, which exhibit again the normal-like shape, are remarkably close and differ most at the maximum point; this difference is 0.00343232.

Example 2.13.2.1

An acceptance sampling plan selects 5 items from a population of 500 items, 16 of which are unacceptable. The lot is accepted if, at most, 2 of the sampled items are unacceptable. Here we compare the exact (hypergeometric) probability with both binomial and Poisson approximations. The hypergeometric probability is

$$P(X \le 2) = \sum_{x=0}^{2} \frac{\binom{16}{x} \cdot \binom{484}{5-x}}{\binom{500}{5}} = 0.99974.$$

The binomial approximation uses $n = 5$ and $p = \frac{16}{500}$ and is equivalent to no replacement, so

$$P(X \le 2) = \sum_{x=0}^{2} \binom{5}{x} \cdot \left(\frac{16}{500}\right)^x \cdot \left(\frac{484}{500}\right)^{5-x} = 0.999688.$$

Finally, the Poisson distribution with $\lambda = 5 \cdot \dfrac{16}{500} = 0.16$ gives

$$P(X \le 2) = \sum_{x=0}^{2} \frac{e^{-0.16} \cdot (0.16)^x}{x!} = 0.999394.$$

The approximations continue to be very good. Actually, the error in the Poisson distribution when approximating the binomial is not easily characterized, but many advise that it is best used when $np \le 5$, a rule that generally works fairly well.

Figure 2.26 compares the binomial with $n = 100$ and $p = 0.03$ to the Poisson distribution with $\lambda = np = 3$. The Poisson approximation is seen to be remarkably good.　　　　　　　　　　　　　　　　　　　　◀

2.14 ▶ The Poisson Process

The Poisson distribution serves as an approximation to the binomial distribution. The binomial distribution models situations where we can observe both the number of successes and the number of failures for a certain number of trials of an experiment.

We now turn our attention to other events that occur in time or space. We may be interested in the following examples: the number of faults in a fixed length of optic cable; the number of customers arriving at a checkout counter in a store; the number of telephone calls received at a telephone switchboard; or the number of messages received at a computer terminal. In each of these examples, we can count the number of occurrences of an event (such as the number of faults in the optic cable), but we cannot count the number of failures. How can such phenomena be modeled?

Consider a continuous interval of length or time or space and suppose that the following are true for the events we wish to observe:

1. The number of events in intervals having no points in common are independent.
2. Consider a short interval of length h. Suppose the probability of exactly one event in this interval is $\lambda \cdot h$ where λ is a constant of proportionality.
3. The probability of more than one event in the interval of length h is 0.

Now divide an interval of unit length into n mutually exclusive parts. By assumption 2], the probability of exactly one event in this interval is $\lambda \cdot \left(\frac{1}{n} \right)$, and so the probability of no events in this interval is $1 - \frac{\lambda}{n}$. Letting X denote the number of events in the unit interval, assumptions **1** and **3** allow us to calculate

$$P(X = x) = \binom{n}{x} \cdot \left(\frac{\lambda}{n} \right)^x \cdot \left(1 - \frac{\lambda}{n} \right)^{n-x}, x = 0, 1, 2, \ldots, n.$$

But we have shown that this can be approximated by a Poisson variable with mean value $n \cdot \frac{\lambda}{n} = \lambda$. So the Poisson distribution can be used in situations for which assumptions **1, 2**, and **3** hold. We see that λ is the expected number of events in a period of time or in an interval of space.

Example 2.14.1

Calls come into a telephone switchboard at a rate of four per minute.

   a. Find the probability of exactly six calls in an interval of two minutes.
   b. Find the probability of at least three calls in three minutes.

Here the interval of interest changes. We present two different solutions.

Solution 1

In part **a** the interval is of length two minutes, so we might suspect that λ, the expected number of events in that interval, is eight. Proceeding on that assumption, and letting X denote the number of calls received in that interval, we have

$$P(X = 6) = \frac{e^{-8} \cdot 8^6}{6!} = 0.122138.$$

In part **b**, the interval is three minutes, and so $\lambda = 12$. We find

$$P(X \geq 3) = 1 - P(X \leq 2) = 1 - \sum_{x=0}^{2} \frac{e^{-12} \cdot 12^x}{x!} = 0.999478.$$

It isn't clear, however, that we can change the interval and retain a Poisson variable.

Solution 2

Let the random variables X_1 and X_2 be defined as follows: Let X_1 denote the number of calls received during the first minute, and let X_2 denote the number of calls received during the second minute.

Since the numbers of calls received during the first and second minutes are independent,

$$P(X_1 + X_2 = 6) = \sum_{x_1=0}^{6} P(X_1 = x_1) \cdot P(X_2 = 6 - x_1)$$

$$= \sum_{x_1=0}^{6} \frac{e^{-4} \cdot 4^{x_1}}{x_1!} \cdot \frac{e^{-4} \cdot 4^{6-x_1}}{(6 - x_1)!}.$$

Multiply and divide by 6! to obtain

$$P(X_1 + X_2 = 6) = \frac{e^{-8}}{6!} \sum_{x_1=0}^{6} \binom{6}{x_1} 4^{x_1} \cdot 4^{6-x_1}$$

and now by the binomial theorem,

$$P(X_1 + X_2 = 6) = \frac{e^{-8}}{6!} (4 + 4)^6 = \frac{e^{-8}}{6!} \cdot 8^6,$$

giving the same result as in solution 1.

Solution 2 indicates that the sum of independent Poisson random variables is again a Poisson random variable. We will return to a discussion of this fact and related facts in Chapter 4, where solution 1 will be completely justified. ◀

Exercises 2.14

1. A Poisson random variable has

$$P(X = 2) = \frac{2}{3} P(X = 1).$$

Find $P(X = 3)$.

2. Deaths in a small city occur at a rate of 5 per week and are known to follow a Poisson distribution.

   a. What is the expected number of deaths in a three-day period?
   b. What is the probability no one dies in a three-day period?
   c. What is the probability that at least 250 people die in 52 weeks?

 3. Traffic accidents at an intersection are assumed to follow a Poisson distribution with 4 accidents expected in a period of one year.

   a. What is the probability of, at most, one accident in a given year?
   b. What is the probability of exactly 3 accidents in 6 months?
   c. It is expected that 2 accidents occur during a year at another intersection. What is the probability that there is a total of at least 3 accidents in a given year at the 2 intersections?

4. The number of typographical errors per page in a book follows a Poisson distribution with parameter $\frac{3}{4}$. What is the probability that there is a total of 10 errors on 10 randomly selected pages in the book?

 5. Twenty percent of the IC chips made in a plant are nonfunctional. Assume that a binomial model is appropriate.

   a. Find the probability that at most 13 nonfunctional chips occur in a sample of 100 chips.
   b. Use the Poisson distribution to approximate the result in part **a**.

6. Let X, the number of hits in a baseball game, be a Poisson variable with parameter α. If the probability of a no-hit game is $\frac{1}{3}$, what is α?

 7. An insurance company has discovered that about 0.1% of the population is involved in a certain type of accident each year. If the 10,000 policy holders of the company are randomly selected from the population, what is the probability that not more than 5 of its clients are involved in such an accident next year?

 8. A study of customers entering a grocery store shows that all the arrivals are Poisson with males entering at an average rate of 3 per minute and females at an average rate of 5 per minute. Find the probability that at least 20 customers enter the store in the next 5 minutes.

 9. Computer programs run on a certain computer are executed during an interval of one minute according to a Poisson process with mean 12. Twenty-five percent of these programs utilize a plotter.

   a. What is the probability there will be a demand for at least 15 programs run in a given minute?
   b. The plotter takes 10 seconds to execute a plot. What is the expected number of seconds the plotter is in use during a given minute?

10. A multiple-choice examination contains 4 choices for each of 100 questions.

    a. Find the exact probability that a student who guesses misses, at most, 4 questions.

    b. Approximate the probability in part **a** using the Poisson distribution.

11. The number of earthquakes of destructive magnitude in California follows a Poisson distribution with one such earthquake expected each year. What is the probability of at least 3 such earthquakes in a six-month period?

12. A quality control inspector follows the following plan in inspecting soccer balls that are produced according to a Poisson process with 4 soccer balls expected each minute. The produced balls fall into a bin that automatically empties at the end of each minute. If the bin collects exactly 3 balls, the inspector takes them out for possible inspection of 10 seconds each. He flips a fair coin for each and inspects them only if a head appears. If the bin should contain 5 balls, he spends 5 seconds inspecting each ball. Otherwise, the inspector does not inspect the output. What is the average amount of time per minute spent in inspecting the soccer balls?

13. Major crimes are reported at an average rate of 5 per night in a given police precinct. The number of these crimes is assumed to follow a Poisson distribution.

    a. What is the probability that on a given night no more than 3 major crimes will be reported?

    b. What is the chance that a full week will pass with no more than 3 major crimes reported on any of the 7 nights?

14. An airline knows that 10% of the people holding reservations for a certain flight will not appear. The plane holds 90 people. Use the Poisson approximation in answering the following questions:

    a. If 95 reservations have been sold, what is the probability that everyone who appears for the flight can be accommodated?

    b. How many reservations should be sold so the probability that the airline can accommodate everyone who appears is at least 0.99?

15. Molecules of a rare gas occur at an average rate of 3 per cubic foot of air and follow a Poisson distribution.

    a. What is the probability that a cubic foot of air contains none of the molecules?

    b. What is the probability that 3 cubic feet of air contain exactly 4 of the molecules?

    c. How much air must be taken as a sample to make the probability at least 0.99 that at least one molecule will be found?

16. A librarian shelves 1000 books per day. If the probability that any particular book is misshelved is 0.001 and if the books are shelved independently of one another,

    a. What is the probability that at most 2 books are misshelved?

    b. Approximate the probability in part **a** using the Poisson distribution.

17. A popular chocolate chip cookie "guarantees" at least 16 chocolate chips per cookie. The actual number of chocolate chips per cookie, however, is a Poisson random variable. What must be the average number of chips per cookie if approximately 95% or more of the cookies are to meet the guarantee?

18. A bakery makes a batch of 1000 chocolate chip cookies and adds n chocolate chips to the batter for each batch and mixes the batter well. Under these assumptions, the number of chocolate chips per cookie should follow a Poisson distribution.

    a. If $n = 4900$, what is the probability that at least 2 chips are in a randomly selected cookie?

    b. If $n = 4900$, what is the number of cookies in each batch that are expected to contain exactly 3 chocolate chips?

    c. FDA regulations declare that at most 1% of cookies labeled "chocolate chip" can fail to contain a single chocolate chip. What is the minimum value for n for the bakery to be within the law?

19. A truck repair shop has facilities for the repair of 3 large trucks per day. The trucks arrive according to a Poisson process with 2 trucks expected per day. If more than 3 trucks arrive, the excess is turned away.

    a. Find the probability that exactly 3 trucks arrive in one day.

    b. Find the probability that trucks are turned away.

    c. Find the probability distribution for X, the number of trucks serviced per day.

    d. Find the expected number of trucks turned away each day.

    e. The shop decides to add facilities so that it can service the trucks arriving during a day about 95% of the time. How many trucks must it be able to service in a day?

20. Calls come into a very busy switchboard at a rate of 6 per minute according to a Poisson process. Unfortunately, some new electronic switching devices work imperfectly and the probability that a received call is switched to the proper extension is only 0.8. It has been observed that the calls are switched independently, however.

    a. If X represents the number of calls correctly switched, find $P(X = k)$ for some one-minute period.

    b. Simplify the result in part **a** and show that X is Poisson with parameter 4.8.

21. Telephone calls coming into a busy switchboard follow a Poisson distribution with 4 calls expected in a one-minute period. The switchboard, however, can answer at most 6 calls in a one-minute interval; any calls exceeding 6 during that period receive a busy signal.

a. Let Y denote the number of calls answered in a one-minute period. Find the probability distribution for Y.

b. Find $E(Y)$.

Chapter Review

This chapter has considered several discrete probability distributions whose importance derives from the fact that they have various applications. Each of the distributions in this chapter arises in one way or another from the binomial distribution.

We began by defining a *random variable* as a real-valued function defined on the points of a sample space. A typical example is throwing two dice and then recording the sum that appears. The sum is a random variable because it is a function, in this case the sum, of the outcomes of the particular sample point that occurs.

If X is a random variable, the *probability distribution function*, or *pdf*, is defined as

$$f(x) = P(X = x).$$

A related function is the *distribution function*, is defined as

$$F(x) = P(X \le x).$$

The distribution function is not often used in this chapter, but has very important applications in the work to come.

Probability distributions are often distinguished and described by the values of their *mean*, μ_x, and their variance, σ_x^2. These are defined as

$$\mu_x = E(X) = \sum_x x \cdot f(x)$$

and

$$\sigma_x^2 = \text{Var}(X) = E(X - \mu_x)^2 = \sum_x (x - \mu_x)^2 f(x),$$

provided, of course, that the sums exist. The variance, σ_x^2, can also be calculated as

$$\sigma_x^2 = E(X^2) - [E(X)]^2.$$

As a (crude) indication that σ actually measures the variation, or dispersion in a random variable, we proved Tchebycheff's Inequality:

$$P(|X - \mu| \le k \cdot \sigma) \ge 1 - \frac{1}{k^2}$$

where k is some positive quantity.

We then turned to some specific discrete probability distributions. Of these, the single most important probability distribution is the *binomial distribution*, whose pdf is given by

$$P(X = x) = \binom{n}{x} p^x q^{n-x}, \quad x = 0, 1, 2, \ldots, n \text{ where } q = 1 - p.$$

This random variable arises from an experiment of n independent trials on each of which the result is one of two outcomes (usually denoted by "success" or "failure"), where p denotes the probability of success and X denotes the total number of successes.

We used a recursion to find that, for the binomial distribution,

$$\mu = n \cdot p$$

and

$$\sigma^2 = n \cdot p \cdot q.$$

We then considered some statistical problems. We first considered the construction of a *confidence interval* when sampling from a binomial distribution with known values of n and p. Frequently, however, p is unknown. We found an approximate 95% confidence interval for p to be

$$P\left(\frac{nX + 2n - 2\sqrt{n^2 X + n^2 - nX^2}}{n^2 + 4n} \leq p \leq \frac{nX + 2n + 2\sqrt{n^2 X + n^2 - nX^2}}{n^2 + 4n} \right) = 0.95$$

where X is the observed number of successes in the binomial process with n trials.

Tests of hypotheses were then considered. We examined tests of $H_o : p = p_0$ against the alternative $H_a : p = p_a$. The two types of error in testing a hypothesis are

$$\alpha = \text{probability of a Type I error}$$
$$= P[H_o \text{ is rejected when it is true}]$$

and

$$\beta = \text{probability of a Type II error}$$
$$= P[H_o \text{ is accepted when it is false}].$$

We considered the effect of the *critical region* – the set of observed values leading to the rejection of the null hypothesis – on the size of β and discussed $1 - \beta$, the *power* of the test. This is the probability that a false H_o is correctly rejected.

We derived the mean and the variance of a *sample proportion* arising from a sample survey. Using these results we found that

$$P\left(p_s - 2 \cdot \sqrt{\frac{p \cdot q}{n}} \leq p \leq p_s + 2 \cdot \sqrt{\frac{p \cdot q}{n}} \right) = 0.95$$

is a 95% confidence interval for the unknown true proportion p based on a sample proportion p_s.

The negative binomial distribution arises when, in a binomial experiment, we wait for, say, the rth success. The probability distribution function is

$$P(X = x) = \binom{x-1}{r-1} \cdot p^r \cdot q^{x-r}, \; x = r, r+1, r+2, \dots$$

with mean $\mu = \frac{r}{p}$ and variance $\sigma^2 = \frac{r \cdot q}{p^2}$. In the special case where $r = 1$, so that we wait for the first binomial success, X is called a *geometric* random variable.

A common situation in which the *hypergeometric* random variable arises is that of *acceptance sampling*. Here a lot, or a collection of a product manufactured over a given period of time, is sampled, but, unlike the binomial distribution, the sampling is done without replacement. If the lot actually contains D unacceptable items and $N - D$ acceptable ones, and if X denotes the number of unacceptable items in a sample of size n, then

$$P(X = x) = \frac{\binom{D}{x} \cdot \binom{N-D}{n-x}}{\binom{N}{n}}, \; x = 0, 1, 2, \dots, Min\{n, D\}.$$

We found that

$$\mu = n \cdot \frac{D}{N}$$

and that

$$\sigma^2 = n \cdot \frac{N-n}{N-1} \cdot \frac{D}{N} \cdot \frac{N-D}{N}.$$

We showed that the hypergeometric random variable is approximated by the binomial random variable when the sample size, n, is small in comparison to the lot size, N. Examples of acceptance sampling were given and we considered two plans for improving the quality of the lot of items sent to the buyer. In one we replaced any unacceptable items in the sample with good ones; in the second, if the sample so indicated, we replaced every unacceptable item in the lot with a good item. Each plan leads to gains with respect to the quality of the outgoing product; under the second plan there is a limit of the percentage of unacceptable products that can be sold. This is known as the *average outgoing quality limit*.

Finally, we considered a *Poisson* random variable, which can be regarded in two ways. We first found the distribution as a limit of the binomial distribution when n is large and p is small. We also considered the *Poisson process* in which events occur over a period of time or space in an independent fashion so that the probability of more than one independent event in a given interval is negligible, and the probability of an event in some interval is proportional to the length of the interval. These assumptions yield the same distribution as the limiting binomial distribution. The Poisson distribution has a variety of applications, many of which were given in the exercises.

In the next chapter we will consider some important continuous probability distributions.

A few typical problems from each section of this chapter are listed here for the convenience of the reader who wishes to review the material.

Problems for Review

Exercises 2.2 # 1, 3, 4, 8, 11
Exercises 2.3 # 1, 2, 5, 6, 7, 11
Exercises 2.5 # 1, 2, 4, 7, 8, 9, 12, 15, 17, 19, 22
Exercises 2.6 # 1, 4, 5, 8, 9
Exercises 2.7 # 1, 4, 6, 7, 10
Exercises 2.8 # 1, 3, 4, 8, 9
Exercises 2.9 # 1, 2, 5, 7, 9, 13
Exercises 2.10 # 1, 3, 5, 6, 8, 10
Exercises 2.11 # 2, 4, 5, 6, 9, 11
Exercises 2.14 # 2, 3, 5, 6, 8, 13, 18, 20

Supplementary Exercises for Chapter 2

1. Calls come into a telephone exchange at a rate of 1.5 per minute. Assuming that the number of calls received follows a Poisson distribution, find the probability that at least 3 calls are received in the next 4 minutes.

2. Twenty percent of the integrated circuit (IC) chips made in a plant are defective. Assume that the chips are produced according to a binomial process.

   a. Find the probability that at most 13 defectives occur in a sample of 100 IC chips.
   b. Approximate the probability in part **a** by a Poisson random variable.

3. A manufacturer of soft drink bottles turns out defectives with probability 0.10. Assume that the bottles are produced according to a binomial process.

   a. Find the probability that there are 4 defective bottles among the next 10 bottles produced.
   b. Find the probability that there are at least 4 defective bottles among the next 10 bottles produced.
   c. How many bottles must be produced to make the probability that at least one bottle among them is defective to be at least 0.95?

4. Earthquakes in a certain part of California occur according to a Poisson process with 3 earthquakes expected each century.

a. What is the probability of exactly 4 earthquakes in a century?

b. What is the probability of at least 2 earthquakes in a 50-year period?

c. Let X be the number of earthquakes in a century. Compare the exact value of $P(\mu - \sigma \leq X \leq \mu + \sigma)$ with the approximation given by Tchebycheff's Inequality.

5. Suppose an event has probability p of occurring and that several independent trials are observed. What value of p maximizes the probability that the first failure occurs on the fifth trial?

6. Suppose that X and Y are independent observations of a Poisson random variable with parameter $\lambda = 1$. Find the probability that the smallest of the 2 observations is 1.

7. A series of trials in which success or failure occurs on each trial has probability of success at the ith trial as $\frac{1}{i+1}$. In 3 trials find the probability of exactly 2 successes.

8. A manufacturer makes a lot of 10 items a day. Two items are drawn (without replacement) and inspected. The lot is accepted if the sample contains, at most, one defective item. Find the probability that a lot containing 3 defective items is accepted.

9. **a.** What is the probability that a poker hand contains exactly 2 aces?

   b. How many poker hands must be selected to make the probability of having at least one hand containing at least 2 aces be at least 0.99?

10. A store sells chocolate donuts at a rate of 16 per hour, the number sold following a Poisson distribution. Find the probability that the store sells at least 3 chocolate donuts in 15 minutes.

11. Five defective transistors are mixed up with 10 good ones. They are inspected one after another until all the good transistors have been found. What is the probability the last good transistor will be found on the 12th test?

12. Errors are known to occur in a digitized message in a communications channel; the probability that an individual bit is incorrectly transmitted is 0.001 and the errors are assumed to be independent.

   a. Find the probability that at most 2 errors occur in a sequence of 10 bits.

   b. Find the mean and variance of the number of errors.

   c. Find the probability of at most 2 errors in a message of 10,000 bits.

13. In a small voting precinct, 100 voters favor candidate A and 80 voters oppose candidate A. What is the probability that a majority of a random sample of 4 voters will oppose candidate A?

14. Customers arrive at a checkout counter in a supermarket at a rate of 20 per half hour, the number following a Poisson distribution. What is the probability that at most 5 customers arrive in a period of 15 minutes?

15. A manufacturer produces items that are good or defective, according to a binomial process where p is the probability an item is defective. Let X denote the number of items produced up to and including the second defective item.

    a. Find an expression for the probability that X is even.

    b. Now suppose that the sixth item is the second defective item produced. What is the most likely value for p?

 16. Thirty percent of the applicants for a position have advanced training in computer programming. Three jobs requiring advanced training are open. Find the probability that the third qualified applicant is found on the fifth interview, supposing that the applicants are interviewed sequentially and at random.

17. A fair die is tossed until a 5 or a 6 appears. Compute the probability that the number of tosses is a multiple of 4.

18. From a lot of 25 items, 5 of which are defective, 4 are chosen at random. Let X be the number of defectives found. Find the probability distribution of X if

    a. the items are chosen with replacement.

    b. the items are chosen without replacement.

    c. In part **a** assume the items are chosen with replacement until a defective item is found. What is the probability that an odd number of drawings is necessary?

 19. Customers arrive at a computer store according to a Poisson process with 5 customers expected per hour. The sales force can accommodate, at most, 10 customers per hour; if more than 10 customers appear in an hour, the excess must be turned away.

    a. What is the probability that customers are turned away in a one-hour period?

    b. Consider two independent one-hour intervals. Let X denote the number of arrivals during the first hour and Y the number of arrivals during the second hour. Find $P(X + Y \leq 8)$.

 20. Fifty chocolate chip cookies are to be made using 150 chocolate chips. The number of chocolate chips per cookie is a Poisson random variable.

    a. What is the probability that a cookie has at least 4 chocolate chips?

    b. How many chocolate chips must be used in order to make the probability that a cookie has at least one chocolate chip be at least 0.90?

 21. A pair of fair dice is rolled 180 times each hour in a dice game at a casino. What is the probability that at least 25 rolls give a sum of 7 during one hour?

22. Telephone calls come into an answering service at an average rate of 3 per hour, the number of calls following a Poisson distribution. During the noon hour only the first 3 calls are answered. What is the expected number of calls answered during the noon hour?

23. A box contains 4 bad and 6 good tubes. The tubes are checked by drawing a tube at random and not replacing it in the box. In how many ways can the fourth bad tube be found on the seventh drawing?

24. A box contains 3 blue and 4 yellow marbles. Marbles are drawn out one at a time, the drawn marbles not being replaced. Drawings are made until all the marbles remaining in the box are of the same color.

    a. Assign probabilities to the sample points and verify that their sum is 1.
    b. What is the probability that only yellow marbles remain in the box when the sampling is finished?

25. A tosses 3 coins which have probability p_A of coming up heads, while B tosses 2 coins which have probability p_B of coming up heads.

    a. Find an expression for the probability that A tosses more heads than B.
    b. Show that the game is fair if the coins are fair.

26. A player pays $\$A$ to play the following game: a coin, loaded to come up heads with probability $\frac{2}{3}$, is tossed 5 times. Let X denote the number of heads. The player wins $\$(X + 1)$ if X is even and wins $\$(X - 1)$ if X is odd. Find A so that the game is fair.

 27. A machine, producing defective parts with probability $\frac{1}{10}$, has produced five parts. Unknown to the operator of the machine, an adjustment to the machine increases this probability to $\frac{1}{5}$. Ten parts are produced after the adjustment. What is the probability the output contains at least 2 defectives? Assume the parts are produced according to a binomial process.

28. Past studies have shown that $\frac{2}{3}$ of professional football players will sustain a permanent injury before retiring. To see if this proportion is true for current players, a sample of 100 retired professional football players showed that 80 of them had sustained permanent injuries. Using $\alpha = 0.05$, test $H_o : p = \frac{2}{3}$ against $H_a : p > \frac{2}{3}$.

29. In exercise 28, find the size of β for the alternative $p = 0.72$.

30. A study of 1200 college students showed that 44% of them said that their political views were similar to those of their parents. Find a 95% confidence interval for the true proportion of college students whose political views are similar to those of their parents.

31. A drug is thought to be effective in 10% of patients with a certain condition. To test this hypothesis, the drug is given to 100 randomly chosen patients with the condition. If 8 or more show some improvement, $H_o : p = 0.10$ is accepted; otherwise, $H_a : p < 0.10$ is accepted. Find the size of the test.

32. Jack thinks that he can guess the correct answer to a multiple-choice question with probability $\frac{1}{2}$. Kaylyn thinks his probability is $\frac{1}{3}$. To decide who is correct, Jack takes a multiple-choice test, guessing the answer to each question. It is decided that Jack will be correct if he answers at least 40 out of 100 questions correctly. Find α and β for this test.

33. A survey of 300 workers showed that 100 are self-employed. Find a 90% confidence interval for the proportion of workers who are self-employed.

34. A management study showed that $\frac{1}{3}$ of American office workers have his or her own office while $\frac{1}{33}$ of Japanese office workers have his or her own office. The study was based on 300 American workers and 300 Japanese workers. Could the difference in these proportions only be apparent and due to sampling variability? (Use 90% confidence intervals.)

35. The Internal Revenue Service says that the chance a United States corporation will have its income tax return audited is 1 in 15. A sample of 75 corporate income tax returns showed that 6 were audited. Do the data support the Internal Revenue Service's claim? Use $\alpha = 0.05$.

36. A survey of 400 children showed that $\frac{1}{8}$ of them were on welfare. Find a 95% confidence interval for the true proportion of children on welfare.

37. How large a sample is necessary to estimate the proportion of people who do not know whose picture is on the $1 bill to within 0.02 with probability 0.90?

38. Three marbles are drawn without replacement from a bag containing 3 white, 3 red, and 5 green marbles. $1 is won for each red selected and $1 is lost for each white selected. No payoff is associated with the green marbles. Let X denote the net winnings from the game. Find the probability distribution function for X.

39. Three fair dice are rolled. You as the bettor are allowed to bet $1 on the occurrence of one of the integers 1, 2, 3, 4, 5, or 6. If you bet on X and X occurs k times ($k = 1, 2, 3$), then you win k; otherwise, you lose the $1 you bet. Let W represent the net winnings per play.

   a. Find the probability distribution for W.
   b. Find $E(W)$.
   c. If you could roll m dice, instead of 3 dice, what would your choice of m be?

40. a. Suppose that X is a Poisson random variable with parameter λ. Find λ if $P(X = 2) = P(X = 3)$.

   b. Show, if X is a Poisson random variable with parameter λ where λ is an integer, that some 2 consecutive values of X have equal probabilities.

 41. Calls come into an office according to a Poisson process with 3 calls expected per hour. Suppose the calls are answered independently with the probability $\frac{3}{4}$ that a call is answered. Find the probability that exactly 4 calls are answered in a one-hour period.

42. Let X be Poisson with parameter λ.

   a. Find a recursion for $P(X = x + 1)$ in terms of $P(X = x)$.

   b. Use the recursion in part **a** to find μ and σ^2.

43. Ten people are wearing tags numbered 1, 2, ..., 10. Three people are chosen at random. What is the probability that the smallest badge number among the three is 5?

► *Chapter 3*

Continuous Random Variables and Probability Distributions

3.1 ▶ Introduction

Discrete random variables were discussed in Chapter 2. However, it is not always possible to describe all the possible outcomes of an experiment with a finite or countably infinite sample space. As an example, consider the wheel shown in Figure 3.1, where the numbers from 0 to 1 have been marked on the outside edge.

The experiment consists of spinning the spinner and recording where the arrow stops. It would be natural here to consider the sample space, S, to be

$$S = \{x | 0 \le x \le 1\}.$$

S is infinite, but not countably infinite.

Now the question arises, "What probability should be put on each of the points in S?" Surely, if the wheel is fair, each point should receive the same probability and the total probability should be 1. What value should that probability be? Suppose, for sake of argument, that a probability of $0.0000000000000000000001 = 10^{-22}$ is put on each point. It is easy to show that the circumference of the wheel contains more than 10^{22} points, so we have used up more than the allotted probability of 1. So we conclude that the only possible assignment of probabilities is

$$P(X = x) = 0 \text{ for any } x \text{ in } S.$$

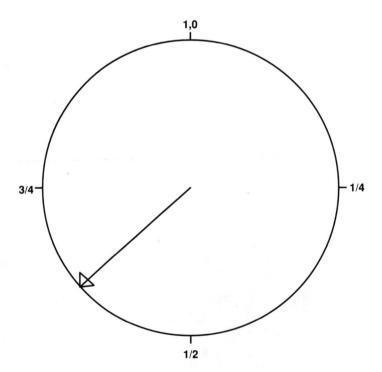

Figure 3.1   The spinner.

Now suppose that the wheel is loaded and that it is three times as likely that the arrow lands in the left half of the wheel than in the right half. We suppose that

$$P\left(X \geq \frac{1}{2}\right) = 3 \cdot P\left(X \leq \frac{1}{2}\right).$$

Again we ask, "What probability should be put on each of the points in S?" Again, because there is still an uncountably infinite number of points in S, the answer is

$$P(X = x) = 0.$$

Definition: If a random variable X takes values on an interval or intervals, then X is said to be a *continuous* random variable.

Of course $P(X = x) = 0$ for *any* continuous random variable X.

So the probability distribution function is not informative in the continuous case, since, for example, we can't distinguish between fair and loaded wheels! The fault, however, lies not in the answer, but in the question. Perhaps we can devise a question whose answer carries more information for us.

Consider now a function, $f(x)$, which we will call a *probability density function*. The abbreviation is again *pdf* (the same abbreviation used for probability distribution function) but the word *density* connotes a continuous distribution. Here are the properties we desire of the new function $f(x)$:

1. $f(x) \geq 0$
2. $\int_{-\infty}^{\infty} f(x)\, dx = 1$
3. $\int_a^b f(x)\, dx = P(a \leq X \leq b)$

These properties are quite analogous to those for a discrete random variable. Property **3** indicates that *areas* under $f(x)$ are probabilities. $f(x)$ must be non-negative, else we encounter negative probabilities, so Property **1** must hold. Property **2** indicates that the total probability on the sample space is 1.

What is $f(x)$ for the fair wheel? Since the circumference of the wheel contains the interval $[0, \frac{1}{4}]$, and because the wheel is a fair one, we would like $P(0 \leq X \leq \frac{1}{4})$ to be $\frac{1}{4}$ so we must have $\int_0^{\frac{1}{4}} f(x)\, dx = \frac{1}{4}$. Many functions have this property. But we would like *any* interval of length $\frac{1}{4}$ to have probability $\frac{1}{4}$. In addition, we would like an interval of length a, say, to have probability a for $0 \leq a \leq 1$. The only function that has this property, in addition to satisfying Properties **1** and **2**, is a *uniform* probability density function:

$$f(x) = \begin{cases} 1, & 0 \leq x \leq 1 \\ 0, & \text{otherwise.} \end{cases}$$

For the loaded wheel, where we want $P(X \geq \frac{1}{2}) = 3P(X \leq \frac{1}{2})$, consider (among many other choices) the function

$$f(x) = \begin{cases} 2x, & 0 \leq x \leq 1 \\ 0, & \text{otherwise.} \end{cases}$$

Then $P(X \geq \frac{1}{2}) = \int_{\frac{1}{2}}^{1} 2x \, dx = \frac{3}{4}$, so that

$$P(X \leq \tfrac{1}{2}) = \tfrac{1}{4}, \quad \text{and} \quad P(X \geq \tfrac{1}{2}) = 3P(X \leq \tfrac{1}{2}).$$

A graph of $f(x)$ is shown in Figure 3.2.

It is also easy to verify that $f(x)$ also satisfies properties **1** and **2** for a probability density function.

We see that $f(x)$, the probability *density* function, distinguishes continuous random variables in an informative way, while the probability *distribution* function (which is useful for discrete random variables) does not. To illustrate this point further, suppose the wheel has been rigged so that it is impossible for the pointer to stop between 0 and $\frac{1}{4}$, while it is still fair for the remainder of the circumference of the wheel. It follows that

$$f(x) = \begin{cases} \frac{4}{3}, & \frac{1}{4} \leq x \leq 1 \\ 0, & \text{otherwise.} \end{cases}$$

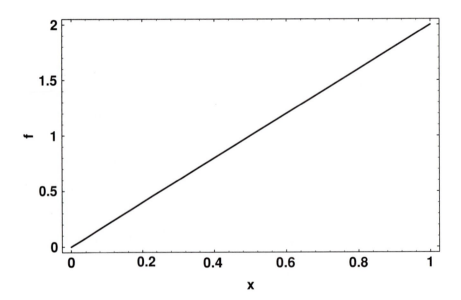

Figure 3.2   Probability density function for the loaded wheel.

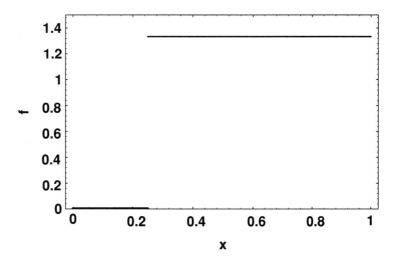

Figure 3.3   Probability density function for the rigged wheel.

This function satisfies all three properties for a probability density function. Its graph is shown in Figure 3.3. For this rigged wheel,

$$P\left(X \geq \frac{1}{2}\right) = \int_{\frac{1}{2}}^{1} \frac{4}{3}\, dx = \frac{2}{3}.$$

It is also useful to define a *cumulative distribution function* (often abbreviated to *distribution function*), defined as

$$F(x) = P(X \leq x) = \int_{-\infty}^{x} f(x)\, dx.$$

We used $F(x)$ in Chapter 2.

The function $F(x)$ accumulates probabilities for a probability density function in exactly the same way $F(x)$ accumulated probabilities in the discrete case. As an example, if

$$f(x) = \begin{cases} 1, & 0 \leq x \leq 1 \\ 0, & \text{otherwise,} \end{cases}$$

then, being careful to distinguish the various regions in which x can be found, we find that

$$F(x) = \begin{cases} \int_{-\infty}^{x} 0\, dx = 0, \ x \leq 0 \\ \int_{-\infty}^{0} 0\, dx + \int_{0}^{x} 1\, dx = x, \ 0 \leq x \leq 1 \\ \int_{-\infty}^{0} 0\, dx + \int_{0}^{1} 1\, dx + \int_{1}^{\infty} 0\, dx = 1, \ x \geq 1. \end{cases}$$

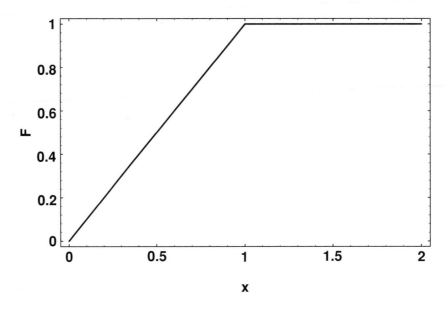

Figure 3.4   Cumulative distribution function for the fair wheel.

A graph of $F(x)$ is shown in Figure 3.4.

It is easy to see that $F(x)$, for any probability density function $f(x)$, has the following properties:

1. $\lim\limits_{x \to -\infty} F(x) = 0$ and $\lim\limits_{x \to \infty} F(x) = 1$,

2. If $a \le b$ then $F(a) \le F(b)$,

3. $P(a \le X \le b) = P(a \le X < b) = P(a < X < b) = F(b) - F(a)$,

and

4. $\dfrac{d[F(x)]}{dx} = f(x)$,

provided the derivative exists.

Mean and Variance

In analogy with the discrete case, the mean and variance of a continuous random variable with probability density function $f(x)$ are defined as

$$E(X) = \mu = \int_{-\infty}^{\infty} x \cdot f(x)\, dx \qquad (3.1)$$

and

$$\text{Var}(X) = \sigma^2 = E(X - \mu)^2 = \int_{-\infty}^{\infty} (x - \mu)^2 \cdot f(x)\, dx \qquad (3.2)$$

provided that the integrals converge.

These definitions are similar to those used for discrete random variables where the values of the random variables were weighted with their probabilities and the results added. It is natural to integrate so that the definitions for the mean and the variance in the continuous case appear to be analogous to their counterparts in the discrete case.

We can expand the definition of Var (X) to find that

$$\text{Var}(X) = \int_{-\infty}^{\infty} (x - \mu)^2 \cdot f(x)\, dx = \int_{-\infty}^{\infty} (x^2 - 2\mu x + \mu^2) \cdot f(x)\, dx$$

$$= \int_{-\infty}^{\infty} x^2 \cdot f(x)\, dx - 2\mu \int_{-\infty}^{\infty} x \cdot f(x)\, dx + \mu^2 \int_{-\infty}^{\infty} f(x)\, dx$$

$$= \int_{-\infty}^{\infty} x^2 \cdot f(x)\, dx - 2\mu^2 + \mu^2,$$

so

$$\text{Var}(X) = E(X^2) - [E(X)]^2.$$

This is the same result we obtained for a discrete random variable.

Other properties of the mean and variance are

$$E(aX + b) = aE(X) + b$$

and

$$\text{Var}(aX + b) = a^2\text{Var}(X)$$

To show these properties, first consider $E(aX + b)$. By definition,

$$E(aX + b) = \int_{-\infty}^{\infty} (ax + b) \cdot f(x)\, dx.$$

Expanding and simplifying the integrals, we find that

$$E(aX + b) = a \int_{-\infty}^{\infty} xf(x)\, dx + b \int_{-\infty}^{\infty} f(x)\, dx,$$

so

$$E(aX + b) = aE(X) + b$$

or

$$E(aX + b) = a \cdot \mu + b,$$

establishing the first property. Now,

$$\text{Var}(aX + b) = E[(aX + b) - (a\mu + b)]^2$$
$$= E[a^2(X - \mu)^2]$$
$$= a^2 E[(X - \mu)^2],$$

so

$$\text{Var}(aX + b) = a^2 \text{Var}(X),$$

establishing the second property.

The definitions of the mean and variance are dependent on the convergence of the integrals involved. To show that this does not always happen, consider the density

$$f(x) = \frac{1}{\pi(1 + x^2)}, \quad -\infty < x < \infty.$$

The fact that

$$\int_{-\infty}^{\infty} f(x)\, dx = \int_{-\infty}^{\infty} \frac{dx}{\pi(1 + x^2)} = \frac{1}{\pi} \text{Arc tan}(x) \Big|_{-\infty}^{\infty} = \frac{1}{\pi}\left[\frac{\pi}{2} - \left(-\frac{\pi}{2}\right)\right] = 1,$$

together with the fact that $f(x) \geq 0$, establishes $f(x)$ as a probability density function. However,

$$E(X) = \int_{-\infty}^{\infty} \frac{x \cdot dx}{\pi(1 + x^2)} = \frac{1}{2\pi} \ln|1 + x^2| \Big|_{-\infty}^{\infty},$$

which does not exist. The random variable X in this case has no variance as well; in fact $E[X^k]$ does not exist for any k. The probability density is called the *Cauchy* density.

We now turn to an example of a better-behaved probability density function.

Example 3.1.1

Given the loaded wheel, for which

$$f(x) = \begin{cases} 2x, & 0 \leq x \leq 1 \\ 0, & \text{otherwise,} \end{cases}$$

we find that

$$F(x) = \begin{cases} 0, & x \leq 0 \\ x^2, & 0 \leq x \leq 1 \\ 1, & x \geq 1. \end{cases}$$

If we want to calculate $P(\frac{1}{2} \le X \le \frac{3}{4})$, we can proceed in two different ways. First,

$$P(\frac{1}{2} \le X \le \frac{3}{4}) = \int_{\frac{1}{2}}^{\frac{3}{4}} 2x \, dx = \frac{5}{16},$$

where $f(x)$ is used in the calculation. We could as easily use $F(x)$:

$$P(\frac{1}{2} \le X \le \frac{3}{4}) = F(\frac{3}{4}) - F(\frac{1}{2}) = \frac{5}{16},$$

giving the same result.

It would appear from this example that $F(x)$ is superfluous, since any probability can be found from a knowledge of $f(x)$ alone (and in fact $f(x)$ is needed to determine $F(x)$!). While this is true, it happens that there are other important uses to which $F(x)$ will be put later and so we introduce the function now. To pique the reader's interest we pose the following question: The loaded wheel above is spun, X being the result. The player then wins $\$3X^2$. If the owner of the wheel wishes to make, on average, $\$0.50$ per play of the game, what is a fair price to charge to play the game? We will answer this question later, making use of $F(x)$, although the reader may be able to answer it now. The function $F(x)$ also plays a leading role in reliability theory, which is considered later in this chapter. ◀

Example 3.1.2

A random variable X has probability density function

$$f(x) = \begin{cases} k \cdot (2 - x), & 0 \le x \le 2 \\ 0, & \text{otherwise.} \end{cases}$$

The constant k, of course, is a special value that makes the total area under the curve 1. It follows that

$$\int_0^2 k \cdot (2 - x) \, dx = 1.$$

It follows from this that $k = \frac{1}{2}$.

Now if we wish to find a conditional probability – for example, $P(X \ge 1 | X \ge \frac{1}{2})$ – first note that the set of values where $X \ge \frac{1}{2}$ does not have area 1, so, as in the discrete case,

$$P(X \ge 1 | X \ge \frac{1}{2}) = \frac{P(X \ge 1 \text{ and } X \ge \frac{1}{2})}{P(X \ge \frac{1}{2})}.$$

This becomes

$$P(X \geq 1 | X \geq \frac{1}{2}) = \frac{P(X \geq 1)}{P(X \geq \frac{1}{2})}.$$

We calculate this conditional probability as $\frac{4}{9}$.   ◄

Before turning to the exercises, we note that there are many important special probability density functions of great interest because they arise in interesting and practical situations. We will consider some of these in detail in the remainder of this chapter.

A Word on Words

We considered, for a discrete random variable, the probability *distribution* function as well as the cumulative *distribution* function. For continuous random variables, the terms probability *density* function and cumulative *distribution* function are terms in common usage.

We will continue to make the distinction here between discrete and continuous random variables by making a distinction in the language we use to refer to them. In part this is because the mathematics useful for discrete random variables is quite different from that for continuous random variables; the language serves to alert us to these distinctions. One would not want to integrate a discrete function nor try to sum a continuous one!

While we will be consistent about this, we will also refer to random variables, either discrete or continuous, as *following* or *having* a certain probability *distribution* function. So we will refer to a random variable as following a binomial distribution, or another random variable as following a Cauchy distribution, although one is discrete and the other is continuous.

Exercises 3.1

1. A loaded wheel has probability density function $f(x) = 3x^2$, $0 \leq x \leq 1$.

   a. Show that $f(x)$ is a probability density function.
   b. Find $P(\frac{1}{2} \leq X \leq \frac{3}{4})$.
   c. Find $P(X \geq \frac{2}{3})$.
   d. Find c so that $P(X \geq c) = \frac{2}{3}$.

2. A random variable X has probability density function

$$f(x) = \begin{cases} k(x^2 - x^3), & 0 \leq x \leq 1 \\ 0, & \text{otherwise.} \end{cases}$$

   a. Find k.
   b. Find $P(X \geq \frac{3}{4})$.

c. Calculate $P(X \geq \frac{3}{4} \mid X \geq \frac{1}{2})$.

3. If
$$f(x) = \begin{cases} k \sin x, & 0 \leq x \leq \pi, \\ 0, & \text{otherwise.} \end{cases}$$

a. Show that $k = \frac{1}{2}$.
b. Calculate $P(X \leq \frac{\pi}{3})$.
c. Find b so that $P(X \leq b) = \frac{1}{3}$.

4. A random variable X has probability density function
$$f(x) = \begin{cases} 4x^3, & 0 \leq x \leq 1, \\ 0, & \text{otherwise.} \end{cases}$$

a. Find the mean μ, and the variance σ^2, for X.
b. Calculate exactly $P(\mu - 2\sigma \leq X \leq \mu + 2\sigma)$ and compare your answer with the result given by Tchebycheff's Inequality.

5. The length of life X, in days, of a heavily used electric motor has probability density function
$$f(x) = \begin{cases} 3e^{-3x}, & x \geq 0. \\ 0, & \text{otherwise.} \end{cases}$$

a. Find the probability that the motor lasts at least $\frac{1}{2}$ of a day, given that it has lasted $\frac{1}{4}$ of a day.
b. Find the mean and variance for X.

6. A random variable X has probability density function
$$f(x) = \begin{cases} kx^2 e^{-x}, & x \geq 0. \\ 0, & \text{otherwise.} \end{cases}$$

a. Find k.
b. Graph $f(x)$.
c. Find μ and σ^2.

7. The distribution function for a random variable X is
$$F(x) = \begin{cases} 0, & x < -4 \\ \frac{1}{8}, & -4 \leq x < -3 \\ \frac{3}{8}, & -3 \leq x < 2 \\ \frac{3}{4}, & 2 \leq x < 5 \\ 1, & x \geq 5. \end{cases}$$

a. Find $P(X = 2)$.

b. Find $P(-3 \le X < 2)$.

8. A continuous random variable X has probability density function

$$f(x) = \begin{cases} k(x - x^2), & 0 < x < 1 \\ 0, & \text{otherwise.} \end{cases}$$

a. Find k.

b. Find the mean and variance of X.

c. Four independent observations of X are made. What is the probability that exactly two of these are greater than $\frac{3}{4}$?

9. Let

$$f(x) = \begin{cases} \frac{1}{2} + \frac{x}{4}, & -2 < x < 0 \\ \frac{1}{2} - \frac{x}{4}, & 0 < x < 2 \\ 0, & \text{otherwise.} \end{cases}$$

a. Show that $f(x)$ is a probability density function.

b. Find $P[|X| < 1]$.

c. Find μ and σ^2.

10. A random variable X has probability density function

$$f(x) = \begin{cases} x, & 0 \le x \le 1 \\ 2 - x, & 1 \le x \le 2 \\ 0, & \text{otherwise.} \end{cases}$$

a. Find $E[X]$.

b. Find $\text{Var}[X]$.

c. Find $F(x)$, being sure to specify this for all values of x.

d. What is the probability that at least two of three independent observations on X are greater than $\frac{1}{2}$?

11. The length of time Y in hours that a student takes to complete an examination is a random variable with

$$g(y) = \begin{cases} cy^2 + y, & 0 \le y \le 1 \\ 0, & \text{otherwise.} \end{cases}$$

a. Find c.

b. Find the cumulative distribution function, $G(y)$.

c. Find an expression for $P(Y > y)$ for any value of y.

12. As a measure of intelligence, mice are timed when going through a maze to reach a reward of food. The time (in seconds) required for any mouse is a random variable Y with probability density function

$$f(y) = \begin{cases} \frac{10}{y^2} & y \geq 10 \\ 0, \text{otherwise.} \end{cases}$$

   a. Show that $f(y)$ has the properties of a probability density function.
   b. Find $P(9 \leq Y \leq 99)$.
   c. Find the probability that a mouse requires at least 15 seconds to traverse the maze if it is known that the mouse requires at least 12 seconds.

13. A continuous random variable has probability density function

$$f(x) = \begin{cases} cx^2, & -3 \leq x \leq 3 \\ 0, & \text{otherwise.} \end{cases}$$

   a. Find the mean and variance of X.
   b. Verify Tchebycheff's Inequality for the case $k = \sqrt{\frac{5}{3}}$.

14. Suppose the distance X between a point target and a shot aimed at the point in a video game is a continuous random variable with probability density function

$$f(x) = \begin{cases} \frac{3}{4}(1 - x^2), & -1 \leq x \leq 1 \\ 0, & \text{otherwise.} \end{cases}$$

   a. Find the mean and variance of X.
   b. Use Tchebycheff's Inequality to give a bound for $P[|X| < \frac{1}{2}]$.

15. If the loaded wheel with $f(x) = 2x$, $0 \leq x \leq 1$, is spun three times, it can be shown that the probability density function for Y, the smallest of the three values obtained, is $g(y) = 6y(1 - y^2)^2$, $0 \leq y \leq 1$. Find the mean and variance for Y.

16. Show that

$$g(y) = \begin{cases} 1260\, y^4 (1 - y)^5, & 0 \leq y \leq 1 \\ 0, & \text{otherwise} \end{cases}$$

   is a probability density function. Then find the mean and variance for Y.

17. Use your computer algebra system to draw a random sample of 100 observations from the distribution

$$f(x) = \begin{cases} 1, & 0 < x < 1 \\ 0, & \text{otherwise.} \end{cases}$$

   The random variable X here is said to follow a *uniform distribution* on the interval $[0, 1]$.

192   *Chapter 3   Continuous Random Variables and Probability Distributions*

a. Enumerate the observations in each of the categories $0 \leq x < 0.1$, $0.1 \leq x < 0.2$, and so on. Do the observations appear to be uniform?

b. We will show in Chapter 4 that if X is uniform on the interval $[0, 1]$ and if $Y = X^2$, then the probability density function for Y is $g(y) = 2y$, $0 \leq y \leq 1$. So the sample in part **a** can be used to simulate a random sample from the loaded wheel discussed in section 3.1. Show a sample from the loaded wheel. Graph the sample values and decide whether or not the sample appears to have been selected from the loaded wheel.

18. Show that

$$g(y) = \begin{cases} \binom{n}{r} \cdot r \cdot y^{r-1} \cdot (1-y)^{n-r}, & 0 \leq y \leq 1 \\ 0, & \text{otherwise} \end{cases},$$

is a probability density function for n, a positive integer, and $r = 1, 2, \ldots, n$.

19. Given

$$f(x) = \begin{cases} \frac{x^2}{2}, & 0 \leq x \leq 1 \\ \frac{3}{4} - \left(x - \frac{3}{2}\right)^2, & 1 \leq x \leq 2 \\ \frac{(x-3)^2}{2}, & 2 \leq x \leq 3 \\ 0, & \text{otherwise.} \end{cases}$$

a. Sketch $f(x)$ and show that it is a probability density function.

b. Find the mean and variance of X.

20. Suppose that X is a random variable with probability distribution function $F(x)$ whose domain is $x \geq 0$.

a. Show that $\int_0^\infty [1 - F(x)] \, dx = E(X)$. (Hint: Write the integral as a double integral and then change the order of integration.)

b. Write an integral involving $F(x)$ whose value is $E(X^2)$.

3.2 ▶ Uniform Distribution

The fair wheel, where $f(x) = 1$, $0 \leq x \leq 1$, is an example of a *uniform* probability density function. In general, if

$$f(x) = \begin{cases} \frac{1}{b-a}, & a \leq x \leq b \\ 0, & \text{otherwise,} \end{cases}$$

then X is said to have a *uniform* probability distribution. This is the continuous analogy of the discrete uniform distribution considered in Chapter 2. A graph is shown in Figure 3.5.

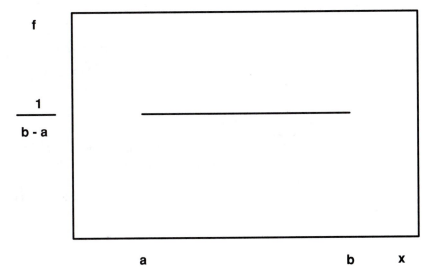

Figure 3.5   Uniform distribution on the interval $[a, b]$.

The mean and variance are calculated as follows:

$$E(X) = \int_a^b \frac{x}{b-a}\, dx = \frac{b+a}{2}$$

and

$$\text{Var}(X) = E(X^2) - [E(X)]^2$$

$$\text{Var}(X) = \int_a^b \frac{x^2}{b-a} dx - \left(\frac{b+a}{2}\right)^2$$

$$= \frac{b^3 - a^3}{3(b-a)} - \left(\frac{b+a}{2}\right)^2$$

$$\text{Var}(X) = \frac{(b-a)^2}{12}.$$

Example 3.2.1

Suppose X is uniform on the interval $[1, 5]$. Then

$$f(x) = \frac{1}{4},\ 1 \le x \le 5.$$

Suppose also that we have an observation that is at least 2. What is the probability the observation is at least 3?

We need

$$P(X \geq 3 \mid X \geq 2) = \frac{P(X \geq 3)}{P(X \geq 2)} = \frac{\frac{1}{2}}{\frac{3}{4}} = \frac{2}{3}. \qquad \blacktriangleleft$$

Example 3.2.2

The wheel in Example 3.1.1 is spun again, the result being X. Now we spin the wheel until an observation greater than the value of X is found, say Y. What is the expected value for Y?

Since the wheel is a fair one, we suppose that Y is uniformly distributed on $(x, 5)$ so that

$$g(y) = \frac{1}{5 - x}, \quad x < y < 5.$$

Then

$$E(Y) = \int_x^5 \frac{y}{5 - x} dy$$

$$= \frac{5^2 - x^2}{2(5 - x)}$$

$$= \frac{5 + x}{2},$$

a natural result since the central value on the interval $(x, 5)$ is $\frac{5+x}{2}$. $\blacktriangleleft$

Exercises 3.2

1. The arrival times of customers at an automobile repair shop are uniformly distributed over the interval from 8 A.M. to 9 A.M. If a customer has not arrived by 8:30, what is the probability he or she will arrive after 8:45?

2. A traffic light is red for 60 seconds, yellow for 10 seconds, and green for 90 seconds. Assuming that arrival times at the light are uniformly distributed, what is the probability of waiting more than 30 seconds?

3. A crude Geiger counter records the number of radioactive particles a substance emits, but often errs in the number of particles recorded. If the error is uniformly distributed on the interval $(-1, 2)$, what is the probability that the counter will underrecord the number of particles emitted?

4. Suppose that X is a random variable uniformly distributed on the interval $(-2, 2)$.

   a. Find $P(\frac{1}{X} < 2)$.
   b. Find $P(\frac{1}{X^2} < 2)$.

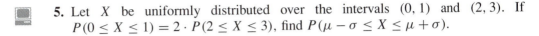 5. Let X be uniformly distributed over the intervals $(0, 1)$ and $(2, 3)$. If $P(0 \le X \le 1) = 2 \cdot P(2 \le X \le 3)$, find $P(\mu - \sigma \le X \le \mu + \sigma)$.

6. The termination of a chemical reaction occurs at a random time T between 6 and 7.5 hours after the start of the experiment. The time follows a uniform distribution.

   a. What is the probability that the reaction lasts at least 6.5 hours and no more than 6.75 hours?
   b. If the reaction is run four independent times, what is the probability that in exactly one of the four replications of the experiment the reaction will last no more than 6.5 hours?

7. Suppose that X is uniformly distributed on $0 < x < 12$. Use Tchebycheff's Inequality to establish a bound on $P(\mu - \frac{3 \cdot \sqrt{3}}{5} \cdot \sigma < X < \mu + \frac{3 \cdot \sqrt{3}}{5} \cdot \sigma)$ and then verify that the bound is correct.

8. Let X be a uniform random variable on the interval $1 < x < b$. Determine b so that

$$\sigma_x^2 = 3\mu_x.$$

9. A random variable X is uniformly distributed on the interval $-1 < x < 1$. Find $P(\frac{1}{4} \le X^2 \le \frac{3}{4})$.

10. Find the probability that at least two of four random observations of a uniform random variable on the interval $[0, 10]$ are greater than 7.

11. Suppose X is a uniform random variable on the interval $[a, a + 2]$. Find a if $P[e^X < 1.765] = \frac{1}{2}$.

▎3.3 ▶ Exponential Distribution

Example 3.3.1

Customers in a checkout line at a supermarket find that the times the checkers take in the checkout process follow the probability density function

$$f(x) = e^{-x}, \ x \ge 0$$

where X is measured in minutes. We see that $f(x) \geq 0$, and that $\int_0^\infty f(x)\,dx = 1$, so $f(x)$ defines a probability density function. Here $f(x)$ is an example of an *exponential probability density function*. What is the probability a customer's checkout time is at least k minutes?

This is

$$P(X \geq k) = \int_k^\infty e^{-x}\,dx = e^{-k}. \tag{3.3}$$

Another calculation yields a somewhat surprising result. Suppose the checkout time has been at least s minutes. What is the probability it will be at least $s + t$ minutes? Using the formula for conditional probability, and Formula 3.3 we have

$$P(X \geq s + t \mid X \geq s) = \frac{P(X \geq s + t)}{P(X \geq s)}$$

$$= \frac{e^{-(s+t)}}{e^{-s}} = e^{-t}$$

so that

$$P(X \geq s + t \mid X \geq s) = P(X \geq t).$$

Consequently, the probability that the customer waits t minutes more, given that the waiting time has been at least s minutes, is the same as the probability that the waiting time is *initially* t minutes! The fact that the customer has been waiting s minutes appears not to affect the future waiting time at all. We call this property of the exponential probability distribution the *memoryless* property. (It can be shown that the exponential probability density function is the only probability density function for which the memoryless property holds. Among discrete distributions, the geometric probability distribution is the only probability distribution function for which the property holds.)

A more general form of the exponential probability density function is

$$f(x) = \lambda e^{-\lambda(x-a)}, \quad x \geq a, \ \lambda > 0.$$

Our checkout-time example is a special case where $\lambda = 1$ and $a = 0$. The graph is shown in Figure 3.6 where a has been taken to be 2.

We note that $f(x) \geq 0$ and that $\int_a^\infty f(x)\,dx = \int_a^\infty \lambda e^{-\lambda(x-a)}\,dx = 1$, so $f(x)$ satisfies the properties of a probability density function. ◄

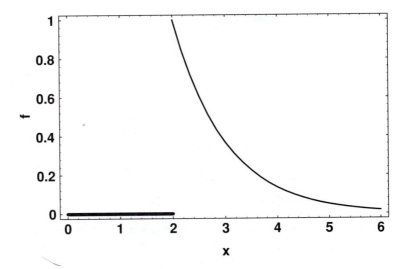

Figure 3.6   An exponential distribution.

3.3.1   Mean and Variance

For $f(x) = \lambda e^{-\lambda(x-a)}$, $x \geq a$, $\lambda > 0$, direct calculation shows that

$$E(X) = \int_a^\infty x \cdot f(x)\,dx = a + \frac{1}{\lambda}$$

and that

$$E(X^2) = \int_a^\infty x^2 \cdot f(x)\,dx = a^2 + \frac{2a}{\lambda} + \frac{2}{\lambda^2}$$

so that

$$\text{Var}(X) = (a^2 + \frac{2a}{\lambda} + \frac{2}{\lambda^2}) - (a + \frac{1}{\lambda})^2$$

so

$$\text{Var}(X) = \frac{1}{\lambda^2}.$$

3.3.2   Distribution Function

For the exponential density $f(x) = \lambda e^{-\lambda(x-a)}$, $x \geq a$, $\lambda > 0$,

$$F(x) = \int_a^x f(x)\,dx = 1 - e^{-\lambda(x-a)}.$$

Example 3.3.2

Refer again to the checkout line and the waiting time density

$$f(x) = e^{-x}, \ x \geq 0.$$

Assume that customers' waiting times are independent. What is the probability that, of the next five customers, at least three will have waiting times in excess of 2 minutes?

There are two random variables here. Let X be the waiting time for an individual customer, and Y be the number of customers who wait at least 2 minutes. Here X is exponential; because the waiting times are independent and $P(X \geq 2)$ is the same for every customer, Y is binomial. Note that while X is continuous, Y is a discrete random variable.

It is easiest to start with X, where $P(X \geq 2)$ determines p in the binomial distribution.

$$P(X \geq 2) = \int_2^\infty e^{-x} \, dx = e^{-2}.$$

Then

$$P(Y \geq 3) = \sum_{y=3}^{5} \binom{5}{y} \cdot \left(e^{-2}\right)^y \cdot \left(1 - e^{-2}\right)^{5-y}.$$

The value of this expression is 0.020028, so the event is not very likely.   ◄

Example 3.3.3

A radioactive source is emitting particles according to a Poisson distribution with 14 particles expected to be emitted per minute. The source is observed until the first particle is emitted. What is the probability density function for this random variable?

Again we have two variables in the problem. If X denotes the number of particles emitted in one minute, then X is Poisson with parameter 14. However, we don't know the time interval until the first particle is emitted. This is also a random variable, which we call Y. Note that Y is a continuous random variable. If y minutes pass before the first particle is emitted, then there must be no emissions in the first y minutes. Since the number of emissions in y minutes is Poisson with parameter $14y$, it follows that

$$P(Y \geq y) = e^{-14y}.$$

We conclude that

$$F(y) = P(Y \leq y) = 1 - e^{-14y}$$

and so

$$f(y) = \frac{dF(y)}{dy} = 14 \cdot e^{-14y}, \ \ y \geq 0.$$

This is an exponential density. In this example, note that X is discrete while Y is continuous.   ◄

Example 3.3.4

As a final example of the exponential density $f(x) = \lambda e^{-\lambda(x-a)}$, $x \geq a$, we check the memoryless property for the more general form of the exponential density.

$$P(X \geq s + t \mid X \geq s) = \frac{P(X \geq s + t)}{P(X \geq s)}$$

$$= \frac{e^{-\lambda(s+t-a)}}{e^{-\lambda(s-a)}} = e^{-\lambda t}$$

so that

$$P(X \geq s + t \mid X \geq s) = P(X \geq a + t).$$

So the memoryless property depends on the value for a.   ◄

3.4 ► Reliability

The *reliability* of a system or of a component in a system refers to the lack of frequency with which failures of the system or component occur. Reliable systems or components fail less frequently than less reliable systems or components. Suppose, for example, that T, the time to failure of a light bulb, has an exponential distribution with expected value 10,000 hours. This gives the probability density function as

$$f(t) = \left(\frac{1}{10000} \right) e^{-t/10000}, \ \ t \geq 0.$$

The reliability, $R(t)$, is defined as

$$R(t) = P(T > t);$$

that is, $R(t)$ gives the probability that the bulb lasts more than t hours. We assume that $R(0) = 1$, and we see that

$$R(t) = P(T > t) = 1 - P(T \leq t) = 1 - F(t)$$

and that

$$-R'(t) = f(t).$$

Because in this case,

$$F(t) = \int_0^t \left(\frac{1}{10000}\right) e^{-t/10000} \, dt = 1 - e^{-t/10000},$$

it follows that

$$R(t) = P(T > t) = e^{-t/10000}.$$

What is the probability that such a bulb lasts at least 2500 hours?

This is $R(2500) = e^{-1/4} = 0.7788$. So although the mean time to failure is 10,000 hours, only about 78% of these bulbs last longer than $\frac{1}{4}$ of the mean lifetime.

If it was crucial that a bulb last 2500 hours, and if this happens with probability 0.95, what should the mean time to failure be? Let this mean time to failure be m. Then

$$e^{-2500/m} = 0.95$$

so

$$m = 48,740 \text{ hours.}$$

Hazard Rate

The *hazard rate* of an item refers to the probability, per unit of time, that an item that has lasted t units of time will last Δt more units of time. We will denote the hazard rate by $h(t)$, so

$$h(t) = \frac{P(t < T < t + \Delta t \mid T > t)}{\Delta t}$$

$$= \frac{F(t + \Delta t) - F(t)}{\Delta t \cdot P(T > t)}.$$

As $\Delta t \to 0$, $h(t)$ approaches

$$h(t) = \frac{f(t)}{1 - F(t)} = \frac{f(t)}{R(t)} = -\frac{R'(t)}{R(t)}.$$

In actuarial work, the hazard rate is called the *force of mortality*. The hazard rate also occurs in econometrics as well as in other fields. In this section we investigate the consequences of a constant hazard rate, λ. Consequences of a non-constant hazard rate will be considered later in this chapter.

Suppose that

$$h(t) = \frac{f(t)}{1 - F(t)} = \lambda$$

where λ is a constant.

$$\frac{f(t)}{1 - F(t)} = -\frac{R'(t)}{R(t)}$$

we have that

$$-\frac{R'(t)}{R(t)} = \lambda.$$

It follows that

$$-\ln[R(t)] = \lambda t + k$$

so that

$$R(t) = c \cdot e^{-\lambda t}.$$

Now $c = R(0)$, and if we suppose that our components begin life at $T = 0$, then $R(0) = 1$ and

$$R(t) = e^{-\lambda t}.$$

Since $f(t) = -R'(t)$, it follows that $f(t) = \lambda e^{-\lambda t}$, $t \geq 0$.

A constant hazard rate then produces an exponential failure law. It is easy to show that an exponential failure law produces a constant hazard rate.

From Example 3.3.3 we conclude that failures occurring according to a Poisson process will also produce an exponential time to failure and hence a constant hazard rate.

Typically, the hazard rate is not constant for components. There is generally a "burn-in" period where the hazard rate may be declining. The hazard rate then usually becomes constant, or nearly so, after which it increases. This produces the "bathtub" function shown in Figure 3.7.

Different hazard rates, although constant, can have surprisingly different consequences. Suppose, for example, that component I has constant hazard rate λ and component II has hazard rate $k \cdot \lambda$ where $k > 0$. Then the corresponding reliability functions are

$$R_I(t) = e^{-\lambda t}$$

while

$$R_{II}(t) = e^{-k\lambda t} = (e^{-\lambda t})^k = [R_I(t)]^k.$$

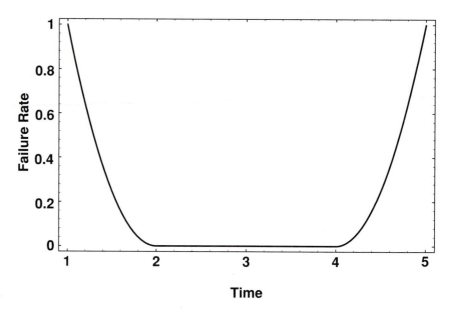

Figure 3.7   A "bathtub" hazard rate function.

So the probability that component II lasts t units or more is the *kth power* of the probability that component I lasts the same time. Since positive powers of probabilities become smaller as the power k increases, component II may rapidly become useless.

Exercises 3.4

1. Let X be an exponential random variable with mean 6.

    a. Find $P(X \geq 4)$.
    b. Find $P(X \geq 4 \mid X \geq 2)$.

 2. The *median* of a probability distribution is the value that is exceeded $\frac{1}{2}$ of the time.

    a. Find the median of an exponential distribution with mean λ.
    b. Find the probability that an observation exceeds λ.

 3. Snowfall in Indiana follows an exponential distribution with mean 15 inches per winter season.

    a. Find the probability that the snowfall will exceed 17 inches next winter.
    b. Find the probability that in four out of the next 5 winters the snowfall will be less than the mean.

4. The length, X, of an international telephone call from a local business follows an exponential distribution with mean 2 minutes. In dollars, the cost of a call of X minutes is $3X^2 - 6X + 2$. Find the expected cost of a telephone call.

5. The lengths of life of batteries in transistor radios follow exponential probability distributions. Radio A takes 2 batteries, each of which has an expected life of 200 hours; radio B uses 4 batteries, but the expected life of each is 400 hours. Radio A works if at least one of its batteries operates; radio B works only if at least 3 of its batteries operate. An expedition needs a radio that will function at least 500 hours. Which radio should be taken, or does it not matter?

6. Accidents at a busy intersection follow a Poisson distribution with 3 accidents expected in a week.

   a. What is the probability that at least 10 days pass between accidents?
   b. It has been 9 days since the last accident. What is the probability that it will be 5 days or more until the next accident?

7. The diameter X of a manufactured part is a random variable whose probability density function is

$$f(x) = \lambda e^{-\lambda x}, \ x > 0.$$

If $X < 1$, the manufacturer realizes a profit of \$3. If $X > 1$, the part must be discarded at a net loss of \$1. The machinery manufacturing the part may be set so that $\lambda = \frac{1}{4}$ or $\lambda = \frac{1}{2}$. Which setting will maximize the manufacturer's expected profit?

8. If X is a random selection from a uniform variable on the interval $(0, 1)$, then the transformation $Y = -\lambda \ln(1 - X)$ is known to produce random selections from an exponential density with mean λ.

   a. Use a uniform random number generator to draw a sample of 200 observations from an exponential density with mean 7.
   b. Draw a histogram of your sample and compare it graphically with the expected exponential density.

9. The hazard rate of an essential component in a rocket engine is .05. Find its reliability at time 125.

10. An exponential process has $R(200) = 0.85$. When is $R = 0.95$?

11. A Poisson process has mean μ. Show that the waiting time for the second occurrence is not exponentially distributed.

12. Find the probability that an item fails before 200 units of time if its hazard rate is 0.008.

13. Suppose that the life length of an automobile is exponential with mean 72,000 miles. What is the expected length of life of automobiles that have lasted 50,000 miles?

14. An electronic device costs $\$K$ to produce. Its length of life X has probability density function

$$f(x) = 0.01e^{-0.01x}, \quad x \geq 0.$$

If the device lasts less than 3 units of time, the item is scrapped and has no value. If the life length is between 3 and 6, the item is sold for $\$S$; if the life length is greater than 6, the item is sold for $\$V$. Let Y be the net profit per item. Find the probability density for Y.

 15. Suppose X is a random variable with probability density function

$$f(x) = 3e^{-3(x-a)}, \quad x \geq 2.$$

a. Show that $a = 2$.
b. Find the cumulative distribution function, $F(x)$.
c. Find $P(X > 5 \mid X > 3)$.
d. If 8 independent observations are made, what is the probability that exactly 6 of them are less than 4?

 16. A lamp contains 3 bulbs, each of which has life length that is exponentially distributed with mean 1000 hours. If the bulbs fail independently, what is the probability that some light emanates from the lamp for at least 1200 hours?

17. According to a kinetic theory, the distance, X, that a molecule travels before colliding with another molecule is described by the probability density function

$$f(x) = \frac{1}{\lambda}e^{-(x/\lambda)}, \quad x > 0, \ \lambda > 0.$$

a. What is the average distance between collisions?
b. Find $P(X > 6 \mid X > 4)$.

3.5 ▶ Normal Distribution

We come now to the most important continuous probability density function, and perhaps the most important probability distribution of any sort – the normal distribution. On several occasions, we have observed its occurrence in graphs from, apparently, widely differing sources: the sums when three or more dice are thrown; the binomial distribution for large values of n; and in the hypergeometric distribution. There are many other examples as well, and several reasons, which will appear here, to call this distribution "normal."

If

$$f(x) = \frac{1}{b\sqrt{2\pi}}\, e^{-(1/2)[(x-a)/b]^2}, \quad -\infty < x < \infty,\ -\infty < a < \infty,\ b > 0, \quad (3.4)$$

we say that X has a *normal* probability distribution. A graph of a normal distribution, where we have chosen $a = 0$ and $b = 1$, appears in Figure 3.8.

The shape of a normal curve is highly dependent on the standard deviation. Figure 3.9 shows some normal curves, each with mean 0, but with different standard deviations. We will show presently that a is the mean value and b is the standard deviation of the normal curve.

We now establish some facts regarding $f(x)$ as defined.

1. $f(x)$ defines a probability density function.

> **PROOF:** $f(x) \geq 0$, and so we must show that $\int_{-\infty}^{\infty} f(x)\, dx = 1$. To do this, let $Z = \frac{X-a}{b}$ in Formula 3.4. We have

$$\int_{-\infty}^{\infty} \frac{1}{b\sqrt{2\pi}}\, e^{-(1/2)[(x-a)/b]^2}\, dx = \int_{-\infty}^{\infty} \frac{1}{\sqrt{2\pi}} e^{-(1/2)z^2}\, dz.$$

Consider the curve $g(x) = \frac{1}{\sqrt{2\pi}}\, e^{-x^2/2}$, $-\infty < x < \infty$ as shown in Figure 3.10.

Let $I = \int_{-\infty}^{\infty} \frac{1}{\sqrt{2\pi}}\, e^{-x^2/2}\, dx$. If the curve is revolved around the y axis, the surface generated is

$$f(x, z) = \frac{1}{\sqrt{2\pi}} e^{-(1/2)(z^2+x^2)}, \quad -\infty < x < \infty,\ -\infty < z < \infty$$

because this surface has circular cross-sections and the proper traces in the coordinate planes. The volume generated is then

$$V = \int_{-\infty}^{\infty}\int_{-\infty}^{\infty} \frac{1}{\sqrt{2\pi}} e^{-(1/2)(z^2+x^2)}\, dz\, dx$$

$$= \frac{1}{\sqrt{2\pi}} (\sqrt{2\pi}\, I)^2 = \sqrt{2\pi}\, I^2.$$

On the other hand, V can be found using cylindrical shells as

$$V = \frac{2\pi}{\sqrt{2\pi}} \int_{0}^{\infty} x \cdot e^{-x^2/2}\, dx = \sqrt{2\pi}.$$

So $I^2 = 1$ and because $I > 0$, $I = 1$. Thus $f(x)$ is a probability density function for $-\infty < a < \infty$ and $b > 0$. ∎

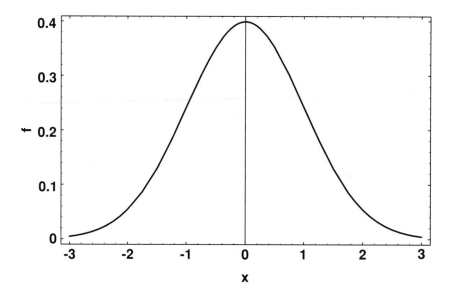

Figure 3.8   Standard normal probability density function.

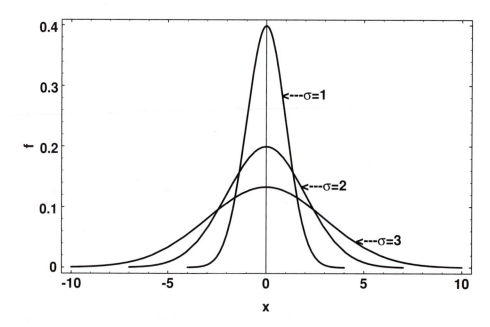

Figure 3.9   Some normal probability density functions.

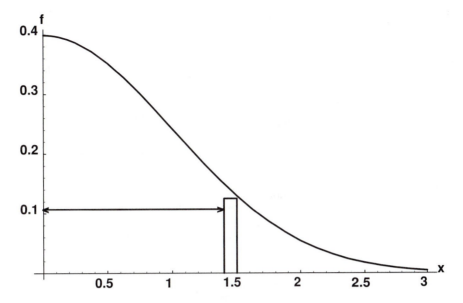

Figure 3.10   Revolving the standard normal curve around the y axis.

2. We now find the mean and variance for X.

$$E(X) = \int_{-\infty}^{\infty} x \cdot f(x)\, dx = \int_{-\infty}^{\infty} x \cdot \frac{1}{b\sqrt{2\pi}}\, e^{-(1/2)[(x-a)/b]^2}\, dx$$

Let $z = \frac{x-a}{b}$ in the integral so that $x = a + bz$, and then

$$E(X) = a \int_{-\infty}^{\infty} \frac{1}{\sqrt{2\pi}} e^{-(1/2)z^2}\, dz + \frac{b}{\sqrt{2\pi}} \int_{-\infty}^{\infty} z\, e^{-(1/2)z^2}\, dz;$$

since the first integral is 1 and the second integral is 0, $E(X) = a$.

To find the variance, first calculate

$$E(X^2) = \int_{-\infty}^{\infty} \frac{x^2}{b\sqrt{2\pi}}\, e^{-(1/2)[(x-a)/b]^2}\, dx.$$

Again let $z = \frac{x-a}{b}$. Then

$$E(X^2) = \int_{-\infty}^{\infty} \frac{(bz+a)^2}{\sqrt{2\pi}}\, e^{-(1/2)z^2}\, dz = b^2 \int_{-\infty}^{\infty} \frac{z^2}{\sqrt{2\pi}}\, e^{-(1/2)z^2}\, dz$$

$$+ 2ab \int_{-\infty}^{\infty} \frac{z}{\sqrt{2\pi}}\, e^{-(1/2)z^2}\, dz + a^2 \int_{-\infty}^{\infty} \frac{1}{\sqrt{2\pi}}\, e^{-(1/2)z^2}\, dz.$$

Since $\frac{1}{\sqrt{2 \cdot \pi}} \int_{-\infty}^{\infty} z \cdot e^{-1/2\,z^2}\, dz = 0$, it follows that

$$E(X^2) = \frac{b^2}{\sqrt{2\pi}} \int_{-\infty}^{\infty} z^2 e^{-1/2\,z^2}\, dz + a^2$$

which simplifies to

$$E(X^2) = b^2 + a^2.$$

We conclude that

$$\mu = a$$

and

$$\sigma^2 = b^2.$$

The probability density function for the normal curve is then usually written as

$$f(x) = \frac{1}{\sigma\sqrt{2\pi}}\, e^{-(1/2)[(x-\mu)/\sigma]^2}, \quad -\infty < x < \infty.$$

We will abbreviate this as

$$X \sim N(\mu, \sigma),$$

where the symbol $\sim$ is read "is distributed as."

   This is our first example of a probability density whose formula involves the standard deviation. The implications of this will be encountered in examples and problems.

3. Finally, we show that if $X \sim N(\mu, \sigma)$ and if $Z = \frac{X-\mu}{\sigma}$, then $Z \sim N(0, 1)$. We call the $N(0, 1)$ curve the *standard* or *unit* normal curve. The statement above indicates that the transformation $Z = \frac{X-\mu}{\sigma}$ can be used on an arbitrary normal curve to produce a standard normal curve.

   To show this, consider the cumulative distribution function for Z, $G(z)$, assuming that $F(x)$ is the cumulative distribution function for X. Then by definition,

$$G(z) = P(Z \le z)$$

$$= P\left(\frac{X - \mu}{\sigma} \le z\right)$$

$$= P(X \le \mu + \sigma \cdot z) = F(\mu + \sigma \cdot z).$$

We now use the fact that

$$g(z) = \frac{dG(z)}{dz}$$

to see that

$$g(z) = \frac{dG(z)}{dz} = \frac{dF(\mu + \sigma \cdot z)}{dz} = \frac{dF(\mu + \sigma \cdot z)}{d(\mu + \sigma \cdot z)} \cdot \frac{d(\mu + \sigma \cdot z)}{dz}$$

by the chain rule. It follows that

$$g(z) = \sigma \cdot f(\mu + \sigma z)$$

$$g(z) = \frac{1}{\sqrt{2\pi}} e^{-1/2\, z^2}, \quad -\infty < z < \infty,$$

which is the standard normal distribution. This indicates that problems involving arbitrary values of μ and σ can be solved using a single, standard, normal curve.

In Chapter 4 we will return to the process by which $g(z)$ was established and apply the same technique to other problems involving functions of random variables. For now, we consider some examples of the normal distribution and problems using it.

Example 3.5.1

Mathematics aptitude scores X on the Scholastic Aptitude Test (SAT) are $N(500, 100)$. Find

a. the probability that an individual's score exceeds 600, and

b. the probability that an individual's score exceeds 600, given that it exceeds 500.

a. Many computer algebra systems will calculate $P(X > 600)$ directly. This will be found to be 0.158655. If a computer algebra system is not available, a table of the standard normal distribution may be used as follows:
The Z transformation here is $Z = \frac{X-500}{100}$, so
$P(X > 600) = P(Z > 1) = 0.158655$ using Table 1 in Appendix 3.

b. Here we need
$$P(X > 600 \mid X > 500) = \frac{P(X > 600)}{P(X > 500)} = \frac{0.158655}{0.500000} = 0.317310. \quad \blacktriangleleft$$

Example 3.5.2

What mathematics SAT score, or greater, can we expect to occur with probability 0.90? We know that $X \sim N(500, 100)$ and we want to find x

so that $P(X \geq x) = 0.90$. So, if $Z = \frac{X-500}{100}$, then

$$P(Z \geq \frac{x-500}{100}) = 0.90,$$

but

$$P(Z \geq -1.287266) = 0.90$$

so

$$\frac{x-500}{100} = -1.287266$$

giving

$$x = 371.$$

◄

Example 3.5.3

From Tchebycheff's Inequality we conclude that the standard deviation is in fact a measure of dispersion for a distribution, because the probability that the interval from $\mu - k\sigma$ to $\mu + k\sigma$ is at least $1 - \frac{1}{k^2}$, a probability that increases as k increases. When the distribution is known, this probability can be determined exactly by integration. We do this now for a standard normal density. Again let $Z = \frac{X-\mu}{\sigma}$.

$$P(\mu - \sigma \leq X \leq \mu + \sigma) = P(-1 \leq Z \leq 1) = 0.6826894921$$
$$P(\mu - 2\sigma \leq X \leq \mu + 2\sigma) = P(-2 \leq Z \leq 2) = 0.9544997361$$
$$P(\mu - 3\sigma \leq X \leq \mu + 3\sigma) = P(-3 \leq Z \leq 3) = 0.9973002039$$

Tchebycheff's Inequality indicates that these probabilities are at least 0, $\frac{3}{4}$, and $\frac{8}{9}$, respectively.

These results, sometimes called the "$\frac{2}{3}$, 95%, 99% Rule," can be very useful in estimating probabilities using the normal curve.

For example, to refer again to the mathematics SAT scores, which are $N(500, 100)$, an estimate for the probability that a student's score is between 400 and 650 may be found by estimating the probability that the corresponding z score is between -1 and 1.50. We know that $\frac{2}{3}$ of the area under the curve is between -1 and 1, and we need to estimate the probability from 1 to 1.50. This can be estimated at $\frac{1}{2}$ the difference between .95 and $\frac{2}{3}$, giving a total estimate of $\frac{2}{3} + (\frac{1}{2})(.95 - \frac{2}{3}) = 0.81$. The exact probability is 0.775. It is a rarity that the answers to probability problems can be estimated in advance of an exact solution.

The occurrence of the normal distribution throughout probability theory is striking. In section 3.6, we explain why the graphs of binomial distributions, considered in Chapter 2, become normal in appearance. ◄

Exercises 3.5

1. IQ scores are known to be normally distributed with mean 100 and standard deviation 10.

   a. Find the probability that an IQ score exceeds 128.
   b. Find the probability that an IQ score is between 90 and 110.

2. The size of a boring in a metal block is normally distributed with mean 3 cm and standard deviation 0.01 cm.

   a. What proportion of the borings have sizes between 2.97 cm and 3.01 cm?
   b. For the borings exceeding 3.005 cm, what proportion exceeds 3.010 cm?

3. Brads labeled $\frac{3}{4}$ in. are normally distributed. Manufacturer I produces brads with mean $\frac{3}{4}$ in. and standard deviation 0.002 in.; manufacturer II produces brads with mean 0.749 in. and standard deviation 0.0018 in.; brads from manufacturer III have mean 0.751 in. and standard deviation 0.0015 in. A builder requires brads in the range $\frac{3}{4} \pm 0.005$ in. From which manufacturer should the brads be purchased?

4. A soft drink machine dispenses cups of a soft drink whose volume is actually a normal random variable with mean 12 oz and standard deviation 0.1 oz.

   a. Find the probability that a cup of the soft drink contains more than 12.2 oz.
   b. Find a volume, v, such that 99% of the time the cups contain at least v oz.

5. Resistors used in an electric circuit have resistances that are normally distributed with mean 0.21 ohms and standard deviation 0.045 ohms. A resistor is acceptable in the circuit if its resistance is at most 0.232 ohms. What percentage of the resistors are acceptable?

6. On May 5, in Indiana, temperatures have been found to be normally distributed with mean $80°$ and standard deviation $8°$. The record temperature on that day is $90°$.

   a. What is the probability that the record of $90°$ will be broken on next May 5?
   b. What is the probability that the record of $90°$ will be broken at least 3 times during the next 5 years on May 5?

7. Sales in a fast food restaurant are normally distributed with mean $42,000 and standard deviation $2000 during a given sales period. During a recent sales period, sales were reported to a local taxing authority to be $37,600. Should the taxing authority be suspicious?

8. Suppose that $X \sim N(\mu, \sigma)$. Find a in terms of μ and σ if

   a. $P(X > a) = 0.90$.

b. $P(X > a) = \frac{1}{3}P(X \le a).$

9. The size of a manufactured part is a normal random variable with mean 100 and variance 25. If the size is between 95 and 110, the parts can be sold at a profit of $50 each. If the size exceeds 110, the part must be reworked, and a net profit of $10 is made per part. A part whose size is less than 95 must be scrapped at a loss of $20. What is the expected profit for this process?

10. Rivets are useful in a device if their diameters are between 0.25 in and 0.38 in. These limits are often called *upper and lower specification limits*. A manufacturer produces rivets that are normally distributed with mean 0.30 in. and standard deviation 0.03 in.

    a. What proportion of the rivets meet specifications?
    b. Suppose the mean of the manufacturing process could be changed, but the manufacturing process is such that the standard deviation cannot be altered. What should the mean of the manufactured rivets be so as to maximize the proportion that meet specifications?

11. Refer to Exercise 10. Suppose that $X \sim N(\mu, \sigma)$ and that upper and lower specification limits are U and L respectively. Show that if σ must be held fixed, then the value of μ that maximizes $P(U \le X \le L)$ is $\frac{U+L}{2}$.

12. Manufacturing processes that produce normally distributed output are often compared by calculating their *process capability indices*. The process capability index for a process with upper and lower specification limits U and L, respectively, is

$$C_p = \frac{U - L}{6\sigma}$$

where the variable X is distributed $N(\mu, \sigma)$.

    What can be said about the process under each of the following conditions?

    a. $C_p = 1$
    b. $C_p < 1$
    c. $C_p > 1$

13. Upper and lower *warning limits* are often established for measurements on manufactured products. Commonly, if $X \sim N(\mu, \sigma)$, these are set at $\mu \pm 1.96\sigma$ so that 5% of the product is outside the warning limits. Discuss the proportion of the product outside the warning limits if the mean of the process increases by one standard deviation.

14. Suppose that $X \sim N(\mu, \sigma)$. Find μ and σ if $P(X > 2) = \frac{2}{3}$ and $P(X > 3) = \frac{1}{3}$.

15. "Forty-pound" bags of cement have weights that are actually $N(39.1, \sqrt{9.4})$.

    a. Find the probability that 2 of 5 randomly selected bags weigh less than 40 lb.

   b. How many bags must be purchased so that the probability that at least $\frac{1}{3}$ of the bags weigh at most 40 lb is at least 0.90?

16. Suppose $X \sim N(0, 1)$. Find

   a. $P\left(|X| < 1.5\right)$.

   b. $P\left(X^2 > 1\right)$.

17. Signals that are either 0's or 1's are sent in a noisy communication circuit. The signal received is the signal sent plus a random variable, ε, that is $N(0, \frac{1}{3})$. If a 0 is sent, the receiver will record a 0 if the signal received is at most a value v; otherwise a 1 is recorded. Find v if the probability that a 1 is recorded when a 0 is actually sent is 0.90.

18. The diameter of a ball bearing is a normally distributed random variable with mean 6 and standard deviation $\frac{1}{2}$.

   a. What is the probability a randomly selected ball bearing has a diameter between 5 and 7?

   b. If a diameter is between 5 and 7, the bearing can be sold for a profit of $1. If the diameter is greater than 7, the bearing may be reworked and sold at a profit of $0.50; otherwise, the bearing must be discarded at a loss of $2. Find the expected value for the profit.

19. Capacitors from a manufacturer are normally distributed with mean 5 μf and standard deviation 0.4 μf. An application requires four capacitors between 4.3 μf and 5.9 μf. If the manufacturer ships 5 randomly selected capacitors, what is the probability that a sufficient number of capacitors will be within specifications?

20. The height, X, that a college high jumper will clear each time she jumps is a normal random variable with mean 6 ft and variance 5.76 in^2.

   a. What is the probability the jumper will clear 6 ft 4 in on a single jump?

   b. What is the greatest height jumped with probability 0.95?

   c. Assuming the jumps are independent, what is the probability that 6 ft 4 in will be cleared on exactly 3 of the next 4 jumps?

21. A Chamber of Commerce advertises that about 16% of the motels in town charge $40 or more for a room and that the average price of a room is $32. Assuming that room rates are approximately normally distributed, what is the variance in the room rates?

22. A commuting student has discovered that her commuting time to school is normally distributed; she has two possible routes for her trip. The travel time by route A has mean 55 min and standard deviation 9 min, while the travel time by route B has mean 60 min and standard deviation 3 min. If the student has, at most, 63 min for the trip, which route should she take?

23. The diameter of an electric cable is normally distributed with mean 0.8 in. and standard deviation 0.02 in.

    a. What is the probability the diameter will exceed 0.81 in.?

    b. The cable is considered defective if the diameter differs from the mean by more than 0.025 in. What is the probability that a randomly selected cable is defective?

    c. Suppose now that the manufacturing process can be altered and that the standard deviation can be changed while keeping the mean at 0.8. If the criterion in part **b** is used, but we want only 10% of the cables to be defective, what value of σ must be met in the manufacturing process?

24. A cathode ray tube for a computer graphics terminal has a fine mesh screen behind the viewing surface, which is under tension produced in manufacturing. The tension readings follow a $N(275, 40)$ distribution, where measurements are in units of mV.

    a. The minimum acceptable tension is 200 mV. What proportion of tubes exceed this limit?

    b. Tension above 375 mV will tear the mesh. Of the acceptable screens, what proportion have tensions of at most 375 mV?

    c. Refer to part **a**. Suppose it is desired to have 99.5% acceptable screens and that a new quality control manager thinks he can reduce σ^2 to an acceptable level. What value of σ^2 must be attained?

25. The life lengths of two electronic devices, at D_1 and D_2, have distributions $N(40, 6)$ and $N(45, 3)$ respectively. If the device is to be used for a 48-hour period, which device should be selected?

26. "One pound" packages of cheese are marketed by a major manufacturer, but the actual weight in pounds of a randomly selected package is a normally distributed random variable with standard deviation 0.02 lb. The packaging machine has a setting allowing the mean value to be varied.

    a. Federal regulations allow for a maximum of 5% short weights (weights below the claim on the label). What should the setting on the machine be?

    b. A package labeled "one pound" sells for $1.50, but costs only $1 to produce. If short-weight packages are not sold, and if the machine's mean setting is that in part **a**, what is the expected profit on 1000 packages of cheese?

3.6 ► Normal Approximation to the Binomial Distribution

Example 3.6.1

A component used in the construction of an electric motor is produced in a factory assembly line. In the past, about 10% of the components have proven unsatisfactory for use in the motor. The situation may be modeled by a binomial process in which p, denoting the probability of an unsatisfactory component, is 0.10. The assembly line produces 500 components per day. If X denotes the number of unsatisfactory components, then the probability distribution function is

$$P(X = x) = \binom{500}{x}(0.10)^x(0.90)^{500-x}, \; x = 0, 1, \ldots, 500.$$

A graph of the distribution is shown in Figure 3.11.

Figure 3.11 is centered on the mean value, $500 \cdot (0.10) = 50$. Note that the possible values of X are from $X = 0$ to $X = 500$ but that the probabilities decrease rapidly, so we show only a small portion of the curve.

The graph in Figure 3.11 certainly appears to be normal. We note, however, that, although the eye may see a normal curve, there are in reality no points on the graph between the X values that are integers, since X can only be an integer. In Figure 3.12 we have used the heights of the binomial curve in Figure 3.11 to produce a histogram. If we consider a particular value of X, say $X = 53$, notice that the base of the bar runs from 52.5 to 53.5 (both impossible values for X!), and that the height of the bar is $P(X = 53)$. Thus the area of the bar at $X = 53$, because the base is of length 1, is $P(X = 53)$. This is the key that allows us to estimate binomial probabilities by the normal curve.

Figure 3.13 shows a normal curve imposed on the histogram of Figure 3.12. What normal curve should be used? It is natural to use a normal curve with mean and variance equal to the mean and variance of the binomial distribution that is being estimated, so we have used $N(500 \cdot 0.10, \sqrt{500 \cdot (0.10) \cdot (0.90)}) = N(50, \sqrt{45})$.

To estimate $P(X = 53)$, we find $P(52.5 \leq X \leq 53.5)$ using the approximation $N(50, \sqrt{45})$; this gives 0.0537716. The exact probability is 0.0524484.

As a final example, consider the probability that the assembly line produces between 36 and 42 unsatisfactory components. This is estimated by $P(35.5 \leq X \leq 42.5)$ where $X \sim N(50, \sqrt{45})$, yielding 0.116449. The exact probability is 0.118181.

When the sum of a large number of binomial probabilities is needed, a computer algebra system might be used to calculate the result exactly, although the computation might well be lengthy. The same computer algebra system would also, more quickly and easily, calculate the relevant normal

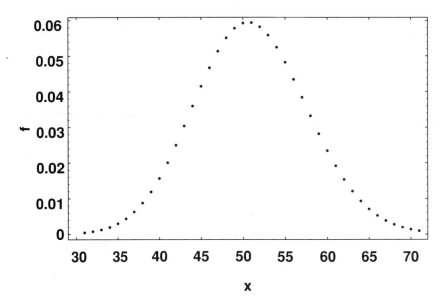

Figure 3.11   Binomial distribution, $n = 500,\ p = 0.10$.

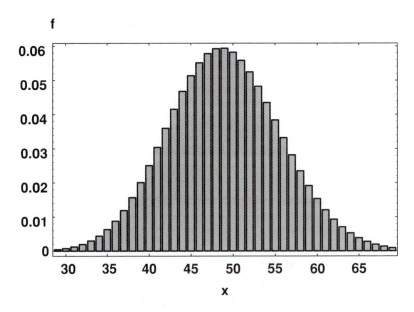

Figure 3.12   A histogram for the binomial distribution, $n = 500,\ p = 0.10$.

probabilities. In any event, whether or not the approximation is used, the approximation of the binomial distribution by the normal distribution is a striking fact. We will justify the approximation more thoroughly when

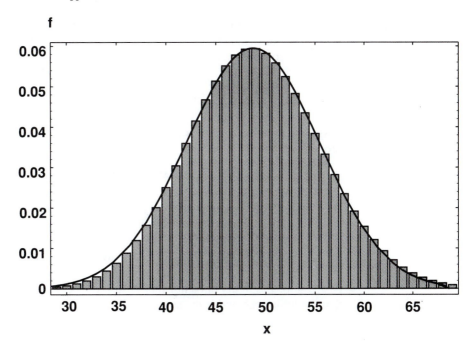

Figure 3.13 Normal curve approximation for the binomial, $n = 500$, $p = 0.10$.

we consider sums of random variables in Chapter 4. We note now that the approximation works well for moderate or large values of n, the quality of the approximation depending somewhat on the value of p.

In using a normal approximation to a binomial distribution, it is well to check the tail probabilities, hoping that these are small so that the approximation is an appropriate one. For example, if we want to approximate $P(9 \leq X \leq 31)$, we should check that the z score for 31.5 exceeds 2.50 and that the z score for 8.5 is less than -2.50 because these scores have about 2% of the curve in each tail. ◄

Exercises 3.6

In solving these problems, find both the exact answer using a binomial distribution as well as the result given by the normal approximation.

1. In 100 tosses of a fair coin, show that 50 heads and 50 tails is the most probable outcome, but that this event has probability of only about 0.08. Show that this compares favorably to the occurrence of at least 58 heads.

2. A manufacturer of components for electric motors has found that about 10% of the production will not meet customer specifications. Find the probability that in a lot of 500 components,

   a. exactly 53 do not meet customer specifications.

   b. between 36 and 42 (inclusive) components do not meet customer specifications.

3. A system of 50 components functions if at least 90% of the components function properly.

   a. Find the probability that the system operates if the probability that a component operates properly is 0.85.

   b. Suppose now that the probability that a component operates properly is *p*. Find *p* if the probability that the system operates properly is 0.95.

4. An acceptance sampling plan accepts a lot if at most 3% of a sample randomly chosen from a very large lot of items does not meet customer specifications. In the past, 2% of the items have not met customer specifications. Find the probability that the lot is accepted if the sample size is

   a. 10.

   b. 100.

   c. 1000.

5. A fair coin is tossed 1000 times. Let X denote the number of heads that occur. Find k so that $P(500 - k \le X \le 500 + k) = 0.90$.

6. An airline finds that 3% of the ticketed passengers do not appear for a certain flight. The plane holds 125 people. How many tickets should be sold if the airline wants to carry all the passengers who show up with probability 0.99?

7. Sam and Joe operate competing minibuses for travel from a central point in a city to the airport. Passengers appear and are equally likely to choose either minibus. During a given time period, 40 passengers appear. How many seats should each minibus have if Sam and Joe each want to accommodate all the passengers who show up for their minibus with probability 0.95?

8. A candidate in an election knows that 52% of the voters will vote for her. What is the probability that, out of 200 voters, she receives at least 50% of the votes?

9. A fair die is rolled 1200 times. Find the probability that at least 210 sixes appear.

10. In 10,000 tosses of a coin, 5150 heads appear. Is the coin loaded?

11. The length of life of a fluorescent fixture has an exponential distribution with expected life length 10,000 hours. Seventy of these bulbs operate in a factory. Find the probability that at most 40 of them last at least 8000 hours.

12. Suppose that X is uniformly distributed on $[0, 10]$.

    a. Find $P(X > 7)$.

    b. Among 4 randomly chosen observations of X, what is the probability that at least 2 of these are greater than 7?

    c. What is the probability that, of 1000 observations, at least 328 are greater than 7?

 13. Two percent of the production of an industrial process is not acceptable for sale. Suppose the company produces 1000 items a day. What is the probability that a day's production contains between 1.4% and 2.2% nonacceptable items?

3.7 ▶ Gamma and Chi-Squared Distributions

In section 3.3 of this chapter we considered the waiting time until the first Poisson event occurred and found that the waiting time followed an exponential distribution. We now want to consider the waiting time for the second Poisson event.

To make matters specific, suppose that the Poisson random variable has parameter λ and that Y is the waiting time for the second event. In y units of time, we expect $\lambda \cdot y$ events. If $Y \geq y$, there is, at most, one event in λy units of time, so

$$P(Y \geq y) = P(X = 0 \text{ or } 1) = \sum_{x=0}^{1} \frac{e^{-\lambda \cdot y}(\lambda \cdot y)^x}{x!}.$$

It follows that

$$F(y) = P(Y \leq y) = 1 - \sum_{x=0}^{1} \frac{e^{-\lambda \cdot y}(\lambda \cdot y)^x}{x!}$$

$$F(y) = 1 - e^{-\lambda \cdot y} - \lambda y e^{-\lambda \cdot y},$$

so

$$f(y) = \frac{dF(y)}{dy} = \lambda^2 y e^{-\lambda \cdot y}, \ y \geq 0.$$

A graph of $f(y)$ is shown in Figure 3.14.

Here $f(y)$ is an example of a more general distribution, called the *gamma* distribution.

Consider now waiting for the rth Poisson event from a Poisson distribution with parameter λ, and let Y denote the waiting time. Then at most $r - 1$ events must occur in y units of time, so

$$1 - F(y) = P(Y \geq y) = \sum_{x=0}^{r-1} \frac{e^{-\lambda \cdot y} \cdot (\lambda y)^x}{x!}.$$

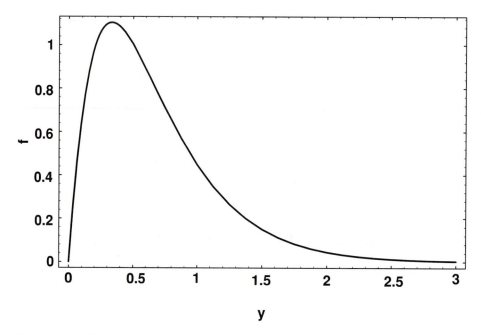

Figure 3.14   Waiting time for the second Poisson event.

It follows that

$$f(y) = \frac{dF(y)}{dy} = e^{-\lambda \cdot y} \sum_{x=0}^{r-1} \frac{\lambda^{x+1} y^x}{x!} - e^{-\lambda \cdot y} \sum_{x=1}^{r-1} \frac{\lambda^x y^{x-1}}{(x-1)!}.$$

This sum collapses, leaving

$$f(y) = \frac{e^{-\lambda y} \lambda^r y^{r-1}}{(r-1)!}, \ \ y \geq 0.$$

Here $f(y)$ defines a *gamma distribution*. The exponential distribution is a special case of $f(y)$ when $r = 1$.

Since $f(y)$ must be a probability density function, it follows that

$$\int_0^\infty \frac{e^{-\lambda y} \lambda^r y^{r-1}}{(r-1)!} \, dy = 1.$$

Now, letting $x = \lambda \cdot y$, it follows that

$$\int_0^\infty e^{-x} x^{r-1} \, dx = (r-1)! \text{ if } r \text{ is a positive integer.}$$

This integral is commonly denoted by $\Gamma(r)$. So

$$\Gamma(r) = \int_0^\infty e^{-x} x^{r-1} \, dx = (r-1)! \text{ if } r \text{ is a positive integer.}$$

Now consider the expected value:

$$E(Y) = \int_0^\infty y \cdot \frac{e^{-\lambda y} \lambda^r y^{r-1}}{(r-1)!} \, dy$$

$$= \frac{r}{\lambda} \int_0^\infty \frac{e^{-\lambda y} \lambda^r y^r}{r!} \, dy$$

so, because $\int_0^\infty e^{-\lambda y} (\lambda y)^r \, dy = r!$,

$$E(Y) = \frac{r}{\lambda} \, .$$

It can also be shown that

$$\mathrm{Var}(Y) = \frac{r}{\lambda^2}.$$

Graphs of $f(y)$ are also easy to produce using a computer algebra system. Figure 3.15 shows $f(y)$ for $r = 7$ and $\lambda = \frac{4}{7}$.

The normal-like appearance of the graph is striking, and so we consider a numerical example to investigate this phenomenon. It will follow from considerations given in Chapter 4 that $f(y)$ does indeed approach a normal distribution.

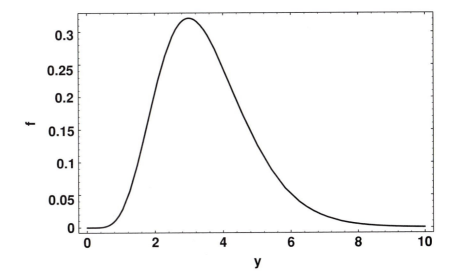

Figure 3.15 A gamma distribution.

Suppose the Poisson process has $\lambda = 2$, and we consider Y to be the waiting time for the seventh occurrence. Then

$$f(y) = \frac{e^{-2y}2^7 y^6}{6!}, \quad y \geq 0.$$

It follows that

$$P(2 \leq Y \leq 5) = \int_2^5 \frac{e^{-2y}2^7 y^6}{6!} \, dy = 0.759185.$$

Using the preceding formulas, $E(Y) = \frac{7}{2}$ and $\text{Var}(Y) = \frac{7}{4}$, so the normal curve approximation uses the normal curve $N(\frac{7}{2}, \frac{\sqrt{7}}{2})$. This gives an approximation of 0.743161. The normal curve approximates this gamma distribution fairly well here.

We return now to the gamma function. We see that if r is a positive integer, $\Gamma(r) = (r-1)!$ so the graph of $\Gamma(r)$ passes through the points $(r+1, r!)$. But $\Gamma(r)$ has values when r is not a positive integer. For example,

$$\Gamma\left(\frac{1}{2}\right) = \int_0^\infty e^{-y} y^{-1/2} \, dy.$$

Letting $y = \frac{z^2}{2}$ in this integral, as well as inserting factors of $\sqrt{2\pi}$, results in

$$\Gamma\left(\frac{1}{2}\right) = \sqrt{2} \cdot \sqrt{2\pi} \int_0^\infty \frac{1}{\sqrt{2\pi}} e^{-(z^2)/2} \, dz.$$

The integral is $\frac{1}{2}$ the area under a standard normal curve and so is $\frac{1}{2}$. So

$$\Gamma\left(\frac{1}{2}\right) = \sqrt{\pi}.$$

Consequently, the gamma distribution is often written as

$$f(y) = \frac{e^{-\lambda y} \lambda^r y^{r-1}}{\Gamma(r)}, \quad y \geq 0.$$

A special case of the gamma distribution occurs when $\lambda = \frac{1}{2}$ and $r = \frac{n}{2}$. The distribution then takes the form

$$f(x) = \frac{e^{-(x/2)} \cdot x^{(n/2)-1}}{2^{n/2}\Gamma(\frac{n}{2})}, \quad x \geq 0.$$

Here X is said to follow a *chi-squared distribution with n degrees of freedom*, which we denote by χ_n^2. The exponent 2 has no particular significance; it is simply part of the notation that is in general use. We will discuss this distribution in greater detail in

Chapter 4, but, because it is a special case of the gamma distribution, and because it has a large variety of practical applications, we show an example of its use now.

First, let's look at some graphs of χ_n^2 for some specific values of n. These graphs, which can be produced by a computer algebra system, are shown in Figure 3.16.

Again we note the approach to normality as n increases. This fact will be established in Chapter 4.

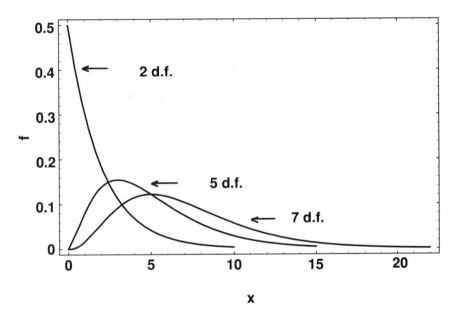

Figure 3.16 Some χ^2 distributions.

Example 3.7.1

A production line produces items that are classified as "Good," "Bad," or "Rework," the latter category indicating items that are not satisfactory on first production but which could be subject to further work and sold as good items. The line, in the past, has been producing 85%, 5%, and 10% in the three categories, respectively. Eight hundred items are produced one day, of which 665 are good, 35 are bad, and 100 need to be reworked. These numbers, of course, are not *exactly* those expected, but the increase in items to be reworked worries the plant management. Has the process in fact changed or is the sample simply the result of random variation?

This is another instance of statistical inference, because we use the sample to draw a conclusion regarding the population, or universe, from which it is selected. Of course we lack an obvious random variable to use in this case. We might begin by computing the expected numbers in the three categories, which are 680, 40, and 80. Let the observed number in the ith category be O_i and the expected number in the ith category be E_i. It can then be shown, although not easily, that

$$\sum_{i=1}^{n} \frac{(O_i - E_i)^2}{E_i}$$

follows a χ_{n-1}^2 distribution where n is the number of categories. In this case we calculate

$$\chi_2^2 = \frac{(665 - 680)^2}{680} + \frac{(35 - 40)^2}{40} + \frac{(100 - 80)^2}{80} = 5.955882353$$

The χ_2^2 curve is an exponential distribution,

$$f(x) = \frac{1}{2} e^{-(x/2)}, \ x \geq 0.$$

This point is quite far out in the right tail of the χ_2^2 distribution, as Figure 3.16 indicates.

It is easy to find that $P(\chi_2^2 > 5.955882353) = 0.0508976$. So we are faced with a decision: if the process has not changed at all, the value of χ_2^2 will exceed that of our sample only about 5% of the time. That is, simple random variation will produce this value of χ_2^2, or an even greater value, about 5% of the time. Because this is fairly small, we would probably conclude that the sample is not simply a consequence of random variation and that the production process had changed. Figure 3.17 shows a more general chi-squared distribution. ◄

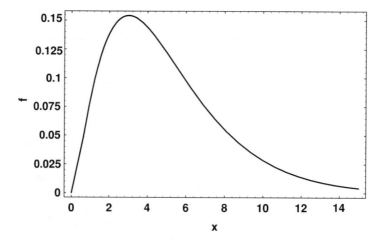

Figure 3.17   A χ^2 distribution.

Exercises 3.7

1. A book publisher finds that the yearly sales X of a textbook (in thousands of books) follow a gamma distribution with $\lambda = 10$ and $r = 5$.

   a. Find the mean and variance for the yearly sales.
   b. Find the probability that the number of books sold in one year is between 200 and 600.
   c. Sketch the probability density function for X.

2. Particles are emitted from a radioactive source with 3 particles expected per minute.

   a. Find the probability density function for the waiting time for the fourth particle to be emitted.
   b. Find the mean and variance for the waiting time for the fourth particle.
   c. Find the probability that at least 20 seconds elapse before the fourth particle is emitted.

3. Weekly sales S, in thousands of dollars, for a small shop follow a gamma distribution with $\lambda = 1$ and $r = 2$.

   a. Sketch the probability density function for S.
   b. Find $P[S > 2 \cdot E(S)]$.
   c. Find $P(S > 1.5 \mid S > 1)$.

4. Yearly snowfall S, in inches, in southern Indiana follows a gamma distribution with $\lambda = \frac{1}{2}$ and $r = 3$.

   a. Find the probability at least 8 inches of snow will fall in a given year.
   b. If 6 inches of snow have fallen in a given year, what is the probability of at least 2 more inches of snow?
   c. Find $P(\mu - \sigma \leq S \leq \mu + \sigma)$.

5. Show, using integration by parts, that $\Gamma(n) = (n - 1)!$ if n is a positive integer.

6. Show that $\Gamma\left(\dfrac{n}{2}\right) = \dfrac{\Gamma(2n)\sqrt{\pi}}{2^{2n-1}\Gamma(n)}$ for $n = 3, 5, 7, 9, \ldots$.

7. Show that $\dbinom{-a}{k} = \dfrac{(-1)^k \Gamma(a + k)}{\Gamma(k + 1)\Gamma(a)}$.

8. If X is a standard normal variable, then it is known that X^2 follows a χ_1^2 distribution. Calculate $P(X^2 < 2)$ in two different ways.

9. A die is tossed 60 times with the following results:

Face	1	2	3	4	5	6
Observations	8	12	9	8	10	13

Is the die fair?

10. Show by direct integration that $E(\chi_n^2) = n$ and that $\text{Var}(\chi_n^2) = 2n$.

11. Phone calls come into a switchboard according to a Poisson process at the rate of five calls per hour. Let Y denote the waiting time for the first call to arrive.

a. Find $P(Y > y)$.
b. Find the probability density function for Y.
c. Find $P(Y \geq 10)$.

3.8 ▶ Weibull Distribution

We considered the reliability of a product and the hazard rate in section 3.3. We showed there that a constant hazard rate produced an exponential time-to-failure law. Now let us consider non-constant hazard rates. A variety of time-to-failure laws are used to produce non-constant hazard rates. As an example we consider a Weibull distribution, because it provides such a variable hazard rate. In addition, a Weibull distribution can be shown to hold when the performance of a system is governed by the least reliable of its components, which is not an uncommon occurrence.

We use the phrase "*a* Weibull distribution" to point out the fact that the distributions about to be described vary widely in appearance and properties and in fact define an entire family of related distributions.

We recall some facts from section 3.4 first. Recall that if $f(t)$ defines a time-to-failure probability distribution, then the reliability function is

$$R(t) = P(T > t) = 1 - P(T \leq t) = 1 - F(t).$$

The hazard rate is

$$h(t) = \frac{f(t)}{R(t)} = -\frac{R'(t)}{R(t)}. \tag{3.5}$$

Now suppose that

$$h(t) = \frac{\alpha}{\beta^\alpha} t^{\alpha-1}, \quad \alpha > 0, \ \beta > 0, \ t \geq 0.$$

Formula 3.5 indicates that

$$-\frac{R'(t)}{R(t)} = \frac{\alpha}{\beta^\alpha} t^{\alpha-1}$$

from which we find

$$R(t) = e^{-(t/\beta)^\alpha} \text{ since } t \geq 0.$$

We also find that

$$f(t) = -\frac{dR(t)}{dt} = \frac{\alpha}{\beta^\alpha} t^{\alpha-1} e^{-(t/\beta)^\alpha}, \ t \geq 0.$$

$f(t)$ describes the *Weibull* family of probability distributions. Varying α and β produces graphs of different shapes, as Figure 3.18 shows.

The reliability functions $R(t)$ also differ widely, as Figure 3.19 shows.

The mean and variance of a Weibull distribution are found using the gamma function. We find, for a Weibull distribution with parameters α and β, that

$$E(T) = \beta \cdot \Gamma\left(\frac{1}{\alpha} + 1\right)$$

and

$$\text{Var}(T) = \beta^2 \cdot \left[\Gamma\left(\frac{2}{\alpha} + 1\right) - \left\{\Gamma\left(\frac{1}{\alpha} + 1\right)\right\}^2 \right].$$

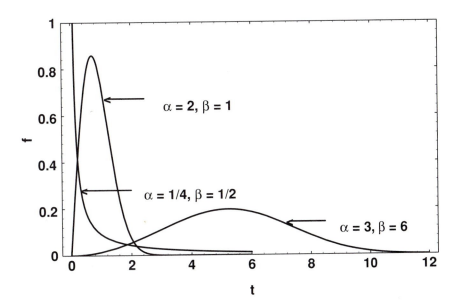

Figure 3.18 Some Weibull Distributions.

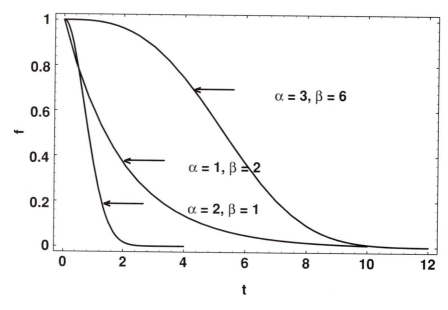

Figure 3.19   Some reliability functions.

Exercises 3.8

1. The lifetime of a part is a Weibull random variable with $\alpha = 2$ and $\beta = 10$ years.

  a. Sketch the probability density function.
  b. Find the probability the part lasts between 3 and 7 years.
  c. Find the probability that a three-year-old part lasts at least 7 years.

2. For the part described in Exercise 1:

  a. If the part carries a 15-year warranty, what percentage of the parts are still good at the end of the warranty period?
  b. What should the warranty be if it is desired to have 99% of the parts still good at the end of the warranty period?

3. The hazard rate for a generator is $10^{-4}t/\text{hr}$.

  a. Find $R(t)$, the reliability function.
  b. Find the expected length of life for the generator.
  c. Find the probability the generator lasts at least 150 hours.

4. A component's life length follows a Weibull distribution with $\alpha = \frac{1}{3}$, $\beta = \frac{1}{27}$.

  a. Plot the probability density function for the life length.
  b. Determine the hazard rate.

    c. Find the probability the component lasts at least 2 hours.

5. One component of a satellite has a hazard rate of $10^{-6}t^2$/hr.

    a. Plot $R(t)$, the reliability function.

    b. Find the probability that the component fails within 100 hours.

6. How many of the components for the satellite in Exercise 5 must be used if we want the probability that at least one lasts at least 100 hours to be 0.99?

7. Find the median of a Weibull distribution with parameters α and β.

8. A Weibull random variable, X, has $\alpha = 4$ and $\beta = 30$. Compare the exact value of $P(20 < X < 30)$ with the normal approximation to that probability.

9. It has been noticed that 56% of a heavily used industrial bulb last at most 10,000 hours. Assuming that the life lengths of these bulbs follow a Weibull distribution with $\beta = 3$, what proportion of the bulbs will last at least 15,000 hours?

Chapter Review

Random variables that can assume any value in an interval or intervals are called *continuous* random variables; they have been the subject of this chapter.

    It is clear that the probability distribution function, $f(x) = P(X = x)$, which was of primary importance in our work with discrete random variables, is of little use when X is continuous, because, in that case, $P(X = x) = 0$ for all values of X, so this function carries no information whatsoever. It is possible, however, to distinguish different continuous random variables by a *probability density function*, $f(x)$, which has the following properties:

1. $f(x) \geq 0$
2. $\int_{-\infty}^{\infty} f(x)\,dx = 1$
3. $\int_{a}^{b} f(x)\,dx = P(a \leq X \leq b)$

We studied several of the most important probability density functions in this chapter.

    The mean and variance of a continuous random variable can be calculated by

$$E(X) = \mu = \int_{-\infty}^{\infty} x \cdot f(x)\,dx$$

and

$$\mathrm{Var}(X) = \sigma^2 = E(X - \mu)^2 = \int_{-\infty}^{\infty} (x - \mu)^2 \cdot f(x)\,dx,$$

provided, of course, that the integrals are convergent.

It is often helpful to use the fact that

$$\sigma^2 = E(X^2) - [E(X)]^2$$

when calculating σ^2.

The first distribution considered was the *uniform* distribution, defined by

$$f(x) = \frac{1}{b-a}, \ a \le x \le b.$$

We found that

$$\mu = \frac{b+a}{2} \text{ and that } \sigma^2 = \frac{(b-a)^2}{12}.$$

The most general form of the *exponential* distribution is

$$f(x) = \lambda e^{-\lambda(x-a)}, \ x \ge a \text{ where } \lambda > 0.$$

A computer algebra program or direct integration shows that

$$E(X) = \int_a^\infty x \cdot f(x)\, dx = a + \tfrac{1}{\lambda}$$

and

$$V(X) = \tfrac{1}{\lambda^2}.$$

An interesting fact is that the waiting time for the first occurrence of a Poisson random variable is an exponential variable.

We then discussed *reliability*, an important modern application of probability theory. We defined the reliability as

$$R(t) = P(T > t)$$

where T is a random variable. The reliability then gives the probability that a component whose lifetime is the random variable T lasts at least t units of time.

The (instantaneous) *hazard rate* is the limit of the probability, per unit of time, that an item that has lasted t units of time will last Δt more units of time. We found that the hazard rate, $h(t)$, is

$$h(t) = \frac{f(t)}{1 - F(t)}$$

where $f(t)$ and $F(t)$ are the probability density and distribution function for T, respectively.

The normal distribution, without doubt the most important continuous distribution of all, was considered next. We showed that its most general form is

$$f(x) = \frac{1}{\sigma\sqrt{2\pi}} \, e^{-(1/2)[(x-\mu)/\sigma]^2}, \quad -\infty < x < \infty.$$

If X has this normal distribution, we write $X \sim N(\mu, \sigma)$.

An important fact is that if $X \sim N(\mu, \sigma)$ and if $Z = \frac{X-\mu}{\sigma}$, then $Z \sim N(0, 1)$, a distribution that is referred to as the *standard* normal distribution. This fact allows a wide variety of normal curve calculations to be carried out using a single normal curve. This is a very unusual circumstance in probability theory, distributions often being highly dependent upon sample size, for example, as we will see in later chapters.

The normal curve arises in a multitude of places; one of its most important applications is that it can be used to approximate a binomial distribution. We discussed the approximation of a binomial variable with parameters n and p by a $N(np, \sqrt{npq})$ curve.

Two distributions whose importance will be highlighted in later chapters are the *gamma* and *chi-squared* distributions. The gamma distribution arises when we wait for the rth Poisson occurrence. Its probability density function is

$$f(y) = \frac{e^{-\lambda y}\lambda^r y^{r-1}}{(r-1)!}, \quad y \geq 0$$

where λ is the parameter in the Poisson distribution.

The chi-squared distribution arises when $\lambda = \frac{1}{2}$ and $r = \frac{n}{2}$.

Finally, we considered the Weibull family of distributions, whose probability density functions are members of the family

$$f(t) = \frac{\alpha}{\beta^\alpha} t^{\alpha-1} e^{-(t/\beta)^\alpha}, \quad \alpha > 0, \ \beta > 0, \ t \geq 0.$$

It is fairly easy to show that

$$E(T) = \beta \cdot \Gamma\left(\frac{1}{\alpha} + 1\right)$$

and

$$\text{Var}(T) = \beta^2 \cdot \left[\Gamma\left(\frac{2}{\alpha} + 1\right) - \left\{\Gamma\left(\frac{1}{\alpha} + 1\right)\right\}^2\right].$$

The Weibull distribution is of importance in reliability theory; several examples were given in the chapter.

Problems for Review

Exercises 3.1 # 2, 3, 4, 5, 7, 10, 14, 18
Exercises 3.2 # 1, 2, 4, 7, 9, 10
Exercises 3.4 # 1, 2, 5, 7, 9, 10, 16
Exercises 3.5 # 1, 3, 5, 8, 9, 10, 15, 16, 17, 19, 23, 26
Exercises 3.6 # 1, 2, 3, 7, 10, 12
Exercises 3.7 # 2, 3, 6, 8, 11
Exercises 3.8 # 1, 3, 4, 6, 7, 9

Supplementary Exercises for Chapter 3

1. A machining operation produces steel shafts having diameters that are normally distributed with mean 1.005 in. and standard deviation 0.01 in. If specifications call for diameters to fall in the interval 1.000 ± 0.02 in., what percentage of the steel shafts will fail to meet specifications?

2. Electric cable is made by two different manufacturers, each of whom claims that the diameters of their cables are normally distributed. The diameters, in inches, from manufacturer I are $N(0.80, 0.02)$, and the diameters from manufacturer II are $N(0.78, 0.03)$. A purchaser needs cable that has diameter less than 0.82 in. Which manufacturer should be used?

3. A buyer requires a supplier to deliver parts that differ from 1.10 by no more than 0.05 units. The parts are distributed according to $N(1.12, 0.03)$. What proportion of the parts do not meet the buyer's specifications?

4. Manufactured parts have lifetimes in hours, X, that are distributed $N(1000, 100)$. If $800 \leq X \leq 1200$, the manufacturer makes a profit of $50 per part. If $X > 1200$, the profit per part is $75. Otherwise, the manufacturer loses $25 per part. What is the expected profit per part?

5. The annual rainfall (in inches) in a certain region is normally distributed with $\mu = 40$, $\sigma = 4$. Assuming rainfalls in different years are independent, what is the probability that in 2 of the next 4 years the rainfall will exceed 50 inches?

6. The weights of oranges in a good year are described by a normal distribution with $\mu = 16$ and $\sigma = 2$ (ounces).

   a. What is the probability that a randomly selected orange has weight in excess of 17 ounces?

   b. Three oranges are selected at random. What is the probability that the weight of exactly one of them exceeds 17 ounces?

   c. How many oranges out of 10,000 are expected to have weights between 15.4 and 17.3 ounces?

7. A sugar refinery has 3 processing plants, all receiving raw sugar in bulk. The amount of sugar in tons that each of the plants can process in a day has an exponential distribution with mean 4.

   a. Find the probability that a given plant processes more than 4 tons in a day.

   b. Find the probability that at least 2 of the plants process more than 4 tons in a day.

8. The length of life in hours, X, of an electronic component has an exponential probability density function with mean 500 hours.

   a. Find the probability that a component lasts at least 900 hours.

   b. Suppose a component has been in operation for 300 hours. What is the probability it will last for another 600 hours?

9. Students in an electrical engineering laboratory measure current in a circuit using an ammeter. Due to several random factors, the measurement, X, follows the probability density function

$$f(x) = 0.025\,x + b, \ 2 < x < 6.$$

   a. Show that $b = 0.15$.

   b. Find the probability that the measurement of the current exceeds 3 amps.

   c. Find $E(X)$.

   d. Find the probability that 3 laboratory partners independently measure the current as less than 4 amps.

10. Let X be a random variable with probability density function

$$f(x) = \begin{cases} k, & 1 \le x \le 2 \\ k(3 - x), & 2 \le x \le 3. \end{cases}$$

   a. Find k.

   b. Calculate $E(X)$.

   c. Find the cumulative distribution function, $F(x)$.

11. The percentage X of antiknock additive in a particular gasoline is a random variable with probability density function

$$f(x) = kx^3(1 - x), \ 0 < x < 1.$$

   a. Show that $k = 20$.

   b. Evaluate $P[X < E(X)]$.

   c. Find $F(x)$.

12. Suppose that $f(x) = 3x^2$, $0 < x < 1$ is the probability density function for some random variable X. Find $P\left(X \ge \frac{1}{2} \mid X \ge \frac{1}{4}\right)$.

13. A point B is chosen at random on a line segment AC of length 10. A right triangle with sides AB and BC is constructed. Determine the probability that the area of the triangle is at least 7 square units.

14. A random variable, X, has probability density function

$$f(x) = \begin{cases} ax, & 0 \le x \le 3 \\ 6a - ax, & 3 \le x \le 6. \end{cases}$$

  a. Show that $a = \dfrac{1}{9}$.
  b. Find $P(X \ge 4)$.

15. Verify Tchebycheff's Inequality for $k = \sqrt{2}$ for the probability density function

$$f(x) = \frac{1}{2}(x+1), \quad -1 < x < 1.$$

16. Suppose X has the distribution function

$$F(x) = \begin{cases} 0, & x < 0 \\ ax - \frac{1}{4}x^2, & 0 \le x < 2 \\ 1, & x \ge 2. \end{cases}$$

  a. Find a.
  b. Find $P(X \ge 1)$.

17. Find the mean and variance of the random variable X whose probability density function is

$$f(x) = \frac{3}{4}(1-x)(x-3), \quad 1 \le x \le 3.$$

18. A random variable X has probability density function

$$f(x) = \begin{cases} 1 + x, & -1 < x < 0 \\ 1 - x, & 0 \le x < 1. \end{cases}$$

  Find $E[X^2 - 2X + 2]$.

19. **a.** Determine k so that $f(x) = kxe^{-x^2}$ is a probability density function for some non-negative random variable X.
  b. Determine $F(x)$ and sketch it.

20. The time (in seconds) a car has to wait for a certain traffic light has probability density function

$$f(x) = \begin{cases} \frac{x}{2500}, & 0 \le x \le 50 \\ \frac{1}{25} - \frac{x}{2500}, & 50 \le x \le 100. \end{cases}$$

  a. What is the probability that the waiting time is between 25 and 75 seconds?

b. If a car has waited 25 seconds, what is the probability that it will wait at least 25 seconds more?

 21. One hundred independent observations are made of the random variable X whose probability density function is

$$f(x) = \begin{cases} x, & 0 \le x \le 1 \\ 2 - x, & 1 \le x \le 2. \end{cases}$$

Find the probability that at least 20 of these observations exceed 1.5.

22. X is a random variable with distribution function

$$F(x) = \begin{cases} 0, & x < -2 \\ \frac{x+2}{4}, & -2 \le x \le 2 \\ 1, & x > 2. \end{cases}$$

Identify the probability density function for X.

23. A random variable X has probability density function

$$f(x) = \begin{cases} 2x, & 0 \le x \le \frac{1}{2} \\ 6 - 6x, & \frac{1}{2} \le x \le 1. \end{cases}$$

Find $F(x)$, being sure to specify this for any value of x.

24. Find the constant c that makes $g(y) = \frac{c}{y}$, $1 \le y \le 2$ a probability density function.

25. Find the mean and variance of the random variable X whose probability density function is

$$f(x) = \begin{cases} 1 - x, & 0 \le x \le 1 \\ x - 1, & 1 \le x \le 2. \end{cases}$$

26. A random variable X has probability density function

$$f(x) = \frac{1}{2x}, \quad \frac{1}{e} < x < e.$$

Two independent observations are made on X. Find the probability that one observation is less than 1 and that the other observation is greater than 1.

27. A random variable X has distribution function

$$F(x) = \begin{cases} 0, & x < -2 \\ \frac{1}{6}, & -2 \le x < -1 \\ \frac{1}{2}, & -1 \le x < 2 \\ 1, & x \ge 2. \end{cases}$$

Find $f(x)$.

28. The probability density function for X, the lifetime in hours of a certain type of electronic device, is given by

$$f(x) = \begin{cases} \frac{10}{x^2}, & x > 10 \\ 0, & x \leq 10. \end{cases}$$

    a. Find $P(X > 20)$.

    b. Find $F(x)$.

29. A random variable T has probability density function

$$g(t) = k(1 + t)^{-2}, \ t \geq 0.$$

Find $P(T \geq 2 \mid T \geq 1)$.

30. A player can win a solitaire card game with probability $\frac{1}{12}$. Find the probability that the player wins at least 10% of 500 games played.

▶ *Chapter 4*

Functions of Random Variables, Generating Functions, and Statistical Applications

▌ 4.1 ▶ Introduction

We now want to expand our applications of statistical inference first encountered in Chapter 2. In particular we want to consider tests of hypotheses and the construction of confidence intervals when continuous random variables are involved; we will also introduce simple linear regression. These considerations have direct bearing on problems of data analysis such as that encountered in the following situation.

A manufacturing process has been producing bearings with mean diameter 2.60 inches; the diameters exhibit some variability around this average value with the standard deviation of the diameters believed to be 0.03 inches. A quality control inspector chooses a random sample of ten bearings and finds their average diameter to be 2.66 inches. Has the process changed?

The quality control inspector has a single observation, namely 2.66 inches, the average of ten observations. This is most commonly the situation: only one sample is available; decisions must be made on the basis of that single sample. Nonetheless we can speculate on what would happen were the sampling to be repeated. In that case, another sample average will most likely occur. In order to decide whether or not 2.66 inches is unusual, we must know the probability distribution of these sample means so that the variation in the mean from sample to sample can be assessed. We can then base a test of the hypothesis that the process mean has not changed on that probability distribution. Confidence intervals can similarly be constructed, but again, the probability distribution of the sample mean must be known.

Determination of the probability distribution here is not particularly easy, so we first need to make some mathematical considerations. This will not only enable us to analyze the example at hand, but will also allow us to solve many other complex problems arising in the analysis of data. We also must investigate the distribution of the sample variance arising from samples drawn from a continuous distribution.

We begin by considering functions of random variables; sums and averages arising from samples are examples of complex functions of sample values. Special functions called *generating functions* provide a particularly powerful technique for solving these problems. While developing these techniques we will solve many interesting problems in probability. Finally we will show several practical statistical problems and their solution, including a statistical process control chart.

▌ 4.2 ▶ Some Examples of Functions of Random Variables

Consider the following problem. An observation X is made from a uniform distribution on the interval [0, 1] and then a square of side X is formed. What is the expected value of its area?

This problem is fairly easily solved.

Since $E(X) = \frac{1}{2}$ and $\text{Var}(X) = E(X^2) - [E(X)]^2 = \frac{1}{12}$, it follows that

$$E(Area) = E(X^2) = \text{Var}(X) + [E(X)]^2 = \frac{1}{12} + \frac{1}{4} = \frac{1}{3}.$$

Other problems of a similar nature, however, may not be quite so easy. As another example, suppose X is an exponential random variable with mean α and we seek $E(\sqrt{X})$. It would be unreasonable to think, for example, that $E(\sqrt{X}) = \sqrt{E(X)}$. This expectation is, after all, an integral, and integrals rarely behave in such simple fashion.

The reader is invited to calculate, or use a computer algebra system, in this example to find that

$$E(\sqrt{X}) = \int_0^\infty \frac{\sqrt{x}}{\alpha} e^{-x/\alpha} \, dx = \frac{1}{2}\sqrt{\pi\alpha}.$$

Another frequently used technique for evaluating the integral we have just encountered is to select a random sample of values from an exponential distribution and then calculate the average value of their square roots. This technique, widely used in problems that prove difficult for analytical techniques, is known as *simulation*. A computer program chose 1000 observations from an exponential distribution with $\alpha = 4$ and then calculated the mean of the square roots of these values. The observed value is 1.800, while the expected value is $\sqrt{\pi} = 1.7725$, so the simulation produced a value quite close to the expected value.

Expectations of many functions of random variables can be carried out by using the probability density function of X directly. In the first example (where X is uniformly distributed on $[0, 1]$ and denotes the length of the side of a square), suppose we want the probability that the area of the square is between $\frac{1}{2}$ and $\frac{3}{4}$. We can calculate

$$P\left(\frac{1}{2} \le X^2 \le \frac{3}{4}\right) = P\left(\frac{1}{\sqrt{2}} \le X \le \frac{\sqrt{3}}{2}\right) = \frac{\sqrt{3}}{2} - \frac{1}{\sqrt{2}} = 0.15892,$$

using the distribution of X directly.

In the second example, supposing X is a random observation from an exponential distribution with mean α, we calculate, for example,

$$P(1 \le \sqrt{X} \le 2) = P(1 \le X \le 4)$$
$$= \int_1^4 (1/\alpha) \cdot e^{-x/\alpha} \cdot dx = e^{-1/\alpha} - e^{-4/\alpha},$$

so often probabilities involving functions of random variables can be found from the distributions of the random variables themselves.

Now suppose that we have two independent observations of a random variable X and we consider the sum of these, $X_1 + X_2$. This is certainly a random variable. How can we calculate $P(X_1 + X_2 \le 2)$ if X_1 and X_2 are, for example, independent observations from the exponential distribution? Clearly, this problem is not as simple as the preceding ones.

It is fortunate that there is another way to look at these problems. It turns out that this other view will solve these problems and has, in addition, considerable implications for the solutions of much more complex problems, solutions that are not easily found in any other way. Our approach will also explain why normality has occurred so frequently in our problems; the reason for this is not simple, as the reader might expect.

The expressions X^2, $\sqrt{X}$, and $X_1 + X_2$ are *functions* of the random variable X. Since X is a random variable, so too are these functions of X; then they have probability distributions. If these probability distributions could be determined, then the problems above, as well as many others, could be solved. So we now consider one method for determining the probability distribution of a function of the random variable X.

4.3 ▶ Probability Distributions of Functions of Random Variables

We begin with an example discussed in Chapter 3.

Example 4.3.1

Suppose that a random variable X has a standard normal distribution, that is, $X \sim N(0, 1)$. Consider the random variable $Y = X^2$, so that Y is a quadratic function of the random variable X. What is the probability density function for Y?

Our answer depends on the simple fact that, when the derivative exists, and where $f(x)$ and $F(x)$ denote the probability density function and distribution function, respectively, then

$$\frac{dF(x)}{dx} = f(x).$$

Let $g(y)$ and $G(y)$ denote the probability density function and the distribution function, respectively, for the random variable Y. Our basic strategy is to find $G(y)$ and then to differentiate it, using the preceding property, to produce $g(y)$. Here

$$G(y) = P(Y \leq y) = P(X^2 \leq y) = P(-\sqrt{y} \leq X \leq \sqrt{y}),$$

so

$$G(y) = F(\sqrt{y}) - F(-\sqrt{y}),$$

by a property of distribution functions. Now we differentiate throughout to find that

$$g(y) = \frac{dG(y)}{dy} = \frac{dF(\sqrt{y}) - dF(-\sqrt{y})}{dy}.$$

Now we must be careful because $\frac{dF(\sqrt{y})}{dy} \neq f(\sqrt{y})$. The problem lies in the fact that the variables in the numerator and denominator are not the same. However, the chain rule comes to our rescue and we find that

$$g(y) = \frac{dF(\sqrt{y})}{d(\sqrt{y})} \cdot \frac{d(\sqrt{y})}{dy} - \frac{dF(-\sqrt{y})}{d(-\sqrt{y})} \cdot \frac{d(-\sqrt{y})}{dy}.$$

This becomes

$$g(y) = \frac{f(\sqrt{y})}{2\sqrt{y}} + \frac{f(-\sqrt{y})}{2\sqrt{y}}. \tag{4.1}$$

But

$$f(y) = \frac{1}{\sqrt{2\pi}} e^{-y^2/2}, \quad -\infty < y < \infty,$$

and

$$f(\sqrt{y}) = f(-\sqrt{y}) = \frac{1}{\sqrt{2\pi}} e^{-y/2}, \quad y > 0,$$

so

$$g(y) = \frac{1}{\sqrt{2\pi y}} e^{-y/2}, \quad y > 0.$$

This is the χ_1^2 variable, first seen in section 3.7. The domain of values for Y is established from that for X: because $-\infty < X < \infty$, then $-\infty < \sqrt{y} < \infty$, or $\sqrt{y} < \infty$, so $y \geq 0$. The same domain is correct for $-\sqrt{y}$.

The calculation that $\int_0^\infty g(y)\, dy = 1$, and the fact that $g(y) \geq 0$, checks our work and shows that $g(y)$ is a probability density function.

This process works well when the derivatives involved can be evaluated which is often the case in the instances that interest us here.

In the previous example, Y is a quadratic function of X and the resulting distribution for Y bears little resemblance to that for X. We expect that a linear function would preserve the shape of the distribution in a sense. We consider a specific example first. ◀

Example 4.3.2

Suppose X is uniform on $[3, 5]$ so that $f(x) = \frac{1}{2}$, $3 \leq x \leq 5$. Let $Y = \frac{X-2}{3}$, a linear function of X. Again we find the distribution function and differentiate it. Here,

$$G(y) = P(Y \leq y) = P(\frac{X-2}{3} \leq y) = P(X \leq 3y + 2) = F(3y + 2).$$

Then

$$g(y) = \frac{dG(y)}{dy} = \frac{dF(3y + 2)}{dy} = \frac{dF(3y + 2)}{d(3y + 2)} \cdot \frac{d(3y + 2)}{dy},$$

so

$$g(y) = f(3y + 2) \cdot 3 = \frac{3}{2}.$$

To establish the domain for y, note that $f(3y + 2) = \frac{1}{2}$ if $3 \leq 3y + 2 \leq 5$, which simplifies to $\frac{1}{3} \leq y \leq 1$, producing the final result:

$$g(y) = \frac{3}{2}, \quad \frac{1}{3} \leq y \leq 1.$$

This is a probability density function because $g(y) \geq 0$ and $\int_{\frac{1}{3}}^{1} g(y)\, dy = 1$.

We observe that the linear transformation $Y = \frac{X-2}{3}$ of X preserves the uniform distribution with which we began.

To consider the problem of a linear transformation in general, suppose that X has probability density function $f(x)$, and that $Y = aX + b$ for some constants a and b provided that $a > 0$. Then

$$G(y) = P(Y \leq y) = P(aX + b \leq y) = P\left(X \leq \frac{y-b}{a}\right) = F\left(\frac{y-b}{a}\right)$$

so

$$g(y) = \frac{dG(y)}{dy} = \frac{dF\left(\frac{y-b}{a}\right)}{dy} = \frac{dF\left(\frac{y-b}{a}\right)}{d\left(\frac{y-b}{a}\right)} \cdot \frac{d\left(\frac{y-b}{a}\right)}{dy}$$

$$g(y) = f\left(\frac{y-b}{a}\right) \cdot \frac{1}{a},$$

showing that the shape of the distribution is preserved under the linear transformation.

If you have not already done so, please note that it is crucial that the variables, denoted by capital letters, be clearly distinguished from their values, denoted by small letters; otherwise, confusion and, most likely, errors will occur. ◄

Example 4.3.3

Consider, one more time, the fair wheel where $f(x) = 1$, for $0 \leq x \leq 1$. Now let's spin the wheel n times, obtaining the random observations $X_1, X_2, \ldots, X_n$. We let Y denote the largest of these, so that

$$Y = \text{Max}\{X_1, X_2, \ldots, X_n\}.$$

Y is clearly a non-trivial random variable. Again we seek the probability density function for Y, $g(y)$.

Note that if the maximum of the $X's$ is at most y, then *each* of the X's must be at most y. So

$$G(y) = P(Y \leq y) = P(\text{Max}\{X_1, X_2, \ldots, X_n\} \leq y)$$
$$= P(X_1 \leq y \text{ and } X_2 \leq y \text{ and} \ldots \text{and } X_n \leq y)$$

Since the X's are independent, it follows that

$$G(y) = [P(X_1 \leq y)] \cdot [P(X_2 \leq y)] \cdots [P(X_n \leq y)].$$

Now, because all the X's have the same probability density function,

$$G(y) = [P(X \leq y)]^n = [F(y)]^n.$$

It follows that

$$g(y) = \frac{dG(y)}{dy} = n[F(y)]^{n-1} \cdot f(y).$$

Since the X's all have the same uniform probability density function, in this case $F(y) = y$ so that $g(y) = ny^{n-1}$, for $0 \leq y \leq 1$.

In the general case, we note that the distribution for Y is dependent on $F(y)$. $F(y)$ is easy in this example, but it could prove intractable, as in the case of a normal variable that has no closed form for its distribution function. In fact the probability distribution of the maximum observation from a random sample of observations from a normal distribution is unknown. ◄

Expectation of a Function of X

In Chapters 1 and 2 we calculated expectations of functions of X using only the probability density function for X. Specifically, we let

$$E[H(X)] = \int_{-\infty}^{\infty} H(x) \cdot f(x)\, dx \tag{4.2}$$

where $H(X)$ is some function of the random variable X and $f(x)$ is the probability density function for the random variable X. We took this as a matter of definition. For example, we wrote that

$$E(X^2) = \int_{-\infty}^{\infty} x^2 \cdot f(x)\, dx.$$

The reader may wonder about this definition. The function $H(X)$ is also a random variable. To find its expectation shouldn't we find its probability density function first and then the expectation of the random variable using that probability density function? That would appear to be a strategy certain of success. Amazingly, it turns out not to be necessary, and Formula 4.2 gives the correct result. Let's see why this is so.

To make matters simple, suppose $Y = H(X)$, and that $H(X)$ is a strictly increasing function of X. (A demonstration similar to that given here can be given for $H(X)$ strictly decreasing.) Then

$$G(y) = P(Y \le y) = P[H(X) \le y] = P[X \le H^{-1}(y)],$$

because $H(X)$ is invertible. This means that

$$G(y) = F[H^{-1}(y)]$$

or

$$g(y) = \frac{dF[H^{-1}(y)]}{dy}$$

so that

$$g(y) = f[H^{-1}(y)]\frac{dH^{-1}(y)}{dy},$$

or

$$g(y) = f(x) \cdot \frac{dx}{dy}$$

where x is expressed in terms of y. This formula can indeed be used in many of our change of variable formulas, but the reader is warned that the function must be strictly increasing or strictly decreasing for this result to work. Now we calculate the expected value:

$$E(Y) = \int_{-\infty}^{\infty} y \cdot g(y)\, dy = \int_{-\infty}^{\infty} H(x) \cdot f(x) \cdot \frac{dx}{dy} \cdot dy = \int_{-\infty}^{\infty} H(x) \cdot f(x)\, dx,$$

showing that our definition of the expectation of a function of X was sound.

Example 4.3.4

Consider the probability density function

$$f(x) = k \cdot x^2,\ 0 < x < 2.$$

We calculate $E(X^2)$ in two ways: first by finding the probability density function for X^2, and secondly without finding that probability density function. k must be determined first. Since

$$\int_0^2 k \cdot x^2\, dx = 1$$

it follows that

$$k \cdot \frac{x^3}{3}\Big|_0^2 = k \cdot \frac{8}{3} = 1$$

so

$$k = \frac{3}{8}.$$

Now consider the transformation $Y = X^2$:

$$G(y) = P(Y \le y) = P(X^2 \le y) = P(X \le \sqrt{y}) = F(\sqrt{y})$$

since X takes on only non-negative values. Now

$$g(y) = \frac{1}{2\sqrt{y}} f(\sqrt{y}),$$

so

$$g(y) = \frac{1}{2\sqrt{y}} \cdot \frac{3}{8} \cdot y = \frac{3}{16} \cdot \sqrt{y},\ 0 < y < 4.$$

Then

$$E(Y) = \frac{3}{16} \cdot \int_0^4 y^{3/2} \cdot dy = \frac{12}{5}.$$

Now we use Formula 4.2 directly:

$$E(Y) = \int_0^2 \frac{3}{8} \cdot x^4 \cdot dx = \frac{12}{5},$$

obtaining the previous result.   ◄

Exercises 4.3

1. Suppose that X is uniformly distributed on the interval $(2, 5)$. Let $Y = 3X - 2$. Find $g(y)$, the probability density function for Y.

2. Suppose that the probability density function for a random variable X is $f(x) = \lambda \cdot e^{-\lambda x}, x \geq 0, \lambda > 0$. Let $Y = 3 - X$. Find $g(y)$, the probability density function for Y.

3. Let X have a uniform distribution on $(0, 1)$. Find the probability density function for $Y = X^2$ and prove that the result is a probability density function.

4. The random variable X has the probability density function $f(x) = 2x$, $0 \leq x \leq 1$.

   a. Let $Y = X^2$ and find the probability density function for Y.
   b. Now suppose that X has the probability density function $f(x)$. What transformation, $Y = H(X)$, will result in Y having a uniform distribution? (Part **a** of this problem may help in discovering the answer.)

5. Suppose that $X \sim N(\mu, \sigma)$, and let $Y = e^X$.

   a. Find the mean and variance of Y.
   b. Find the probability density function for Y. The result is called the *lognormal* probability density function because the logarithm of the variable is $N(\mu, \sigma)$.

6. Random variable X has probability density function $f(x) = 4x(1 - x^2)$, $0 \leq x \leq 1$. Find $E(X^2)$ in two ways:

   a. without finding the probability density function of $Y = X^2$.
   b. using the probability density function of $Y = X^2$.

7. If X has a Weibull distribution with parameters α and β, show that the variable $Y = \left(\frac{X}{\alpha}\right)^\beta$ is an exponential variable with mean 1.

8. The *folded normal* distribution is the distribution of $|X|$ where $X \sim N(0, \sigma)$.

   a. Find the probability density function for a folded normal variable.
   b. Find $E(|X|)$.

9. Find the probability density function for $Y = X^3$ where X has an exponential distribution with mean value 1.

10. A circle is drawn by choosing a radius from the uniform distribution on the interval $(0, 1)$. Find the probability density function for the area of the circle.

11. Suppose that X is a uniform random variable on the interval $(-1, 1)$. Find the probability density function for the variable $Y = \sin(X)$.

12. Find the probability density function for $Y = e^X$ where X is uniformly distributed on $[0, 1]$.

13. Random variable X has probability density function

$$f(x) = \frac{1}{(1 + x)^2}, \ x \geq 0.$$

   a. Find the probability density function for $Y = \sqrt{X}$.
   b. Show that $P(0 \leq Y \leq b) = 1 - \frac{1}{1+b^2}$, where $b \geq 0$.

14. A random variable X has the probability density function

$$f(x) = \frac{x + 1}{2}, \ -1 \leq x \leq 1.$$

Find $E(X^2)$

   a. by first finding the probability density function for $Y = X^2$.
   b. without using the probability density function for $Y = X^2$.

15. A fluctuating electric current, I, is a uniformly distributed random variable on the interval $[9, 11]$. If this current flows through a 2-ohm resistor, the power is $Y = 2I^2$. Find $E(Y)$ by first finding the probability density function for the power.

16. A random variable X has the probability density function $f(x) = \frac{1}{x^2}, \ x \geq 1$. Find the probability density function for $Y = 1 - \frac{1}{X}$ and prove that your result is a probability density function.

17. Independent observations $X_1, X_2, X_3, \ldots, X_n$ are taken from the exponential distribution $f(x) = \lambda e^{-\lambda x}$ where $x > 0$ and $\lambda > 0$. Find the probability density function for $Y = \text{Min}(X_1, X_2, X_3, \ldots, X_n)$.

18. In triangle ABC, angle ABC is $\pi/2$, $|AB| = 1$, and angle BAC (in radians) is a random variable uniformly distributed on the interval $[0, \frac{\pi}{3}]$. Find the expected length of side BC.

19. Find the probability density function for $Y = X^2$ if X has the probability density function

$$f(x) = \begin{cases} \frac{x}{4} + \frac{1}{2}, & -2 \le x \le 0 \\ -\frac{x}{4} + \frac{1}{2}, & 0 \le x \le 2. \end{cases}$$

20. Find the probability density function for $Y = -\ln X$ if X is uniformly distributed on $(0, 1)$.

21. Is $E\left(\frac{1}{X}\right) = \frac{1}{E(X)}$ if $f(x) = e^{-x}$, $x \ge 0$?

22. Find $g(y)$, the probability density function for $Y = X^2$ if X is uniformly distributed on $(-1, 2)$.

23. Computers commonly produce random numbers that are uniform on the interval $(0, 1)$. These can often be used to simulate random selections from other probability distributions in the following way. Suppose we wish a function of the uniform variables to have a given probability distribution function, say $g(y)$. Then, if $G(y)$ is invertible, consider the transformation $Y = G^{-1}(X)$. Then,

$$P(Y \le y) = P[G^{-1}(X) \le y] = P[X \le G(y)] = G(y)$$

because X is a uniform random variable, showing that Y has the required probability density function.

   a. Find a function Y of a uniform $(0, 1)$ random variable so that Y is uniform on (a, b).
   b. Find a function Y of a uniform $(0, 1)$ random variable so that Y has an exponential distribution with expected value $1/\lambda$.

24. Show that $E[H(X)] = \int_{-\infty}^{\infty} H(x) \cdot f(x) \cdot |\frac{dx}{dy}| \, dy$, where $f(x)$ is the probability density function of X, if $H(X)$ is a strictly decreasing function of X.

25. Show, without using the probability density function of $\sqrt{X}$, that $E(\sqrt{X}) = \frac{1}{2}\sqrt{\alpha\pi}$ if X is an exponential random variable with mean α. (Hint: The variance of a $N(0, 1)$ variable is 1.)

26. Show that if $f(x) = \frac{1}{\pi} \cdot \frac{1}{1+x^2}$, $-\infty < x < \infty$, and if $Y = \frac{1}{X}$, then
$g(y) = \frac{1}{\pi} \cdot \frac{1}{1+y^2}$, $-\infty < y < \infty$.

27. An area is lighted by lamps whose length of life is exponential with mean 8000 hours. It is very important that some light be available in the area for 20,000 hours. How many lamps should be installed?

28. Random variable X has a Cauchy distribution, that is

$$f(x) = \frac{1}{\pi(1 + x^2)}, \quad -\infty < x < \infty.$$

Let $Y = \frac{1}{1+X^2}$.

a. Show that the probability density function of Y is

$$g(y) = \frac{1}{\pi\sqrt{y(1-y)}}, \quad 0 < y < 1.$$

b. Show that the distribution function for Y is

$$F(y) = \frac{2}{\pi}\arcsin(\sqrt{y}), \quad 0 < y < 1.$$

c. Find $E(Y)$ and $\mathrm{Var}(Y)$.

4.4 ▶ Sums of Random Variables I

Random variables can often be regarded as sums of other random variables. For example, if a coin is tossed and X, the number of heads that appear, is recorded (X can only be 0 or 1), and subsequently the coin is tossed again and Y, the number of heads that appear, is recorded, then clearly $X + Y$ denotes the total number of heads that appear. So the total number of heads when two coins are tossed can be regarded as a sum of two individual (and in this case independent) random variables. Clearly we expect $X + Y$ to be a binomial random variable with $n = 2$. We can extend this argument to n tosses; the sum is then a binomial random variable. In Chapter 1 we encountered the random variable denoting the sum when two dice are thrown, so we have actually considered sums before.

Now we intend to study the behavior of sums of random variables primarily because the results are interesting, and because the consequences have extensive implications to some problems in statistics. In this section we start with some interesting results and examples.

Example 4.4.1

In the example just cited, X is a random variable that takes on the values 1 or 0 with probabilities p and $1 - p$, respectively. Y has a similar distribution, and because X and Y are independent, the distribution of $X + Y$ can be found by considering all the possibilities:

$$P(X + Y = 0) = P(X = 0) \cdot P(Y = 0) = (1 - p)^2$$
$$P(X + Y = 1) = P(X = 0) \cdot P(Y = 1) + P(X = 1) \cdot P(Y = 0)$$
$$= 2p(1 - p)$$
$$P(X + Y = 2) = P(X = 1) \cdot P(Y = 1) = p^2.$$

Recall that the individual variables X and Y are often called *Bernoulli* random variables; their sum, as the preceding calculation shows, is a binomial random variable with $n = 2$. Since the Bernoulli random variable can be regarded as a binomial variable with $n = 1$, we see that in this case the sum of two independent binomial variables is also binomial. This raises the question, "Is the sum of two independent binomial random variables in general always binomial?"

To answer this question, let's proceed with a calculation. Suppose X is binomial (n, p) and Y is binomial (m, p) and let $Z = X + Y$. The event $Z = z$ can arise in several mutually exclusive ways: $X = 0$ and $Y = z$; $X = 1$ and $Y = z - 1$; and so on until $X = z$ and $Y = 0$. So, assuming also that X and Y are independent,

$$P(Z = z) = \sum_{k=0}^{z} P(X = k) \cdot P(Y = z - k),$$

or

$$P(Z = z) = \sum_{k=0}^{z} \binom{n}{k} p^k (1 - p)^{n-k} \cdot \binom{m}{z - k} p^{z-k} (1 - p)^{m-(z-k)}.$$

This can be simplified to

$$P(Z = z) = p^z (1 - p)^{n+m-z} \sum_{k=0}^{z} \binom{n}{k} \binom{m}{z - k}.$$

But we recognize $\sum_{k=0}^{z} \binom{n}{k} \binom{m}{z-k}$ from the hypergeometric distribution as $\binom{n+m}{z}$. So

$$P(Z = z) = \binom{n + m}{z} p^z (1 - p)^{n+m-z}, \quad z = 0, 1, 2, \ldots, n + m,$$

a binomial distribution with parameters $n + m$ and p. This establishes the fact that sums of independent binomial random variables are also binomial.

We note here, because $E(X) = np$ and $\text{Var}(X) = np(1 - p)$ (with similar results for Y), that $E(Z) = (n + m)p$ and $\text{Var}(Z) = (n + m)p(1 - p)$. We summarize these results as follows:

$$E(X + Y) = E(X) + E(Y)$$

and

$$\text{Var}(X + Y) = \text{Var}(X) + \text{Var}(Y),$$

since X and Y are independent. As we will see later, the assumption of independence is a crucial one here.

This example shows that the sum of independent binomial random variables is again binomial. Occasionally random variables are *reproductive* in the sense that their sums are distributed in the same way as the summands, but this is not always the case. In fact it is not the case with binomials if the probability of success at any trial for the random variable X differs from the probability of success at any trial for the random variable Y. We turn now to an example where the probability distribution of the sum is not of the same form as the summands. ◄

Example 4.4.2

Suppose X and Y are each discrete uniform random variables; that is,

$$P(X = x) = \frac{1}{n}, \quad x = 1, 2, \ldots, n$$

with an identical distribution for Y. What happens if we add two randomly chosen observations? We investigate the probability distribution of the sum, $Z = X + Y$, assuming that X and Y are independent.

The special case $n = 4$ may be instructive. Then if we wanted to find, for example, $P(Z = 6)$, we could work out all the possibilities:

$$P(Z = 6) = P(X = 2) \cdot P(Y = 4) + P(X = 3) \cdot P(Y = 3)$$
$$+ P(X = 4) \cdot P(Y = 2)$$

$$= \left(\tfrac{1}{4}\right) \cdot \left(\tfrac{1}{4}\right) + \left(\tfrac{1}{4}\right) \cdot \left(\tfrac{1}{4}\right) + \left(\tfrac{1}{4}\right) \cdot \left(\tfrac{1}{4}\right) = \tfrac{3}{16}.$$

Proceeding in a similar way for other values of z, we find

$$P(Z = z) = \begin{cases} \frac{1}{16}, & z = 2 \\ \frac{2}{16}, & z = 3 \\ \frac{3}{16}, & z = 4 \\ \frac{4}{16}, & z = 5 \\ \frac{3}{16}, & z = 6 \\ \frac{2}{16}, & z = 7 \\ \frac{1}{16}, & z = 8. \end{cases}$$

This result can also be summarized as

$$P(Z = z) = \begin{cases} \frac{z-1}{16}, & z = 2, 3, 4, 5 \\ \frac{9-z}{16}, & z = 6, 7, 8. \end{cases}$$

A graph of this is shown in Figure 4.1.

The sum is certainly not uniform. It is not clear what might happen when we increase the number of summands. We might conjecture that, as we add more independent uniform random variables, the sums become normal. This is in fact the case, but we need to develop some techniques before we can consider that circumstance and verify the normality. We will begin to do that in the next section. ◄

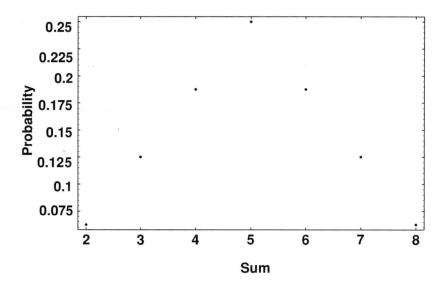

Figure 4.1   Sum of two independent discrete uniform variables, $n = 4$

Exercises 4.4

1. Show that the sum of two independent Poisson variables with parameters λ_x and λ_y, respectively, has a Poisson distribution with parameter $\lambda_x + \lambda_y$.

2. Let X and Y be independent geometric random variables so that $P(X = x) = (1 - p)^{x-1} \cdot p, \; x = 1, 2, 3, \ldots,$ with a similar distribution for Y. Show, if X and Y are independent, that $X + Y$ has a negative binomial distribution.

3. Find the probability distribution for $X + Y + Z$ where $X, Y,$ and Z each have a discrete uniform distribution on the integers $1, 2, 3, 4$.

4. Let X denote a Bernoulli random variable – i.e., $P(X = 1) = p$ and $P(X = 0) = 1 - p$ – and let Y be a binomial random variable with parameters n and p. Show that $X + Y$ is binomial with parameters $n + 1$ and p.

5. A coin, loaded so as to come up heads with probability $\frac{2}{3}$, is thrown until a head appears, then a fair coin is thrown until a head appears.

   a. Find the probability distribution for Z, the total number of tosses necessary.

   b. Find the expected value for Z from the probability distribution for Z.

6. Phone calls come into an office according to a Poisson distribution with four calls expected in an interval of two minutes. The calls are answered according to a binomial process with $p = \frac{1}{2}$. Find the probability that exactly three calls are answered in a two-minute period.

7. Generalize Exercise 6: Consider Poisson events, in a given interval of time, with parameter λ, which are recorded according to a binomial process with parameter p. Show that the number of events recorded in the interval of time is Poisson with parameter λp.

4.5 ▶ Generating Functions

At this point we have found the probability distribution functions of sums of random variables by working out the probabilities for each possible value of the sum. This technique of course cannot be carried out when the summands are continuous or when the number of summands is large.

We consider now another technique that will make some complex problems involving sums of either discrete or continuous random variables tractable. We start with the discrete case in this section.

Example 4.5.1

Consider throwing a fair die once, and this function:

$$G(t) = \frac{1}{6}(t + t^2 + t^3 + t^4 + t^5 + t^6).$$

If X is the random variable denoting the face showing on the die, then we observe that the coefficient of t^k in $G(t)$ is the probability that X equals k, $P(X = k)$. For example, $P(X = 3)$ is the coefficient of t^3, which is $\frac{1}{6}$. Since $G(t)$ has this property, it is called a *probability generating function.*

If X is a random variable taking values on the non-negative integers, then any function of the form

$$\sum_{k=0}^{\infty} t^k \cdot P(X = k)$$

is called a *probability generating function.*

Note that in $G(t)$ we could easily load the die by altering the coefficients of the powers of t to reflect the different probabilities with which the faces appear. For example, the function

$$H(t) = \frac{1}{10}t + \frac{1}{5}t^2 + \frac{1}{10}t^3 + \frac{1}{5}t^4 + \frac{1}{5}t^5 + \frac{1}{5}t^6$$

generates probabilities on a die loaded so that faces numbered 1 and 3 appear with probability $\frac{1}{10}$, while each of the other faces appears with probability $\frac{1}{5}$.

Probability generating functions are of great importance in probability; they provide neat summaries of probability distributions and have other remarkable properties, as we will see.

Continuing our example, if we square $G(t)$ we find that

$$G^2(t) = \frac{1}{36}(t^2 + 2t^3 + 3t^4 + 4t^5 + 5t^6 + 6t^7 + 5t^8 + 4t^9 + 3t^{10} + 2t^{11} + t^{12})$$

$G^2(t)$ is also a probability generating function – its coefficients are the probabilities of the sums when two dice are thrown. In general, $G^n(t)$ generates probabilities

$$P(X_1 + X_2 + X_3 + \cdots + X_n = k)$$

where X_i is the face showing on die i, $i = 1, 2, \ldots, n$.

We may use this fact to find, for example, the probability that when four fair dice are thrown a sum of 17 is obtained. We would find this to be a very difficult problem if we were constrained to write out all the possibilities for which the sum is 17. Since $G(t)$ can be written as

$$G(t) = \frac{t(1 - t^6)}{6(1 - t)},$$

it follows that

$$G^4(t) = \frac{t^4(1 - t^6)^4}{6^4(1 - t)^4}.$$

This reduces our problem to finding the coefficient of t^{13} in $(1 - t^6)^4 \cdot (1 - t)^{-4}$. Expanding this by the binomial theorem, and ignoring the division by 6^4 for the moment, we see that the coefficient we seek is the coefficient of t^{13} in

$$\left[1 - \binom{4}{1} t^6 + \binom{4}{2} t^{12} + \cdots \right] \left[1 + \binom{-4}{1}(-t) + \binom{-4}{2}(-t)^2 + \cdots \right].$$

So the coefficient of t^{13} is

$$\binom{-4}{13}(-1)^{13} - \binom{4}{1}\binom{-4}{7}(-1)^7 + \binom{4}{2}\binom{-4}{1}(-1)$$

$$= \binom{16}{3} - 4\binom{10}{3} + 6\binom{4}{3} = 104.$$

Therefore the probability that we seek is $\frac{104}{6^4} = \frac{13}{162}$.

This process is certainly an improvement on that of counting all the possibilities, a technique that clearly becomes impossible when the number of dice is large.

A computer algebra system allows us to find

$$G^4(t) = \left[\frac{1}{6}(t + t^2 + t^3 + t^4 + t^5 + t^6)\right]^4$$

giving directly the following table of probabilities for the sums on four fair dice:

Sum	Probability	Sum	Probability
4	$\frac{1}{1296}$	15	$\frac{35}{324}$
5	$\frac{1}{324}$	16	$\frac{125}{1296}$
6	$\frac{5}{648}$	17	$\frac{13}{162}$
7	$\frac{5}{324}$	18	$\frac{5}{81}$
8	$\frac{35}{1296}$	19	$\frac{7}{162}$
9	$\frac{7}{162}$	20	$\frac{35}{1296}$
10	$\frac{5}{81}$	21	$\frac{5}{324}$
11	$\frac{13}{162}$	22	$\frac{5}{648}$
12	$\frac{125}{1296}$	23	$\frac{1}{324}$
13	$\frac{35}{324}$	24	$\frac{1}{1296}$
14	$\frac{73}{648}$		

While the computer can give us high powers of $G(t)$, it can also give us great insight into the problem. Consider a graph of the coefficients of $G(t)$ as shown in Figure 4.2.

Now consider $G^2(t)$ whose coefficients are shown in Figure 4.3.

A graph of the coefficients in $G^4(t)$ is shown in Figure 4.4.

Finally, Figure 4.5 shows the probabilities for sums on 12 fair dice.

This is probably enough to convince the reader that normality, once again, is involved. The probability distribution for the sums on 12 fair dice is in fact remarkably close to a normal curve. We find, for example, that

$$P(36 \leq \text{Sum} \leq 48) = 0.724753, \text{ exactly,}$$

while the normal curve gives 0.728101, a very good approximation.

Before the normality can be explained analytically we must consider some more characteristics of probability generating functions. We will examine these in the next section. ◄

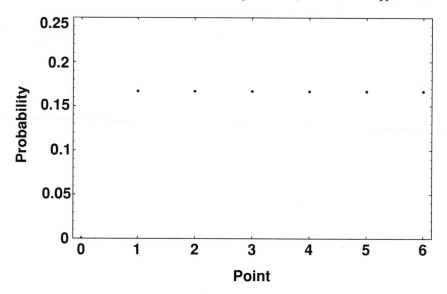

Figure 4.2   Probabilities for one fair die.

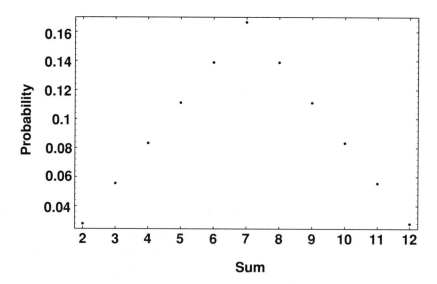

Figure 4.3   Probabilities for sums on two fair dice.

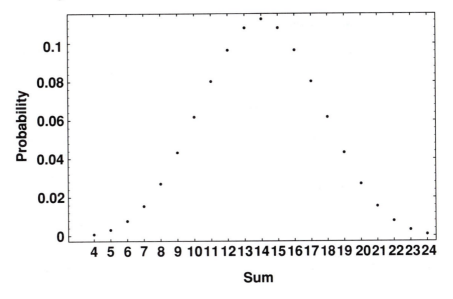

Figure 4.4   Probabilities for sums on four fair dice.

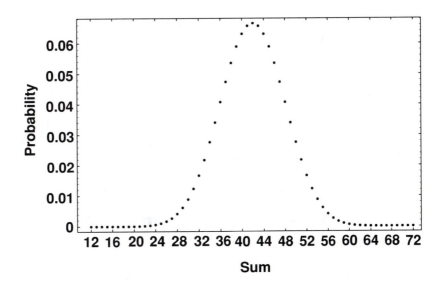

Figure 4.5   Probabilities for sums on 12 fair dice.

Exercises 4.5

1. Show that the function $(1 + t)^n$ generates the binomial coefficients $\binom{n}{x}$, $x = 0, 1, 2, \ldots, n$.

2. What sequence is generated by $(1 - 4t)^{-1/2}$?

3. Consider the set $\{a, b, c\}$. What does the function $(1 + at)(1 + bt)(1 + ct)$ generate?

4. Find a function that generates the sequence $0^2, 1^2, 2^2, 3^2, \ldots$.

5. Find a function that generates the sequence $\frac{1}{1 \cdot 2}, \frac{1}{2 \cdot 3}, \frac{1}{3 \cdot 4}, \frac{1}{4 \cdot 5}, \ldots$.

6. A die is loaded so that the probability that a face appears is proportional to the face. If the die is thrown 5 times, find the probability that the sum obtained is 17.

7. Suppose that the probability generating function for the random variable X is $P_X(t)$. Find an expression for the probability generating function for

a. $X + k$ where k is a constant

b. $k \cdot X$ where k is a constant.

8. Verify in Example 4.5.1 that if X is the sum on 12 fair dice, then $P(36 \leq X \leq 48) = 0.724753$.

9. A fair die and a die loaded so that the probability that a face appears is proportional to the face are thrown. Find the probability distribution for the sums appearing on the uppermost faces.

4.6 ▶ Some Properties of Generating Functions

Let's first explain why the products of generating functions generate probabilities associated with sums of random variables.

Suppose $A(t)$ and $B(t)$ are probability generating functions for random variables X and Y, respectively, where X and Y are defined on the set of non-negative integers or some subset of them, and let

$$A(t) = a_0 + a_1 t + a_2 t^2 + \cdots$$

and

$$B(t) = b_0 + b_1 t + b_2 t^2 + \cdots.$$

Then

$$A(t) \cdot B(t) = a_0 b_0 + (a_0 b_1 + a_1 b_0)t + (a_0 b_2 + a_1 b_1 + a_2 b_0)t^2 + \cdots,$$

so the coefficient of t^k is

$$\sum_{i=0}^{k} a_i b_{k-i} = \sum_{i=0}^{k} P(X = i) \cdot P(Y = k - i) = P(X + Y = k).$$

This explains why we could find powers of $G(t)$ in Example 4.5.1 and generate probabilities associated with throwing more than one die.

Since $E(t^X) = \sum_{i=0}^{\infty} t^i P(X = i)$, it follows that a probability generating function, say $P_X(t)$, can be regarded as an expectation

$$P_X(t) = E(t^X) = \sum_{i=0}^{\infty} t^i P(X = i)$$

for a random variable X.

For example, $G(t) = \sum_{i=1}^{6} t^i \cdot P(X = i) = \sum_{i=1}^{6} t^i \cdot \frac{1}{6}$. Note that if $t = 1$, then

$$P_X(1) = \sum_{i=0}^{\infty} P(X = i) = 1.$$

Also,

$$\frac{dP_X(t)}{dt} = \frac{d}{dt} \sum_{i=0}^{\infty} t^i \cdot P(X = i) = \sum_{i=0}^{\infty} \frac{d(t^i)}{dt} P(X = i),$$

from which it follows that

$$P_X'(t) = \sum_{i=0}^{\infty} i \cdot t^{i-1} \cdot P(X = i).$$
So

$$P_X'(1) = \sum_{i=0}^{\infty} i \cdot P(X = i) = E(X).$$

In addition,

$$P_X''(t) = \sum_{i=0}^{\infty} i \cdot (i - 1) \cdot t^{i-2} \cdot P(X = i),$$

so

$$P_X''(1) = E[X \cdot (X - 1)],$$

with similar results holding for higher-order derivatives.

Since $E(X^2) = E[X \cdot (X - 1)] + E(X)$, it follows that

$$\text{Var}(X) = E[X \cdot (X - 1)] + E(X) - [E(X)]^2$$

or

$$\text{Var}(X) = P_X''(1) + P_X'(1) - [P_X'(1)]^2.$$

As an example, consider throwing a single die and let

$$G(t) = \frac{1}{6}(t + t^2 + t^3 + t^4 + t^5 + t^6).$$

Then

$$G'(t) = \frac{1}{6}(1 + 2t + 3t^2 + 4t^3 + 5t^4 + 6t^5).$$

So

$$G'(1) = \frac{1}{6}(1 + 2 + 3 + 4 + 5 + 6) = \frac{7}{2}$$

giving $E(X)$ and

$$G''(t) = \frac{1}{6}(2 + 6t + 12t^2 + 20t^3 + 30t^4).$$

So

$$G''(1) = \frac{1}{6}(2 + 6 + 12 + 20 + 30) = \frac{70}{6}.$$

It follows that

$$\text{Var}(X) = P_X''(1) + P_X'(1) - [P_X'(1)]^2$$

$$\text{Var}(X) = \frac{70}{6} + \frac{7}{2} - \left(\frac{7}{2}\right)^2 = \frac{35}{12}.$$

4.7 ▶ Probability Generating Functions for Some Specific Probability Distributions

Probability generating functions for the binomial and geometric random variables are particularly useful, so we derive their probability generating functions in this section.

Binomial Distribution

For the binomial distribution,

$$P_X(t) = E(t^X) = \sum_{x=0}^{n} t^x \binom{n}{x} p^x q^{n-x}.$$

This sum can be written as

$$P_X(t) = \sum_{x=0}^{n} \binom{n}{x} (tp)^x q^{n-x}.$$

The binomial theorem shows that $P_X(t) = (q + pt)^n$. It is easy to check that

$$P'_X(t) = np(q + pt)^{n-1}$$

so that, because $p + q = 1$ $P'_X(1) = np$ as expected. Also,

$$P''_X(t) = n(n-1)p^2(q+pt)^{n-2},$$

so

$$E[(X(X-n)] = P''_X(1) = n(n-1)p^2,$$

from which it follows that $\text{Var}(X) = n(n-1)p^2 + np - (np)^2 = np - np^2 = npq$.

  Now we show, using probability generating functions, that the sum of independent binomial random variables is binomial. Suppose that X and Y are independent binomial variables with the probability generating functions $P_X(t) = (q + pt)^{n_x}$ and $P_Y(t) = (q + pt)^{n_y}$, respectively. If $Z = X + Y$, then the probability generating function for Z is

$$P_Z(t) = P_X(t) \cdot P_Y(t)$$

$$P_Z(t) = (q + pt)^{n_x} \cdot (q + pt)^{n_y} = (q + pt)^{n_x + n_y}.$$

  Assuming that the probability generating functions are unique – that is, assuming that a probability generating function can arise from one and only one probability distribution function – this shows that Z is binomial with parameters $n_x + n_y$ and p.

  The preceding derivation, done in one line, shows the power of the probability generating function technique; the reader can compare this with the derivation in Example 4.4.1.

  It should be pointed out, however, as may have occurred to the reader, that the fact that sums of binomials, with the same probabilities of success at any trial, are binomial is hardly surprising. If we have a series of n_x binomial trials and we record X, the number of successes, and follow this by a series of n_y trials recording Y successes, it is obvious, because the trials are independent, that we have $X + Y$ successes in $n_x + n_y$ trials. The fact that we paused somewhere in the experiment to record the number of successes so far and then continued has nothing to do with the entire series of trials.

This raises the question of pausing in the series and changing the probability of success at that point. Now the resulting distribution is not at all obvious. Such trials are, confusingly perhaps, called *Poisson's trials*. This problem can be considered using generating functions.

Poisson's Trials

As an example, suppose we toss a fair coin 20 times followed by 10 tosses of a coin loaded so that the probability of a head is $\frac{1}{3}$. What is the probability of exactly 15 heads resulting?

Using probability generating functions, we see that we need the coefficient of t^{15} in $(\frac{1}{2} + \frac{1}{2}t)^{20} \cdot (\frac{2}{3} + \frac{1}{3}t)^{10}$. This is

$$\sum_{x=5}^{15} \binom{20}{x}\left(\frac{1}{2}\right)^x \left(\frac{1}{2}\right)^{20-x} \binom{10}{15-x}\left(\frac{2}{3}\right)^{x-5}\left(\frac{1}{3}\right)^{15-x}$$

$$= \frac{156031933}{1289945088} = 0.12096.$$

A computer algebra system will give this result as well as all the other coefficients in $\left(\frac{1}{2} + \frac{1}{2}t\right)^{20} \cdot \left(\frac{2}{3} + \frac{1}{3}t\right)^{10}$ directly, so it is of immense value in problems of this sort. A graph of these coefficients is remarkably normal, as shown in Figure 4.6.

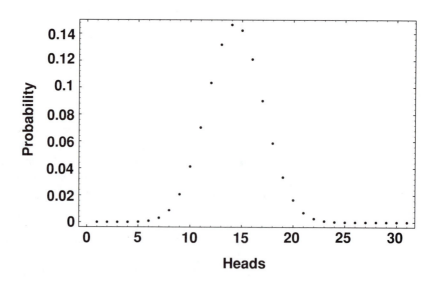

Figure 4.6   Probabilities for the total number of heads when a fair coin is tossed 20 times followed by 10 tosses of a loaded coin with $p = \frac{1}{3}$.

If X is the number of heads in the first series and Y is the number of heads in the second series, it is still true that $E(X + Y) = E(X) + E(Y)$. In this example, $E(X + Y) = 20 \cdot \frac{1}{2} + 10 \cdot \frac{1}{3} = \frac{40}{3}$ and, because the tosses are independent,

$$\text{Var}(X + Y) = \text{Var}(X) + \text{Var}(Y)$$

$$= 20 \cdot \frac{1}{2} \cdot \frac{1}{2} + 10 \cdot \frac{1}{3} \cdot \frac{2}{3} = \frac{65}{9}.$$

We would expect a normal curve with these parameters to fit the distribution of $X + Y$ fairly well.

Example 4.7.1

A series of n binomial trials with probability p is conducted, and is followed by a series of m trials with probability $\frac{x}{n}$, where x is the number of successes in the first series of trials. Let Y denote the number of successes in the second series of trials. Then

$$E(X + Y) = E(X) + E(Y)$$

$$= n \cdot p + m \cdot \frac{E(X)}{n} = p \cdot (n + m).$$

The variance of the sum is another matter, however, since the second series of trials is very clearly dependent on the first series and, because of this, $\text{Var}(X + Y) \neq \text{Var}(X) + \text{Var}(Y)$. General calculations of this sort will be considered in Chapter 5 when we discuss sample spaces with two or more random variables defined on them.

For now, consider, as an example of this, a first series comprised of five trials with probability of success 1/2, followed by a series of three trials. What is the probability of exactly four successes in the entire experiment? We find that

$$P(X + Y = 4) = \sum_{x=1}^{4} \binom{5}{x} \left(\frac{1}{2}\right)^x \left(\frac{1}{2}\right)^{5-x} \binom{3}{4-x} \left(\frac{x}{5}\right)^{4-x} \left(1 - \frac{x}{5}\right)^{x-1},$$

or

$$P(X + Y = 4) = \frac{73}{400}.$$

Again, a computer algebra system is of great use in doing the calculations.   ◀

Geometric Distribution

The waiting time X for the first occurrence of a binomial random variable with parameter p has the probability distribution function

$$P(X = x) = q^{x-1}p, \ x = 1, 2, \ldots,$$

so $P_X(t) = \sum_{x=1} t^x \cdot q^{x-1} p = \frac{p}{q} \sum_{x=1} (tq)^x = \frac{pt}{1-qt}$, provided that $|qt| < 1$. Since $0 < q < 1$, and we are interested only when $t = 1$, the restriction is not important for us.

Using $P_X(t)$, we find that $P_X'(1) = E(X) = 1/p$, and that $\text{Var}(X) = q/p^2$.

The variable X here denotes the waiting time for the first binomial success. When we wait for, say, the rth success, the negative binomial distribution arises. Since a negative binomial variable is the sum of geometric variables, it follows, if X is now the waiting time for the rth binomial success, that

$$P_X(t) = \left(\frac{pt}{1 - qt} \right)^r = \frac{p^r t^r}{(1 - qt)^r}.$$

$P_X(t)$ can be used to show that the negative binomial distribution has mean r/p and variance rq/p^2. This is left as an exercise for the reader.

Collecting Premiums in Cereal Boxes

The manufacturer of your favorite breakfast cereal, in an effort to urge you to buy more cereal, encloses a toy or a premium in each box. How many boxes must you buy in order to collect all the premiums? This problem is often called the *coupon collector's problem* in the literature on probability theory. Of course we can't be *certain* to collect all the premiums, given finite resources, but we could think about the average, or expected number, of boxes to be purchased.

To make matters specific, suppose there are six premiums. The first box gives us a premium we didn't have before. The probability that the next box will not duplicate the premium we already have is $\frac{5}{6}$. This waiting time for the next premium not already collected is a geometric random variable, with probability $\frac{5}{6}$. The expected waiting time for the second premium is then $\frac{1}{(5/6)}$. Now we have two premiums, so the probability that the next box contains a new premium is $\frac{4}{6}$. This is again a geometric variable and our expected waiting time for collecting the third premium is $\frac{1}{(4/6)}$. This process continues. Since the expectation of a sum is the sum of the expectations of the summands, and if we let X denote the total number of boxes purchased in order to secure all the premiums, we conclude that

$$E(X) = 1 + \frac{1}{\frac{5}{6}} + \frac{1}{\frac{4}{6}} + \frac{1}{\frac{3}{6}} + \frac{1}{\frac{2}{6}} + \frac{1}{\frac{1}{6}}$$

$$E(X) = 1 + \frac{6}{5} + \frac{6}{4} + \frac{6}{3} + \frac{6}{2} + \frac{6}{1}$$

$$E(X) = 1 + 1.2 + 1.5 + 2 + 3 + 6 = 14.7 \text{ boxes.}$$

Clearly the cereal company knows what it's doing! An exercise will ask the reader to show that the variance of X is 38.99, so unlucky cereal eaters could be in for buying many more boxes than the expectation would indicate.

This is an example of a series of trials, analogous to Poisson's trials, in which the probabilities vary. Since the total number of trials X can be regarded as a sum of geometric variables (plus 1 for the first box), and because the probability generating function for a geometric variable is $\frac{qt}{1-pt}$, the probability generating function of X is

$$P_X(t) = \frac{\frac{5}{6}t}{1 - \frac{1}{6}t} \cdot \frac{\frac{4}{6}t}{1 - \frac{2}{6}t} \cdot \frac{\frac{3}{6}t}{1 - \frac{3}{6}t} \cdot \frac{\frac{2}{6}t}{1 - \frac{4}{6}t} \cdot \frac{\frac{1}{6}t}{1 - \frac{5}{6}t}.$$

This can be written as

$$P_X(t) = \frac{5!t^5}{(6 - t)(6 - 2t)(6 - 3t)(6 - 4t)(6 - 5t)}.$$

The first few terms in a power series expansion of $P_X(t)$ are

$$P_X(t) = \frac{5t^5}{324} + \frac{25t^6}{648} + \frac{175t^7}{2916} + \frac{875t^8}{11664} + \frac{11585t^9}{139968} + \frac{875t^{10}}{10368} + \frac{616825t^{11}}{7558272} + \cdots.$$

Probabilities can be found from $P_X(t)$, but not at all easily without a computer algebra system. The preceding series shows that the probability that it takes nine boxes in total to collect all six premiums is $875/11664 = 0.075$.

A graph of the probability distribution function is shown in Figure 4.7. The probabilities shown there are the probabilities that it takes n boxes to collect all six premiums.

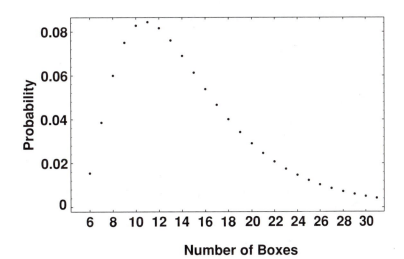

Figure 4.7   Probabilities for the cereal box problem.

Exercises 4.7

1. **a.** Find the probability generating function for a Poisson random variable with parameter λ.

   b. Use the generating function in part **a** to find the mean and variance of a Poisson random variable.

2. Use probability generating functions to show that the sum of independent Poisson variables, with parameters λ_x and λ_y, respectively, has a Poisson distribution with parameter $\lambda_x + \lambda_y$.

3. A discrete random variable X has probability distribution function $f(x) = k/2^x$, $x = 0, 1, 2, 3, 4$.

   a. Find k.

   b. Find $P_X(t)$, the probability generating function.

   c. Use $P_X(t)$ to find the mean and variance of X.

4. Use the probability generating function to find the mean and variance of a negative binomial variable with parameters r and p.

5. A fair coin is tossed 8 times followed by 12 tosses of a coin loaded so as to come up heads with probability $\frac{3}{4}$. What is the probability that

   a. exactly 10 heads occur?

   b. at least 10 heads occur?

6. Use the probability generating function of a Bernoulli random variable to show that the sum of independent Bernoulli variables is a binomial random variable.

7. A random variable has probability distribution function

$$f(x) = \frac{1}{e \cdot x!}, \quad x = 0, 1, 2, 3, \dots .$$

   a. Find the probability generating function for X, $P_X(t)$.

   b. Use $P_X(t)$ to find the mean and variance of X.

8. Suppose a series of 10 binomial trials with probability $\frac{1}{2}$ of success is conducted, giving x successes. These trials are followed by 8 binomial trials with probability $\frac{x}{10}$ of success. Find the probability of exactly 6 successes in the entire series.

9. Verify that the variance of X is 38.99 in the cereal box problem.

10. Suppose X and Y are independent geometric variables with parameters p_1 and p_2 respectively.

    a. Find the probability generating function for $X + Y$.

b. Use the probability generating function to find $P(X + Y = k)$ and then verify your result by calculating the probability directly.

4.8 ► Moment Generating Functions

Another generating function that is commonly used in probability theory is the *moment generating function*. For a random variable X, this function generates the *moments*, $E(X^k)$, for the probability distribution of X. If $k = 1$, the moment becomes $E(X)$, or the mean of the distribution. If $k = 2$, then the moment is $E(X^2)$, which we use in calculating the variance.

The word *moment* has a physical connotation. If we think of the probability distribution as being a very thin piece of material of area 1, then $E(X)$ is the same as the center of gravity of the material, and $E(X^2)$ is used in calculating the moment of inertia. Hence the name *moment* for these quantities that we use to describe probability distributions.

The extent to which we are successful in using the moments to describe probability distributions may be judged from certain considerations. If we were to specify $E(X)$ as a value for a probability distribution, this would certainly constrain the set of random variables X under consideration, but we could still be considering an infinite set of variables. Were we to specify $E(X^2)$ as well, this would narrow the set of possible random variables. A value for $E(X^3)$ further narrows the set. For the examples we will consider, we ask the reader to accept the fact that, were *all* the moments specified, X would be determined uniquely.

Now let's see how this fact can be used. We begin with a definition of the moment generating function.

Definition: The *moment generating function* of a random variable X is

$$M[X; t] = E[e^{tX}],$$

provided that the expectation exists.

It follows that

$$M[X; t] = \sum_x e^{tx} \cdot P(X = x), \quad \text{if } X \text{ is discrete,}$$

and

$$M[X; t] = \int_{-\infty}^{\infty} e^{tx} \cdot f(x)\, dx, \quad \text{if } X \text{ is continuous}$$

provided the sum or integral exists, and where $f(x)$ is the probability density function of X.

First we show that $M[X; t]$ does in fact generate moments. Consider the continuous case so that

$$M[X; t] = \int_{-\infty}^{\infty} e^{tx} \cdot f(x)\, dx.$$

Expanding e^{tx} in a power series, we have

$$M[X; t] = \int_{-\infty}^{\infty} (1 + t\,x + \frac{t^2 x^2}{2!} + \frac{t^3 x^3}{3!} + \cdots) \cdot f(x)\, dx.$$

Use the fact that the integral of a sum is the sum of the integrals and factor out all the powers of t to find that

$$M[X; t] = \int_{-\infty}^{\infty} f(x)\, dx + t \int_{-\infty}^{\infty} x \cdot f(x)\, dx$$

$$+ \frac{t^2}{2!} \cdot \int_{-\infty}^{\infty} x^2 f(x)\, dx + \frac{t^2}{3!} \cdot \int_{-\infty}^{\infty} x^3 f(x)\, dx + \cdots$$

so

$$M[X; t] = 1 + t \cdot E(X) + \frac{t^2}{2!} \cdot E(X^2) + \frac{t^3}{3!} \cdot E(X^3) + \cdots$$

provided that the series converges.

$M[X; t]$ generates moments in the sense that *the coefficient of $\frac{t^k}{k!}$ is $E(X^k)$.*

We used the derivatives of the probability generating function, $P_X(t)$, to calculate $E(X)$, $E[X(X - 1)]$, $E[X(X - 1)(X - 2)]$, ..., quantities that are often called *factorial moments*. The moments defined above could be calculated from them. We did that on several occasions to find the variance.

The derivatives of $M[X; t]$ also have some significance. Since

$$M[X; t] = 1 + t \cdot E(X) + \frac{t^2}{2!} \cdot E(X^2) + \frac{t^3}{3!} \cdot E(X^3) + \cdots,$$

$$M'[X; t] = \frac{dM[X; t]}{dt} = E(X) + t \cdot E(X^2) + \frac{t^2}{2!} \cdot E(X^3) + \cdots$$

and

$$M''[X; t] = \frac{d^2 M[X; t]}{dt^2} = E(X^2) + t \cdot E(X^3) + \frac{t^2}{2!} \cdot E(X^4) + \cdots.$$

so it is evident that $M'[X; 0] = E(X)$, and $M''[X; 0] = E(X^2)$. There are then two methods for calculating moments – either a series expansion or by the derivatives of $M[X; t]$. There are in practice very few examples where each method is feasible;

generally one method works well while the other presents difficulties. We turn now to some examples.

Example 4.8.1

For the uniform random variable, $f(x) = 1$, for $0 \le x \le 1$. The moment generating function is then

$$M[X; t] = \int_0^1 1 \cdot e^{tx} \, dx = \frac{1}{t} e^{tx} \Big|_0^1 = \frac{1}{t}(e^t - 1).$$

In this instance it is easy to express $M[X; t]$ in a power series. Using the power series for e^t we find that

$$M[X; t] = 1 + \frac{t}{2!} + \frac{t^2}{3!} + \frac{t^3}{4!} + \cdots,$$

so $E(X^k) = \frac{1}{k+1}$.

However, this is a fact that is much more easily found directly:

$$E(X^k) = \int_0^1 x^k \, dx = \frac{1}{k+1}.$$

The moment generating function does little here but provide a very difficult way in which to find the moments. This is almost always the case. Moment generating functions are rarely used to generate moments; it is almost always easier to proceed by definition. What then is the use of the moment generating function? The answer is that we use it almost exclusively to establish the distributions of functions of random variables and the distributions of sums of random variables, basing our conclusions on the fact that moment generating functions are unique; that is, only one distribution has a given moment generating function. We will return to this point later.

For now, continuing with the example, we found that

$$M[X; t] = \frac{1}{t}(e^t - 1).$$

If we differentiate this,

$$M'[X; t] = \frac{te^t - e^t + 1}{t^2}.$$

As $t \to 0$, we use L'Hôspital's Rule to find that

$$M'[X; t] \to \frac{1}{2},$$

so the process yields the correct result. This is without doubt the most difficult way in which to establish the fact that the mean of a uniform random variable on the interval $(0, 1)$ is $\frac{1}{2}$!

Clearly we have other purposes in mind; the fact is that the moment generating function is an extremely powerful tool. Using it, facts that are very difficult to establish in other ways can be easily established.

We continue with further examples since the generating functions themselves are of importance.   ◄

Example 4.8.2

Consider the exponential distribution $f(x) = e^{-x}$, $x \geq 0$. We calculate the moment generating function:

$$M[X; t] = \int_{-\infty}^{\infty} e^{tx} \cdot f(x) \, dx = \int_{0}^{\infty} e^{tx} \cdot e^{-x} \, dx.$$

This can be simplified to

$$M[X; t] = \int_{0}^{\infty} e^{-(1-t)x} \, dx = \frac{-1}{1-t} e^{-(1-t)x} \bigg|_{0}^{\infty} = \frac{1}{1-t} \text{ if } t < 1.$$

Again, the power series is easy to find:

$$M[X; t] = 1 + t + t^2 + t^3 + \cdots,$$

establishing the fact that $E[X^k] = k!$, for k a positive integer. This is a nice way to establish the fact that

$$\int_{0}^{\infty} x^k e^{-x} \, dx = k!$$

which arose earlier when the gamma distribution was considered.

The reader may also want to show that the moment generating function for $f(x) = \lambda e^{-\lambda x}$, $x > 0$, is

$$M(X; t) = \frac{\lambda}{\lambda - t}.$$ ◄

Example 4.8.3

The moment generating function for a normal random variable is by far our most important result, as will be seen later. Here we use a standard normal distribution:

$$M[X; t] = \frac{1}{\sqrt{2\pi}} \int_{-\infty}^{\infty} e^{tx} \cdot e^{-x^2/2} \, dx.$$

The simplification of this integral takes some manipulation. Consider the exponent:

$$tx - \frac{x^2}{2} = -\frac{1}{2}(x^2 - 2tx + t^2) + \frac{t^2}{2} = -\frac{1}{2}(x - t)^2 + \frac{t^2}{2}$$

by completing the square. This means that the generating function can be written as

$$M[X; t] = e^{t^2/2} \int_{-\infty}^{\infty} \frac{1}{\sqrt{2\pi}} e^{-\frac{1}{2}(x-t)^2} \, dx.$$

The integral is 1 because it represents the area beneath a normal curve with mean t and variance 1. It follows that

$$M[X; t] = e^{t^2/2}.$$

We can also find a power series for this generating function as

$$M[X; t] = 1 + \frac{t^2}{2} + \frac{1}{2!}\left(\frac{t^2}{2}\right)^2 + \frac{1}{3!}\left(\frac{t^2}{2}\right)^3 + \cdots .$$

It follows that

$$E(X^k) = 0 \text{ if } k \text{ is odd}$$

and

$$E(X^{2k}) = \frac{(2k)!}{k!2^k} \text{ for } k = 1, 2, 3, \ldots .$$

Moment generating functions for other commonly occurring distributions will be established in the exercises. ◀

Exercises 4.8

1. Show that if $P_X(t)$ is the probability generating function for a random variable X, then

$$M[X; t] = P_X(e^t).$$

2. **a.** Find the moment generating function for a binomial random variable with parameters n and p.
   b. Take the case $n = 2$ and expand the moment generating function to find $E(X^2)$ from this expansion.
   c. Use the moment generating function to find the mean and variance of a binomial random variable.

3. **a.** Find the moment generating function for a Poisson random variable with parameter λ.
   b. Use the moment generating function to find the mean and variance of a Poisson random variable.

4. Find the moment generating function for an exponential random variable with mean λ.

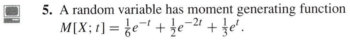

5. A random variable has moment generating function
   $M[X; t] = \frac{1}{6}e^{-t} + \frac{1}{2}e^{-2t} + \frac{1}{3}e^{t}$.

   a. Find the mean and variance of X.
   b. Find the first 5 terms in the power series expansion of $M[X; t]$.

6. A random variable X has probability distribution function $f(x) = k \cdot \left(\frac{1}{2}\right)^x$, $x = 1, 2, 3, \ldots$.

   a. Find k.
   b. Find the moment generating function and show the first 5 terms in its power series expansion.
   c. Find the mean and variance of X from the moment generating function in 2 ways.

7. **a.** Find the moment generating function for a gamma distribution.
   b. Use the moment generating function to find the mean and variance for the gamma distribution.

8. **a.** Find the moment generating function for a χ_n^2 random variable.
   b. Use the moment generating function to find the mean and variance of a χ_n^2 random variable.

9. A random variable X has the probability density function

$$f(x) = \begin{cases} \frac{1}{3} & -2 < x < -1 \\ k & 1 < x < 4. \end{cases}$$

Find the moment generating function for X.

 10. Find $E[X^4]$ for a random variable whose moment generating function is $e^{t^2/2}$.

11. Random variable X has moment generating function $M[X; t] = e^{t^2/2}$.

    a. Find $P(-1 \le X \le 1)$.
    b. Find $P(X^2 < 2)$.

12. A random variable X has the probability density function

$$f(x) = \begin{cases} \frac{1}{2}e^{-x} & x > 0 \\ \frac{1}{2}e^{x} & x < 0. \end{cases}$$

    a. Show that $M[X; t] = (1 - t^2)^{-1}$.
    b. Find the mean and variance of X.

13. Suppose that X is a uniformly distributed random variable on $[2, 3]$.

    a. Find the moment generating function for X.
    b. Expand $M[X; t]$ in an infinite series and from this series find μ_x and σ_x^2.

14. Find the moment generating function for X if $f(x) = 2x, \ 0 < x < 1$. Then use the moment generating function to find μ_x and σ_x^2.

 15. The moment generating function for a random variable X is $(\frac{2}{3} + \frac{1}{3}e^t)^5$.

    a. Find the mean and variance of X.
    b. What is the probability distribution for X?

16. The moment generating function for a random variable X is e^{t^2}. Find the mean and variance of X.

4.9 ▶ Properties of Moment Generating Functions

A primary use of the moment generating function is in determining the distributions of functions of random variables. It happens that the moment generating function for linear functions of X is easily related to the moment generating function for X.

THEOREM:

   a. $M[cX; t] = M[X; ct]$.
   b. $M[X + c; t] = e^{ct} M[X; t]$ where c is a constant.

PROOF:

   a. $M[cX; t] = E(e^{(cX)t}) = E(e^{X(ct)}) = M[X; ct]$.
   b. $M[X + c; t] = E[e^{(X+c)t}] = E[e^{Xt} \cdot e^{ct}] = e^{ct} E[e^{Xt}] = e^{ct} M[X; t]$. ∎

So multiplying the variable by a constant simply multiplies t by the constant in the generating function; adding a constant multiplies the generating function by e^{ct}.

Example 4.9.1

We use the preceding theorem to find the moment generating function for a $N(\mu, \sigma)$ random variable from the generating function for a $N(0, 1)$ random variable. Let $Z = \frac{X-\mu}{\sigma}$. Since $M[Z; t] = e^{t^2/2}$, it follows that

$$M\left[\frac{X - \mu}{\sigma}; t\right] = M\left[X - \mu; \frac{t}{\sigma}\right] = e^{-(\mu t/\sigma)} M\left[X; \frac{t}{\sigma}\right],$$

from which it follows that

$$M\left[X; \frac{t}{\sigma}\right] = e^{(t^2/2)+(\mu t/\sigma)}.$$

We conclude from this that

$$M[X; t] = e^{\mu t + (\sigma^2 t^2/2)}.$$

Therefore, if a random variable X has, for example, $M[X; t] = e^{(3t/4)+(t^2/3)}$ then X is normal with mean $\frac{3}{4}$ and variance $\frac{2}{3}$.

A remarkable fact, and one we use frequently here, is that, if X and Y are independent,

$$M[X + Y; t] = M[X; t] \cdot M[Y; t].$$

To indicate why this is true, we start with $M[X + Y; t] = E[e^{(X+Y)t}] = E[e^{tX} \cdot e^{tY}]$, which is the expectation of the product of two functions. If we can show that the expectation of the product is the product of the expectations, then

$$M[X + Y; t] = E[e^{tX} \cdot e^{tY}] = E[e^{tX}] \cdot E[e^{tY}] = M[X; t] \cdot M[Y; t].$$

As a partial explanation of the fact that the expectation of a product of independent random variables is the product of the expectations, consider X and Y as discrete independent random variables. Then

$$E(X \cdot Y) = \sum_x \sum_y x \cdot y \cdot P(X = x \text{ and } Y = y).$$

But if X and Y are independent, then

$$P(X = x \text{ and } Y = y) = P(X = x) \cdot P(Y = y)$$

and so

$$E[X \cdot Y] = \sum_x \sum_y x \cdot y \cdot P(X = x \text{ and } Y = y)$$

$$= \sum_x \sum_y x \cdot y \cdot P(X = x) \cdot P(Y = y)$$

$$= \sum_x x \cdot P(X = x) \cdot \sum_y y \cdot P(Y = y) = E(X) \cdot E(Y).$$

So it is plausible that the expectation of the product of independent random variables is the product of their expectations, and we accept the fact that

$$M[X + Y; t] = M[X; t] \cdot M[Y; t]$$

if X and Y are independent.

We will return to this point in Chapter 5 when we consider bivariate probability distributions. For now we will make use of the result to establish some surprising results. ◄

4.10 ► Sums of Random Variables II

We have used the facts that $E(X + Y) = E(X) + E(Y)$ and, if X and Y are independent, that $\text{Var}(X + Y) = \text{Var}(X) + \text{Var}(Y)$. However, these facts do not establish the distribution of $X + Y$. We now turn to determining the distribution of the sums of two or more independent random variables. Our solution here will show the power and usefulness of the moment generating function. We will return to this subject in Chapter 5 where we can demonstrate another procedure for finding the probability distribution of sums of random variables. Here we use the fact that the moment generating function for a sum of independently distributed random variables is the product of the individual generating functions.

Example 4.10.1 (Sums of Normal Random Variables)

It is probably not surprising to find that sums of independent normal variables are also normal. The proof of this is now easy: If X and Y are independent normal variables,

$$M[X + Y; t] = M[X; t] \cdot M[Y; t].$$

The exponent in the product on the right in this equation is

$$\mu_x t + \frac{t^2 \sigma_x^2}{2} + \mu_y t + \frac{t^2 \sigma_y^2}{2}.$$

This can be rearranged as

$$(\mu_x + \mu_y)t + \frac{t^2(\sigma_x^2 + \sigma_y^2)}{2},$$

showing that $X + Y \sim N[\mu_x + \mu_y, \ \sigma_x^2 + \sigma_y^2]$. Note that the mean and variance of the sum can be established in other ways. This argument establishes the *normality*, which otherwise would be very difficult to show.

However, the big surprise is that sums of non-normal variables also become normal. We will explain this fully in section 4.11, but the reader may note that this may explain the frequency with which we have seen the normal distribution up to this point. For the moment we continue with another example.   ◀

Example 4.10.2 (Sums of Exponential Random Variables)

We begin with a decidedly non-normal random variable, namely an exponential variable where we take the mean to be 1. So

$$f(x) = e^{-x}, \ x \geq 0.$$

We know that

$$M[X; t] = (1 - t)^{-1}.$$

It follows that the moment generating function of the sum of two independent exponential random variables is

$$M[X + Y; t] = (1 - t)^{-2}.$$

This, however, is the moment generating function of $f(x) = xe^{-x}, \ x \geq 0$. The graph of this distribution is shown in Figure 4.8.

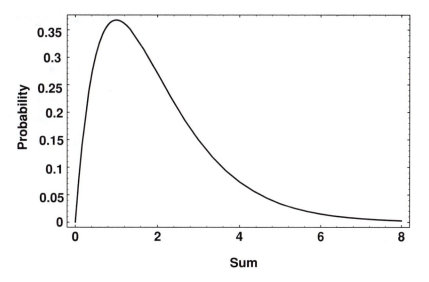

Figure 4.8   Sum of two independent exponential random variables.

Now consider the sum of three independent exponential random variables. The moment generating function is $M[X + Y + Z; t] = (1 - t)^{-3}$. A computer algebra system, or otherwise, shows that this is the moment generating function for $f(x) = \frac{x^2}{2} e^{-x}, \ x \geq 0$. Figure 4.9 shows a graph of this distribution.

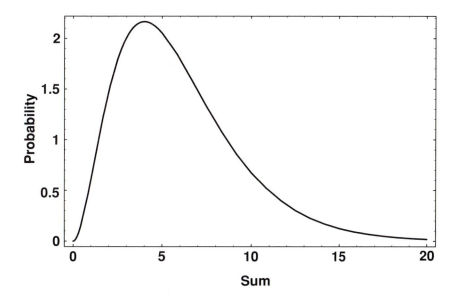

Figure 4.9   Sum of three independent exponential variables.

We now strongly suspect the occurrence of normality if we add more variables. We know that

$$M[(X_1 + X_2 + X_3 + \cdots + X_n); t] = (1 - t)^{-n}.$$

This is the moment generating function for the gamma distribution, $f(x) = \frac{1}{\Gamma[n]} x^{n-1} e^{-x}$, $x \geq 0$. Since the mean and variance of each of the X_i's above is 1, we can consider $X = \sum_{i=1}^{n} X_i$ and $Z = \frac{X-n}{\sqrt{n}}$. Then,

$$M[Z; t] = M\left[\frac{X - n}{\sqrt{n}}; t\right] = M\left[X - n; \frac{t}{\sqrt{n}}\right] = e^{-t\sqrt{n}}\left(1 - \frac{t}{\sqrt{n}}\right)^{-n}.$$

The behavior of this is most easily found using a computer algebra system. We expand $M[Z; t]$ and let $n \to \infty$. We find that

$$M[Z; t] \to e^{t^2/2},$$

showing that Z approaches the standard normal distribution. Some details of this calculation are given in Appendix One. This establishes the fact that the sums of independent exponential variables approach a normal distribution.   ◄

In the beginning of this chapter we indicated that the sums showing on n dice – the sums of independent discrete uniform random variables – became normal, although we lacked the techniques for proving this at that time. The proof of this will not be shown here, but we note that the process followed in Example 4.10.2 will work in this case. Now we know that the distribution of sums of independent exponential variables also approaches the normal distribution. The fact that the distribution of the sum of widely different summands approaches the normal distribution is perhaps one of the most surprising facts in mathematics. The fact that normality should occur for a wide range of random variables is investigated in the next section.

Exercises 4.10

1. The moment generating function for a random variable X is $M[X; t] = \frac{5}{5-t}$.

    a. Find the mean and variance of X.

    b. Identify the probability distribution for X.

2. Random variable X has $M[X; t] = \left(\frac{2}{3} + \frac{1}{3}e^t\right)^6$.

    a. Find the mean and variance of X.

    b. Identify the probability distribution for X.

3. Find the mean and variance for the random variable whose moment generating function is $M(Z; t) = (1 - 2t)^{-5}$.

4. Find the moment generating function for the exponential random variable whose probability density function is $f(x) = 2e^{-2x}$, $x \geq 0$.

5. Suppose $X \sim N(36, \sqrt{10})$ and $Y \sim N(15, \sqrt{6})$. If X and Y are independent, find $P(X + Y \geq 43)$.

6. A random variable X has probability density function $f(x) = xe^{-x}$, $x \geq 0$.

   a. Find the moment generating function, $M[X; t]$.
   b. Use $M[X; t]$ to find μ_x and σ_x^2.
   c. Find a formula for $E(X^k)$.

7. Find the variance of a random variable whose moment generating function is $M[X; t] = (1 - t)^{-1}$.

8. Explain why the function $2 + \frac{1}{1-t}$ cannot be the moment generating function for any random variable.

9. What is the probability distribution for a random variable whose moment generating function is

$$M[X; t] = e^{-\lambda(1 - e^t + t)} ?$$

10. Identify the random variable whose moment generating function is

$$M[X; t] = \left(\frac{1}{2}\right)^{16} \cdot e^{-4t} \cdot (1 + e^{t/2})^{16}.$$

11. Show that the sum of 2 independent binomial random variables with parameters n and p, and m and p, respectively, is binomial with parameters $n + m$ and p.

12. Show that the sum of independent Poisson random variables with parameters λ_x and λ_y, respectively, is Poisson with parameter $\lambda_x + \lambda_y$.

13. Show that the sum of independent χ^2 random variables is a χ^2 random variable.

14. Show that a Poisson variable with parameter λ becomes normal as $\lambda \to \infty$. (Hint: Find the limit of $M\left[\frac{X - \lambda}{\sqrt{\lambda}}; t\right]$.)

15. a. If X is uniformly distributed on $[0, 1]$, show that $M[X; t] = \frac{1}{t}(e^t - 1)$.
   b. Suppose that X_1 and X_2 are independent observations from the uniform distribution in part **a**. Find $M[X_1 + X_2 ; t]$.

c. Let Y be a random variable with the probability density function

$$g(y) = \begin{cases} y, & 0 < y < 1 \\ 2 - y, & 1 < y < 2. \end{cases}$$

Find the moment generating function for Y.

d. What can be concluded from parts **b** and **c** above?

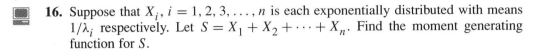 **16.** Suppose that X_i, $i = 1, 2, 3, \ldots, n$ is each exponentially distributed with means $1/\lambda_i$ respectively. Let $S = X_1 + X_2 + \cdots + X_n$. Find the moment generating function for S.

17. Suppose that X and Y are independent random variables with probability density functions

$$f(x) = 1, \ 0 \le x \le 1 \quad \text{and} \quad g(y) = 1, \ -1 \le y \le 0.$$

Find $M[X + Y; t]$.

18. If $M[X; t] = e^{-\lambda(1-e^t)}$, what is $M[15 - 3X; t]$?

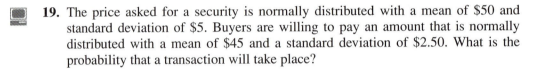 **19.** The price asked for a security is normally distributed with a mean of \$50 and standard deviation of \$5. Buyers are willing to pay an amount that is normally distributed with a mean of \$45 and a standard deviation of \$2.50. What is the probability that a transaction will take place?

20. A rod is made up of five sections. A study of the individual sections shows that the end sections have mean lengths of 1.001 inches, and the three middle sections have mean lengths of 1.999 inches. The standard deviation of the length of each section is 0.004 inches. If random assembly is employed,

a. what will be the average length of the assembled rods?

b. what will be the standard deviation of the assembled lengths?

c. what is the probability that the assembled rod will have length in excess of 8.002 inches?

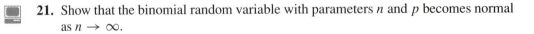 **21.** Show that the binomial random variable with parameters n and p becomes normal as $n \to \infty$.

22. Two independent observations, X and Y, are selected from a probability distribution with $f(x) = 2x$, $0 \le x \le 1$.

a. Find the moment generating function for the sum $Z = X + Y$.

b. Find $E(Z^4)$.

23. Let S denote the sum of r independent exponential random variables, each with expectation $1/\alpha$. Show that $2\alpha S$ has a χ^2_{2r} distribution.

24. Show that the moment generating function for the gamma distribution

$$f(x) = \frac{1}{\Gamma[n]} x^{n-1} e^{-x}, \; x \geq 0$$

is $(1-t)^{-n}$.

▌4.11 ▶ **The Central Limit Theorem**

We have had numerous examples of sums that approach normality as the number of summands increases. We now want to consider means, which are multiples of sums, of random variables and consider the limiting distribution of such averages.

THEOREM:
   If $\overline{X}$ denotes the mean of n observations of a random variable X with mean μ and variance σ^2, then the limiting distribution of $\frac{\overline{X}-\mu}{\frac{\sigma}{\sqrt{n}}}$ is $N(0, 1)$ provided that X has a moment generating function. ∎

   The theorem indicates that the probability distribution of the random variable $\frac{\overline{X}-\mu}{\frac{\sigma}{\sqrt{n}}}$ approaches the $N(0, 1)$ probability distribution. The result is known as the *central limit theorem*. Actually there is a class of theorems known as central limit theorems in probability, but since this is the only one we will consider, we will refer to it uniquely and call it the central limit theorem. We now indicate a proof.
   Since we presume that X has a moment generating function, let this moment generating function be

$$M[X; t] = 1 + \mu_1 t + \mu_2 \frac{t^2}{2!} + \mu_3 \frac{t^3}{3!} + \cdots$$

where, for convenience, μ_k denotes $E(X^k)$. Now

$$M[\overline{X}; t] = M\left[\frac{\sum_{i=1}^{n} X_i}{n}; t\right] = \left(M\left[X; \frac{t}{n}\right]\right)^n$$

so

$$\log(M[\overline{X}; t]) = n \log \left(M\left[X; \frac{t}{n}\right]\right).$$

Using the series expansion $\log(1 + x) = x - \frac{x^2}{2} + \frac{x^3}{3} - \cdots$ where $|x| < 1$, we find

$$\log\left(M\left[X;\frac{t}{n}\right]\right) = \left(\mu_1\frac{t}{n} + \mu_2\frac{t^2}{2!n^2} + \mu_3\frac{t^3}{3!n^3} + \cdots\right)$$

$$- \frac{1}{2}\left(\mu_1\frac{t}{n} + \mu_2\frac{t^2}{2!n^2} + \mu_3\frac{t^3}{3!n^3} + \cdots\right)^2$$

$$+ \frac{1}{3}\left(\mu_1\frac{t}{n} + \mu_2\frac{t^2}{2!n^2} + \mu_3\frac{t^3}{3!n^3} + \cdots\right)^3 - \cdots.$$

So $n \cdot \log\left(M\left[X;\frac{t}{n}\right]\right) = \log(M[\overline{X};t])$ simplifies to $\log(M[\overline{X};t]) = \mu_1 t + \sigma^2\frac{t^2}{2n}$ plus terms that approach 0 as $n \to \infty$.

This shows that $M[\overline{X};t] \to e^{\mu t + (\sigma^2/n)(t^2/2)}$, the moment generating function for a normal curve with mean μ and variance $\frac{\sigma^2}{n}$.

This explains many of the normal-like graphs we have encountered previously. If the variables are sums, or means, of variables with moment generating functions (as all of ours have been), we expect normality as the number of summands increases. This is exactly what has happened. This phenomenon was encountered in section 4.10 where we examined sums of independent exponential random variables and found that these approach a normal probability distribution.

This also explains why the normal curve can be regarded as something that is normal in the sense that it is usual or expected. It can, in fact, be generated from almost any probability distribution by taking sums or forming averages.

We will show some very important statistical applications of this result in the remaining sections in this chapter.

Example 4.11.1

We noticed in Chapter 2 that the graphs of the binomial distributions we considered became normal-like as the number of trials increased. We also promised a full explanation of the fact that binomial curves with large values for n can be approximated by normal curves. We will do this by using the technique just described, namely by finding the limiting behavior of the moment generating function.

Let X be a binomial random variable with parameters n and p. Then the moment generating function of X is

$$M[X;t] = (q + pe^t)^n.$$

As usual we let $Z = \frac{X-\mu}{\sigma}$ so that

$$M[Z;t] = e^{-\mu t/\sigma} \cdot (q + pe^{t/\sigma})^n.$$

So

$$\log(M[Z;t]) = -\frac{\mu t}{\sigma} + n \log \left[q + p \left(1 + \frac{t}{\sigma} + \frac{t^2}{2!\sigma^2} + \frac{t^3}{3!\sigma^3} + \cdots \right) \right]$$

$$= -\frac{\mu t}{\sigma} + n \log \left[1 + p \cdot \frac{t}{\sigma} + p \cdot \frac{t^2}{2!\sigma^2} + p \cdot \frac{t^3}{3!\sigma^3} + \cdots \right]$$

$$= -\frac{\mu t}{\sigma} + n \left(p \cdot \frac{t}{\sigma} + p \cdot \frac{t^2}{2!\sigma^2} + p \cdot \frac{t^3}{3!\sigma^3} + \cdots \right)$$

$$- \frac{n}{2} \left(p \cdot \frac{t}{\sigma} + p \cdot \frac{t^2}{2!\sigma^2} + p \cdot \frac{t^3}{3!\sigma^3} + \cdots \right)^2 + \cdots .$$

Now, using the facts that $\mu = n \cdot p$ and that $\sigma^2 = n \cdot p \cdot q$, we find that

$$\log(M[Z;t]) = \frac{t^2}{2} + \text{ terms that approach } 0 \text{ as } n \to \infty.$$

It follows that $M[Z;t]$ approaches the moment generating function of the standard normal random variable.

This justifies our use of the normal distribution in approximating the binomial distribution for large values of n, although computer algebra systems allow us to compute binomial probabilities exactly for values of n that occur in most practical cases.

The central limit theorem has wide application to the statistical analysis of data. We will show some of the statistical applications of this result in the remaining sections of this chapter.   ◀

Exercises 4.11

1. Show that, if X is normal with mean μ and variance σ^2, then $\overline{X}$ is normal with mean μ and variance $\frac{\sigma^2}{n}$, where $\overline{X}$ is the mean of n of the X's.

2. Use the central limit theorem to approximate the probability that the sum on 12 fair dice is 38 and then compare the approximation to the exact value.

3. Using the central limit theorem, approximate the probability that the sum of 8 observations taken from an exponential distribution with mean 2 exceeds 5.

4. Light fixtures in a warehouse contain bulbs whose life lengths are exponential with a mean of 720 hours. When a light burns out, it is immediately replaced with a new bulb.

   a. What is the probability that 3 bulbs last at least 2000 hours?

b. If we want the probability that the bulbs on hand will last at least 3500 hours with probability 0.95, how many bulbs should be stocked?

5. Suppose that X is a random variable with an unknown mean μ, but its variance is known to be 100. How many observations of X must be taken so that the probability that $\overline{X}$ is within 2 units of μ is 0.99?

6. The components in a system are known to have $R(1000 \text{ hours}) = 0.91$, where R denotes the reliability function.

a. Approximate the reliability of a system of 100 such components if at least 70 of the components must function at least 1000 hours.

b. How many components must be installed in the system if at least 90 components must last at least 1000 hours with probability 0.98?

7. Articles are shipped in lots of 1000 items. It is known that the probability that an item is defective is 0.04. Presume that the production process follows the assumptions of the binomial distribution.

a. Approximate the probability that in 100 lots, the average number of defectives is less than 39.5.

b. Now suppose we would like the probability that the average number of defectives in 100 lots is less than 39.5 to be 0.10. What should the size of each lot be?

8. Traffic accidents at an intersection follow a Poisson distribution with 40 accidents expected per year. Approximate the probability of, at most, 55 accidents in a given year at that intersection.

9. An elevator can carry a maximum of 1575 pounds. What is the probability that 10 people will overload the elevator if their weights are random selections from $N(150, 10)$?

10. A class has 200 graduates. Assume each graduate invites 2 guests who attend, independently, with probability 0.8. How many seats for guests should be provided at commencement if the class desires to be 99% confident of seating everyone?

11. A machine turns out precision bolts whose lengths may be regarded as a normal random variable with mean 6 and variance 0.0036. To check on whether or not the machine is in control, 36 bolts are randomly selected from each day's production. The machine is considered to be under control if the mean of these lengths falls between 5.970 and 6.015. What is the probability that a sample will fail to meet this criterion, even though the machine is under control?

12. At a local discount store, service times at the checkout counter are observed to be normally distributed with mean 3.5 minutes and variance 1.44 min^2.

a. Find the probability that a customer takes more than 5 minutes to check out.

    b. A customer has been checking out for 3 minutes. What is the probability that it will take at least 5 minutes for the entire process?

    c. What is the probability that the next 6 customers check out in a total of 20 minutes or less?

13. One hundred bolts are packed in a box. The weight of a bolt has mean 1 oz and standard deviation 0.1 oz. Approximate the probability that a box weighs more than 102 oz.

14. A candy maker produces mints that have a label weight of 20.4 grams, but the actual distribution of the weights has $\mu = 21.37$ g, and $\sigma^2 = 0.16$ g^2. Let $\overline{X}$ be the mean weight of a sample of 36 units. Find $P(21.21 \leq \overline{X} \leq 21.45)$.

15. Let X be a normal random variable with mean 1 and variance 16.

    a. What is the probability that an observation is within 2 units of the mean?

    b. What is the probability that the mean of 4 observations is within 2 units of the mean?

16. Civil engineers believe that W, the weight (in units of 1000 pounds) that the span of a bridge can withstand without structural damage resulting, is 78.5. Suppose that the weight (again in units of 1000 pounds) is a random variable with mean 3 and standard deviation 0.3. How many cars can be allowed on the bridge span for the probability that structural damage will not occur to be 0.99?

17. In Example 2.3.1.2 we remarked that a deviation of more than 0.11 in the average result when a fair die is thrown 1000 times is highly unlikely. Show that this is true.

4.12 ▶ Weak Law of Large Numbers

    If we have a manufacturing process that is producing a defective item with probability p, we have an intuitive notion that we can discover the value of p, at least within a given range, if we observe the production process long enough. If we have a distribution with unknown mean μ, we have a similar belief, namely that we can determine μ, again with a given accuracy, if we take a large enough sample and compute the mean of the sample. It is reasonable to believe that the mean of this sample ought to be close to μ. These ideas are correct, and here we examine mathematical demonstrations of them.

    We might even refer to these results as a law of averages, although the literature of probability generally refers to these results as the *Weak Law of Large Numbers*.

    We consider the second problem first. Suppose that $X_1, X_2, X_3, \ldots, X_n$ is a random sample from some distribution with finite mean and variance, say $E(X_i) = \mu$ and $\text{Var}(X_i) = \sigma^2$ for $i = 1, 2, \ldots, n$. By the central limit theorem, $E(\overline{X}) = \mu$ and $\text{Var}(\overline{X}) = \sigma^2/n$.

We can apply Tchebycheff's Inequality to find that

$$P\left[\, |\,\overline{X} - \mu\,| \le k \cdot \frac{\sigma}{\sqrt{n}} \,\right] \ge 1 - \frac{1}{k^2} \ \text{for some } k > 0.$$

Now let $\varepsilon = k \cdot \frac{\sigma}{\sqrt{n}}$; so that $k = \varepsilon \cdot \frac{\sqrt{n}}{\sigma}$ so that the inequality above becomes

$$P[\, |\,\overline{X} - \mu\,| \le \varepsilon\,] \ge 1 - \frac{\sigma^2}{n \cdot \varepsilon^2}.$$

As $n \to \infty$, $P[\, |\,\overline{X} - \mu\,| \le \varepsilon\,] \to 1$ even when ε is arbitrarily small. So the probability that $\overline{X}$ and μ are arbitrarily close approaches 1. This is a verification of our conjecture that a sample mean can be made arbitrarily close to the population mean as the sample size increases.

For the first conjecture, let p_s denote the sample proportion of defective items chosen from a manufacturing process that is producing defective items with probability p. We know that $E(p_s) = p$ and that $\text{Var}(p_s) = \frac{p \cdot q}{n}$ where n is the sample size. Again applying Tchebycheff's Inequality, we find that

$$P\left[\, |p_s - p| \le k \cdot \sqrt{\frac{p \cdot q}{n}} \,\right] \ge 1 - \frac{1}{k^2} \ \text{for some } k > 0.$$

If $\varepsilon = k \cdot \sqrt{\frac{p \cdot q}{n}}$, the inequality becomes

$$P[\, |p_s - p| \le \varepsilon\,] \ge 1 - \frac{p \cdot q}{n \cdot \varepsilon^2}.$$

So p_s and p can be made arbitrarily close as n becomes large with probability approaching 1.

A computer simulation provides some concrete evidence of the above statement. Figure 4.10 shows the result of 100 samples of size 100 each, drawn from a binomial population with $p = 0.38$. The horizontal axis shows the number of samples, and the vertical axis displays the cumulative ratio of successes to the total number of trials. While the initial values exhibit fairly large variation, the later ratios are very close to 0.38, as we expect.

The convergence in statements such as $P[\, |\,\overline{X} - \mu\,| \le \varepsilon\,] \to 1$, indicating that a sequence of means approaches a population value with probability 1, is referred to as *convergence in probability*. It differs from the convergence usually encountered in calculus; that convergence is normally pointwise.

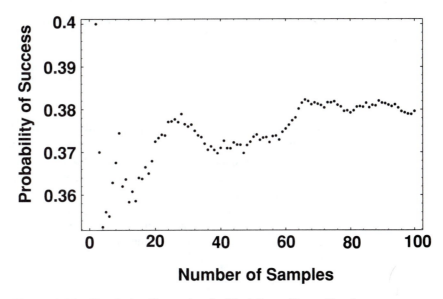

Figure 4.10   Simulation illustrating the Weak Law of Large Numbers.

4.13 ▶ Sampling Distribution of the Sample Variance

The remainder of this chapter will be devoted to data analysis, so we now turn to some statistical applications of the theory presented to this point. In particular we want to investigate hypothesis tests, some confidence intervals, and the analysis of data arising in many practical situations. We will also examine the theory of least squares as it applies to fitting a linear function to data.

The central limit theorem indicates that the probability distribution of sample means drawn from a variety of populations is approximately normal even for samples of moderate size. The probability distributions of other quantities calculated from samples (usually referred to as *statistics*) do not have such simple distributions and, in addition, are often seriously affected by the type of probability distribution from which the samples come.

In this section we determine the probability distribution of the *sample variance*. Other statistics will become important to us, and we will consider their probability distributions when they arise. It is worth considering the sample variance by itself first.

First we define the sample variance for a sample $x_1, x_2, \ldots, x_n$ as

$$s^2 = \frac{1}{n-1} \sum_{i=1}^{n} (x_i - \overline{x})^2$$

where $\bar{x}$ is the mean of the sample. The formula may also be written as

$$s^2 = \frac{n\sum_{i=1}^{n} x_i^2 - (\sum_{i=1}^{n} x_i)^2}{n(n-1)}.$$

Clearly there is some relationship between s^2 and σ^2. The divisor of $n-1$ may be puzzling, but, as we will presently show, this is chosen so that $E(s^2) = \sigma^2$. Since $E(s^2) = \sigma^2$, s^2 is called an *unbiased estimator* of σ^2. If a divisor of n had been chosen, the expected value of the sample variances thus calculated would not be the population value σ^2.

Now let's look at a specific example. Consider all the possible samples of size 3, chosen without replacement, from the discrete uniform distribution on the set of integers $\{1, 2, \ldots, 20\}$. We calculate the *sample variance* for each sample. Each sample variance is calculated using

$$s^2 = \frac{1}{2}\sum_{i=1}^{3}(x_i - \bar{x})^2 \text{ where } \bar{x} = \frac{x_1 + x_2 + x_3}{3}.$$

The probability distribution of s^2 in part is as follows. Permutations of the samples have been ignored.

s^2	1	$\frac{7}{3}$	4	$\frac{13}{3}$	$\frac{19}{3}$	$\cdots$	$\frac{301}{3}$	$\frac{307}{3}$	$\frac{313}{3}$	109	$\frac{343}{3}$
$1140 \cdot$ Prob.	18	34	16	32	30	$\cdots$	2	4	2	2	2

The complete distribution is easy to work out with the aid of a computer algebra system. There are 83 possible values for s^2. A graph of the distribution of these values is shown in Figure 4.11.

The graph indicates that large values of the variance are unusual. The graph also indicates that the probability distribution of s^2 is probably not normal. However, the sample size is quite small, so we can't draw any definite conclusions here.

We do see that the probability distribution shown in Figure 4.12 strongly resembles that suggested by Figure 4.11.

As another example, 500 samples of size 5 each were selected from a standard normal distribution. The sample variance for each of these samples was computed; the results are shown in the histogram in Figure 4.13. We see a distribution with a long tail, which resembles the probability distribution shown in Figure 4.14. Figure 4.14 is in fact a graph of the probability distribution of a chi-squared distribution with 4 degrees of freedom. We will now show that this is the probability distribution of a function of the sample variance s^2.

The sample variance s^2 is a very complex random variable, because it involves $\bar{x}$, which, in addition to the sample values themselves, varies from sample to sample. To narrow our focus, suppose that the sample comes from a normal distribution $N(\mu, \sigma)$. Note that no such distributional restriction was necessary in discussing the distribution of the sample mean. The restriction to normality is common among functions of the

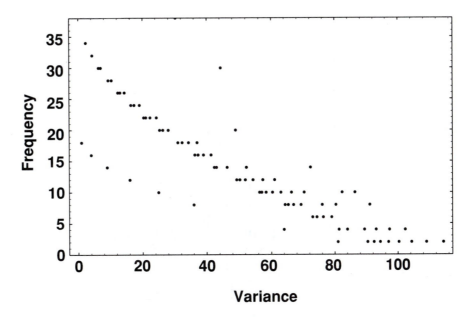

Figure 4.11   Sampling distribution of the sample variance.

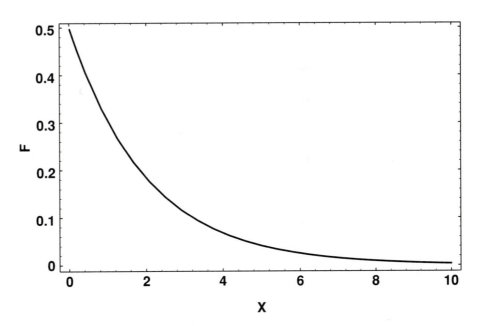

Figure 4.12   A probability distribution suggested by Figure 4.11.

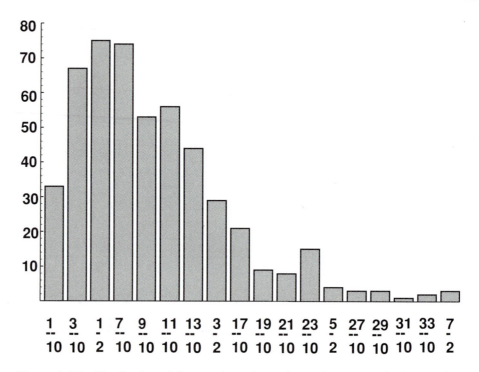

Figure 4.13   Distribution of the sample variance chosen from a standard normal distribution.

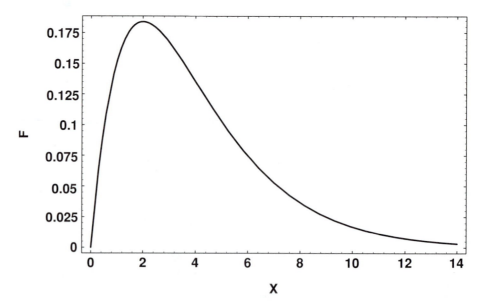

Figure 4.14   Chi-squared distribution with 4 degrees of freedom.

sample values other than the sample mean, and, although much is known when this restriction is lifted, we cannot discuss this in this book.

We now present a fairly plausible derivation of the distribution of a function of the sample variance provided that the sample is chosen from a $N(\mu, \sigma)$ distribution. From the definition of the sample variance, we can write

$$\frac{(n-1)s^2}{\sigma^2} = \sum_{i=1}^{n} \frac{(x_i - \overline{x})^2}{\sigma^2}.$$

The sum in the numerator can be written as

$$\sum_{i=1}^{n} (x_i - \overline{x})^2 = \sum_{i=1}^{n} [(x_i - \mu) - (\overline{x} - \mu)]^2.$$

This in turn simplifies to

$$\sum_{i=1}^{n} (x_i - \overline{x})^2 = \sum_{i=1}^{n} (x_i - \mu)^2 - n(\overline{x} - \mu)^2,$$

so

$$\sum_{i=1}^{n} \frac{(x_i - \overline{x})^2}{\sigma^2} = \sum_{i=1}^{n} \frac{(x_i - \mu)^2}{\sigma^2} - \frac{(\overline{x} - \mu)^2}{\sigma^2/n},$$

or

$$\frac{(n-1)s^2}{\sigma^2} + \frac{(\overline{x} - \mu)^2}{\sigma^2/n} = \sum_{i=1}^{n} \frac{(x_i - \mu)^2}{\sigma^2}.$$

It can be shown, in sampling from a normal population, that $\overline{X}$ and s^2 are independent. This fact is far from being intuitively obvious; its proof is beyond the scope of this book, but a proof can be found in Hogg and Craig [14]. Using this fact of independence it follows that

$$M\left[\frac{(n-1)s^2}{\sigma^2}; t\right] \cdot M\left[\frac{(\overline{x} - \mu)^2}{\sigma^2/n}; t\right] = M\left[\sum_{i=1}^{n} \frac{(x_i - \mu)^2}{\sigma^2}; t\right]$$

where $M[X; t]$ denotes the moment generating function. Now $\sum_{i=1}^{n} \frac{(x_i - \mu)^2}{\sigma^2}$ is the sum of squares of $N(0, 1)$ variables and hence has a chi-squared distribution with n degrees of freedom. Also, $\frac{(\overline{x} - \mu)^2}{\sigma^2/n} = \left(\frac{\overline{x} - \mu}{\sigma/\sqrt{n}}\right)^2$ is the square of a single $N(0, 1)$ variable and so has a chi-squared distribution with 1 degree of freedom. Therefore, using the moment generating function for the chi-squared random variable, we have

$$M\left[\frac{(n-1)s^2}{\sigma^2};t\right] \cdot (1-2t)^{-1/2} = (1-2t)^{-n/2}$$

or

$$M\left[\frac{(n-1)s^2}{\sigma^2};t\right] = (1-2t)^{-(n-1)/2},$$

indicating that $\frac{(n-1)s^2}{\sigma^2}$ has a χ^2_{n-1} distribution.

Since it can be shown that $E(\chi^2_{n-1}) = n-1$, it follows that $E\left[\frac{(n-1)s^2}{\sigma^2}\right] = n-1$, from which it follows that

$$E(s^2) = \sigma^2,$$

showing that s^2 is an unbiased estimator for σ^2.

It is also true that $\text{Var}(\chi^2_{n-1}) = 2(n-1)$, so

$$Var\left[\frac{(n-1)s^2}{\sigma^2}\right] = \frac{(n-1)^2}{\sigma^4}\text{Var}(s^2) = 2(n-1)$$

or

$$\text{Var}(s^2) = \frac{2\sigma^4}{n-1}.$$

This shows that the sample variance is very variable, a multiple of the fourth power of the standard deviation. The variability of the sample variance was noted in the early part of this section, and this result verifies that observation.

Also in the early part of this section, we considered the sampling distribution of the sample variance when we took samples of size 3 from the uniform distribution on the integers $\{1, 2, 3, \ldots, 20\}$. The graph in Figure 4.11 resembles that in Figure 4.12, which in reality is a chi-squared distribution with 2 degrees of freedom. Figure 4.11, while at first appearing to be somewhat chaotic, is in reality remarkable because the sampling is certainly not done from a normal distribution with mean 0 and variance 1. This indicates that the sampling distribution of the sample variance may be somewhat *robust*; that is, insensitive to deviations from the assumptions used to derive it.

Example 4.13.1

Samples of size 5 are drawn from a normal population with mean 20 and variance 300. A 95% confidence interval for the sample variance s^2 is found by using the χ^2_4 curve whose graph is shown in Figure 4.14. A table of values for some chi-squared distributions can be found in Appendix 3.

The normal distribution has a point of symmetry, which is often used in calculations. The chi-squared distribution, however, has no point of symmetry and so tables must be used to find both upper and lower significance points. We find, for example, that

$$P(0.2972011 \le \chi_4^2 \le 10.0255) = 0.95$$

so

$$P(0.2972011 \le \frac{4s^2}{300} \le 10.0255) = 0.95$$

or

$$P(22.290 \le s^2 \le 751.9125) = 0.95,$$

a very large range for s^2. It is approximately true that

$$P(4.721 \le s \le 27.42) = 0.95$$

by taking square roots in the confidence interval for s^2. The exact distribution for s could be found by finding the probability distribution of the square root of a χ^2 distribution, but the above interval is a good approximation. We will consider the exact distribution in section 4.17.

Other 95% confidence intervals are possible. Another example is

$$P(0.48442 \le \chi_4^2 \le 11.1433) = 0.95$$

which leads to the interval

$$P(36.3315 \le s^2 \le 835.7475) = 0.95.$$

There are many other possibilities that can be found easily with the aid of a computer algebra system because tables give very restricted choices for the chi-squared values needed. Note that the preceding two 95% confidence intervals have unequal lengths. This is due to the lack of symmetry of the chi-squared distribution. ◄

Exercises 4.13

1. A sample of five "six-hour" VCR tapes had actual lengths (in minutes) of 366, 339, 364, 356, and 379. Find a 95% confidence interval for σ^2, assuming that the lengths are $N(\mu, \sigma)$.

2. It is crucial that the variance of a measurement of the length of a piston rod be no greater than 1 square unit. A sample gave the following lengths (which have been coded for convenience): $-3, 6, -7, 8, 4, 0, 2, 12, -8$. Find a one-sided 99% confidence interval for the true variance of the length measurements.

3. Suppose $X \sim N(\mu, \sigma)$ where μ is known. Find a 95% two-sided confidence interval for σ^2 based on a random sample of size n.

4. Suppose that $\{X_1, X_2, \ldots, X_{2n}\}$ is a random sample from a distribution with $E[X] = 0$ and $Var[X] = \sigma^2$. Find k if

$$E[k \cdot \{(X_1 - X_2)^2 + (X_3 - X_4)^2 + (X_5 - X_6)^2 + \cdots + (X_{2n-1} - X_{2n})^2\}] = \sigma^2.$$

5. A random sample of n observations from a $N(\mu, \sigma)$ has $s^2 = 42$ and produced a two-sided 95% confidence interval for σ^2 of length 100. Find n.

6. Six readings on the amount of calcium in drinking water gave $s^2 = 0.0285$. Find a 90% confidence interval for σ^2.

7. A random sample of 12 observations is taken from a normal population with variance 100. Find the probability that the sample variance is between 50 and 240.

8. A random sample of 12 shearing pins is taken in a study of the Rockwell hardness of the head of a pin. Measurements of the Rockwell hardness were made for each of the 12, giving a sample average of 50 with a sample standard deviation of 2. Find a 90% confidence interval for the true variance of the Rockwell hardness. What assumptions must be made for your analysis to be correct?

9. A study of the fracture toughness of base plate of 10% nickel steel gave $s^2 = 5.04$ based on a sample of 22 observations. Assuming the sample comes from a normal population, construct a 99% confidence interval for σ^2, the true variance.

4.14 ▶ Hypothesis Tests and Confidence Intervals for a Single Mean

We are now prepared to return to the structure of hypothesis testing considered in Chapter 2 and show some applications of the preceding theory to the statistical analysis of data. Only the binomial distribution was available to us in Chapter 2. Now we have not only continuous distributions but the central limit theorem as well, which is the basis for much of our analysis. We begin with an example.

Example 4.14.1

A manufacturer of steel has measured the hardness of the steel produced and found that the hardness X has had in the past a mean value of 2200 lb with a known standard deviation of 4591.84 lb. It is desired to detect any significant shift in the mean value, and for this purpose samples of 25 pieces of the steel are taken periodically and the mean strength $\overline{X}$ of the sample is found. The manufacturer is willing to have the probability of a

Type I error no greater than 0.05. When should the manufacturer decide that the steel no longer has mean hardness 2200 lb?

In this case, because it is desired to detect deviations either greater than or less than 2200 lb, we take as null and alternative hypotheses

$$H_o : \mu = 2200$$
$$H_a : \mu \neq 2200.$$

The central limit theorem tells us that

$$\frac{\overline{X} - \mu}{\frac{\sigma}{\sqrt{n}}}$$

is approximately a $N(0, 1)$ variable.

Since the alternative hypothesis is *two-sided* – that is, it is comprised of the two *one-sided* hypotheses $\mu > 2200$ and $\mu < 2200$ – we take a two-sided rejection region, $\{\overline{X} > k\} \cup \{\overline{X} < h\}$. Since $\alpha = 0.05$, we find k and h such that

$$P[\overline{X} > k] = P[\overline{X} < h] = 0.025$$

so that

$$\frac{k - 2200}{\sqrt{\frac{21085000}{25}}} = 1.96 \text{ and } \frac{h - 2200}{\sqrt{\frac{21085000}{25}}} = -1.96.$$

These equations give $k = 4000$ and $h = 400$, approximately. So H_o is accepted if $400 \leq \overline{X} \leq 4000$.

The size of the Type II error, β, is a function of the specific alternative hypothesis. In this case if the alternative is $H_a : \mu = 2600$, for example, then

$$\beta = P\left[400 < \overline{X} < 4000 \mid \mu = 2600\right]$$

$$= P\left[\frac{400 - 2600}{\sqrt{\frac{21085000}{25}}} \leq z \leq \frac{4000 - 2600}{\sqrt{\frac{21085000}{25}}}\right]$$

$$= P[-2.39555 \leq z \leq 1.5244]$$

$$= 0.927998 \,,$$

so the test is not particularly sensitive to this alternative.   ◄

Confidence Intervals, σ Known

Suppose that $\overline{X}$ is the mean of a sample of n observations selected from a population with known standard deviation σ. By the central limit theorem for a given α we can find z so that

$$P\left(-z \leq \frac{\overline{X} - \mu}{\frac{\sigma}{\sqrt{n}}} \leq z\right) = 1 - \alpha.$$

These inequalities can in turn be solved for μ, producing $(1 - \alpha)\%$ confidence intervals

$$\overline{X} - z \cdot \frac{\sigma}{\sqrt{n}} \leq \mu \leq \overline{X} + z \cdot \frac{\sigma}{\sqrt{n}}.$$

Each of these confidence intervals has length $2 \cdot z \cdot \frac{\sigma}{\sqrt{n}}$.

Example 4.14.2

A sample of ten observations from a normal distribution with $\sigma = 6$ gave a sample mean $\overline{X} = 28.45$. A 90% confidence interval for the unknown mean μ of the population is

$$28.45 - 1.645 \cdot \frac{6}{\sqrt{10}} \leq \mu \leq 28.45 + 1.645 \cdot \frac{6}{\sqrt{10}}$$

or

$$25.3288 \leq \mu \leq 31.5712.$$

◀

Example 4.14.3

How large a sample must be selected from a normal distribution with standard deviation 12 in order to estimate μ to within 2 units with probability 0.95?

In this case, $\frac{1}{2}$ the length of a 95% confidence interval is 2. So,

$$z \cdot \frac{\sigma}{\sqrt{n}} = 1.96 \cdot \frac{12}{\sqrt{n}} = 2$$

so

$$n = \left(\frac{1.96 \cdot 12}{2}\right)^2.$$

Therefore a sample size of $n = 139$ is sufficient.

◀

Student's *t* Distribution

In the previous example, it was assumed that σ^2 was known. What if this parameter is also unknown? This of course is the most commonly encountered situation in practice; that is, neither μ nor σ is known when the sampling is done. Although we will not prove it here, the following theorem is useful if the sampling is done from a normal population.

THEOREM:

The ratio of a standard normal random variable and the square root of an independent chi-squared random variable divided by n, its degrees of freedom, follows Student's *t* distribution with n degrees of freedom. Symbolically,

$$\frac{N(0, 1)}{\sqrt{\chi_n^2/n}} = t_n.$$

A proof can be found in Hogg and Craig [14]. ∎

How is this of help here? We know that $\frac{\overline{X}-\mu}{\sigma/\sqrt{n}}$ is approximately normal by the central limit theorem, and we know from the previous section that if the sampling is done from a normal population, then $\frac{(n-1)s^2}{\sigma^2}$ is a chi-squared random variable with $n - 1$ degrees of freedom. So

$$\frac{\dfrac{\overline{X} - \mu}{\sigma/\sqrt{n}}}{\sqrt{\dfrac{(n - 1)s^2}{\sigma^2}/(n - 1)}} = \frac{\overline{X} - \mu}{s/\sqrt{n}} = t_{n-1}.$$

The sample provides all the information we need to calculate t. Student's *t* distribution (which was discovered by W. G. Gossett, who wrote using the pseudonym "Student") becomes normal-like as the sample size increases, but differs significantly from the normal distribution for small samples. Several *t* distributions are shown in Figure 4.15. A table of critical values for various *t* distributions can be found in Appendix 3.

Now tests of hypotheses can be carried out and confidence intervals can be calculated if the sampling is from a normal distribution with unknown variance, as the following example indicates.

Example 4.14.4

Tests on a ball bearing manufactured in a day's run in a plant show the following diameters (which have been coded for convenience): 8, 7, 3, 5, 9, 4, 10, 2, 6, 7. The sample gives $\overline{x} = 6.1$ and $s^2 = 203/30$. If we wish to

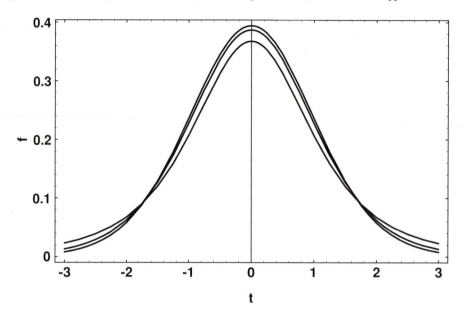

Figure 4.15   Student's t distributions for 3, 8, and 20 degrees of freedom.

test $H_o : \mu = 7$ against the alternative $H_a : \mu \neq 7$ with $\alpha = 0.05$, we find that

$$t_9 = \frac{6.1 - 7}{\sqrt{(203/30)/10}} = -1.09.$$

A table of t values can be found in Appendix 3. The critical values for t_9 are ± 2.26, so the hypothesis is accepted.

Confidence intervals for μ can also be constructed. Using the sample data, we have

$$P\left[-2.26 \leq \frac{\overline{X} - \mu}{s/\sqrt{n}} \leq 2.26\right] = 0.95$$

so

$$P\left[-2.26 \leq \frac{6.1 - \mu}{\sqrt{(203/30)/10}} \leq 2.26\right] = 0.95$$

which simplifies as

$$P[4.5697 \leq \mu \leq 7.63023] = 0.95.$$

The 95% confidence interval is also the acceptance region for a hypothesis tested at the 5% level. Recall that, when σ is known, the confidence intervals arising from separate samples all have the same length. This was shown previously. If, however, σ is unknown, then the confidence intervals will have varying widths as well as various central values. Some possible 95% confidence intervals are shown in Figure 4.16. ◄

p Values

We have always given the α or *significance* value when constructing a test of a hypothesis. These values of α have an arbitrary appearance, to say the least. Who is to say that this significance level should be 0.05 or 0.01, or some other value? How does one decide what value to choose?

These are often troublesome questions for an experimenter. The acceptance or rejection of a hypothesis is of course completely dependent upon the choice of the significance level. Another way to report the result of a test would be to report the *smallest value of α at which the test results would be significant*. This is called the *p value* for the test. We give some examples.

Example 4.14.5

In Example 4.14.4, we found that the sample of size 10 gave $\overline{x} = 6.1$ and $s^2 = 203/30$. This in turn produced $t_9 = -1.09$, and the hypothesis was accepted because α had been chosen as 5%.

However, we can use tables or a computer algebra system to find that

$$P(t_9 < -1.09) = 0.152018.$$

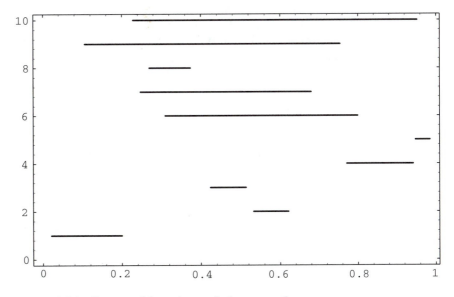

Figure 4.16 Some confidence intervals for μ, σ unknown.

This means that the observed value for t would be in the rejection region if half the α value were less than 0.152018.

Since the test in this case is two-sided, we report the p value as twice the above value, or 0.304036.

The person interpreting the test results can decide if this value suggests that the results are significant or not. Undoubtedly the decision in this case would be that the result is not significant, although this p value would be of value and interest in many studies.   ◄

Example 4.14.6

Suppose we revise Example 4.14.1 as follows. Assume the hypotheses are

$$H_o : \mu = 2200$$
$$H_a : \mu > 2200,$$

and that a sample of size 25 gave a sample mean of $\bar{x} = 3945$. Since we know in this case that $\sigma^2 = 21{,}085{,}000$, we find that

$$z = \frac{3945 - 2200}{\sqrt{\dfrac{21085000}{25}}} = 1.90011$$

and

$$P(Z > 1.90011) = 0.0287094.$$

so this is the p value for this test. If the significance level is greater than 0.0287094, then the result is significant; otherwise, it is not. Many computer statistical packages now report p values together with other test results.   ◄

Exercises 4.14

1. Test runs with an experimental engine showed it operated, respectively, for 24, 28, 21, 23, 32, and 22 minutes with one gallon of fuel.

   a. Is this evidence at the 1% significance level that $H_o : \mu = 29$ should be accepted against $H_a : \mu < 29$?

   b. Find the p value for the test.

2. Machines used in producing a particular brand of yarn are given periodic checks to help insure stable quality. A machine has been set so it is expected that strands of yarn it produces will have breaking strength $\mu = 19.50$ oz, with a standard deviation of 1.80 oz. A random sample of 12 pieces of yarn has a mean of 18.46 oz. Assume that the standard deviation remains constant over a fairly wide range of values for μ.

a. Test $H_o : \mu = 19.50$ against $H_a : \mu \neq 19.50$ at the 5% significance level. Find the p value for the test.

b. Now suppose that σ is also unknown and that the sample standard deviation is 1.80. Test the hypothesis in part **a** again. Are any additional assumptions needed?

c. Under the conditions in part **a**, find β for the alternative $H_a : \mu = 19.70$.

3. "One-quarter-inch" rivets are produced by a machine that is checked periodically by taking a random sample of 10 rivets and measuring their diameters. It is thought that the wear-off factor in the machine will eventually cause it to produce rivets with diameters that are less than 1/4 in. Assume that the variance of the diameters is known to be $(0.0015)^2$.

a. Describe the critical region, in terms of $\overline{X}$, for a test at the 1% level of significance for $H_o : \mu = 0.25$ against the alternative $H_a : \mu < 0.25$.

b. What is the power of the test at $\mu = 0.2490$?

c. Now suppose we wish to test $H_o : \mu = 0.25$ against $H_a : \mu = 0.2490$ with $\alpha = .01$ and so that the power of the test is 0.99. What sample size is necessary to achieve this?

4. A manufacturer of light bulbs claims that the life of the bulbs is normally distributed with mean 800 hr and standard deviation 40 hr. Before buying a large lot, a buyer tests 30 of the bulbs and finds an average life of 789 hr.

a. Test the hypothesis $H_o : \mu = 800$ against the alternative $H_a : \mu < 800$ using a test of size 5%.

b. Find the probability of a Type II error for the alternative $H_a : \mu = 790$.

c. Find the p value for the test.

5. A sample of size 16 from a distribution whose variance is known to be 900 is used to test $H_o : \mu = 350$ against the alternative $H_a : \mu > 350$, using the critical region $\overline{X} > 365$.

a. What is α for this test?

b. Find β for the alternative $H_a : \mu = 372.50$.

6. A manufacturer of sports equipment has developed a new synthetic fishing line that he claims has a mean breaking strength of 8 kg. To test $H_o : \mu = 8$ against the alternative $H_a : \mu \neq 8$, a sample of 50 lines is tested; the sample has a mean breaking strength of 7.8 kg.

a. If σ is assumed to be 0.5 kg and $\alpha = 5\%$, is the manufacturer's claim supported by the sample?

b. Find β for the above test for the alternative $H_a : \mu = 7.7$.

c. Find the p value for the test.

7. For a certain species of fish, a sample of measurements for DDT is 5, 10, 8, 7, 4, 9, and 13 parts per million.

   a. Find a range of values of μ_o for which the hypothesis $H_a : \mu = \mu_o$ would be accepted at the 5% level.

   b. Find a 95% confidence interval for σ^2, the true variance of the measurements.

8. The time to repair breakdowns of an office copying machine is claimed by the manufacturer to have a mean of 93 minutes. To test this claim, 23 breakdowns of a model were observed, resulting in a mean repair time of 98.8 min and a standard deviation of 26.6 min.

   a. Test $H_o : \mu = 93$ against the alternative $H_a : \mu > 93$ with $\alpha = 5\%$ and state your conclusions.

   b. Supposing that $\sigma^2 = 625$, find β for the alternative $H_a : \mu = 95$.

   c. Find the p value for the test.

9. A firm produces metal wheels. The mean diameter of these wheels should be 4 inches. Because of other factors as well as chance variation, the diameters of the wheels vary with standard deviation 0.05 in. A test is conducted on 50 randomly selected wheels.

   a. Find a test with $\alpha = 0.01$ for testing $H_o : \mu = 4$ against the alternative $H_a : \mu \neq 4$.

   b. If the sample average is 3.97, what decision is made?

   c. Calculate β for the alternative $H_a : \mu = 3.99$.

10. A tensile test was performed to determine the strength of a new adhesive for a metal-to-glass assembly. The data are 16, 14, 19, 18, 19, 20, 15, 18, 17, 18. Test $H_o : \mu = 19$ against the alternative $H_a : \mu < 19$ with $\alpha = 5\%$,

    a. if σ^2 is known to be 2.

    b. if σ^2 is unknown.

11. The activation times for an automatic sprinkler system are a subject of study by the system's manufacturer. A sample of activation times is 27, 41, 22, 27, 23, 35, 30, 33, 24, 27, 28, 22, and 24 seconds. The design of the system calls for its activation in, at most, 25 s. Do the data contradict the validity of this design specification?

12. The breaking strengths of cables produced by a manufacturer have mean 1800 lb. It is claimed that a new manufacturing process will increase the mean breaking strength of the cables. To test this hypothesis, a sample of 30 cables manufactured using the new process is tested, giving $\overline{X} = 1850$ and $s = 100$.

    a. If $\alpha = 0.05$, what conclusion can be drawn regarding the new process?

    b. Find the p value for the test.

13. A sample of 80 observations is taken from a population with known standard deviation 56 to test $H_o : \mu \leq 300$ against the alternative $H_a : \mu > 300$, giving $\overline{X} = 310$.

    a. Find the minimum value of α so that H_o would be rejected by the sample.

b. Assuming that the critical region is $\overline{X} > 310$, find β for the alternative $\mu = 315$.

14. A contractor must have cement with a compressive strength of at least 5000 kilograms per square centimeter (kg/cm^2). He knows that the standard deviation of the compressive strengths is 120. In order to test $H_o : \mu \geq 5000$ against the alternative $H_a : \mu < 5000$, a random sample of 4 pieces of cement is tested.

   a. If the average compressive strength of the sample is 4870 kg/cm^2, is the concrete acceptable? Use $\alpha = 0.01$.

   b. The contractor must be 95% certain that the compressive strength is not less than 4800 kg/cm^2. How large a sample should be taken to insure this?

15. The assembly time in a plant is a normal random variable with mean 18.5 seconds and standard deviation 2.4 seconds.

   a. A random sample of 10 assembly times gave $\overline{X} = 19.6$. Is this evidence that $H_o : \mu = 18.5$ should be rejected in favor of the alternative $H_a : \mu > 18.5$ if $\alpha = 5\%$?

   b. Find the probability that H_o is accepted if $\mu = 19$.

   c. It is very important that the assembly time not exceed 20 seconds. How large a sample is necessary to reject $H_o : \mu = 18.5$ with probability 0.95 if $\mu = 20$?

16. A lot of rolls of paper is acceptable for making bags for grocery stores if its true mean breaking strength is not less than 40 lb. It is known from past experience that $\sigma = 2.5$ lb. A sample of 20 is chosen.

   a. Find the critical region for testing the hypothesis $H_o : \mu = 40$ against the alternative $H_a : \mu < 40$ at the 5% level of significance.

   b. Find the probability of accepting H_o if in fact $\mu = 40.5$ lb.

   c. If σ were unknown and a sample of 20 gave $\overline{X} = 39$ lb and $s = 2.4$ lb, would H_o be accepted with $\alpha = 5\%$?

17. The drying time of a particular brand and type of paint is known to be normally distributed with $\mu = 75$ min and $\sigma = 9.4$ min. In an attempt to improve the drying time, a new additive has been developed. Use of the additive in 100 test samples of the paint gave an average drying time of 68.5 min. We wish to test $H_o : \mu = 75$ against the alternative $H_a : \mu < 75$.

   a. Find the critical region if $\alpha = 5\%$.

   b. Does the experimental evidence indicate that the additive improves drying time?

   c. What is the probability that H_o will be rejected if in fact $\mu = 72$ min?

   d. Find the p value for the test.

18. The breaking strength of a fiber used in manufacturing cloth is required to be not less than 160 pounds per square inch (psi). Past evidence indicates that $\sigma = 3$ psi.

A random sample of 4 specimens is tested and the average breaking strength is found to be 158 psi.

a. Test $H_o : \mu = 160$ against a suitable alternative using $\alpha = 5\%$.
b. Find β for the alternative $\mu = 157$.

19. An engineer is investigating the wear characteristics of a particular type of radial automobile tire used by the company fleet of cars. A random sample of 16 tires is selected and each tire used until the wear bars appear. The sample gave $\bar{x} = 41,116$ miles and $s^2 = 1,814,786$ miles^2.

a. Find a so that $P(\mu > a) = 0.95$.
b. Find a 90% confidence interval for σ^2.
c. Answer part **a** assuming that the sample size is 43 with $\bar{x}$ and s^2 as before.

20. The diameter of steel rods produced by a subcontractor is known to have standard deviation 2 cm, and, in order to meet specifications, must have $\mu = 12$.

a. If the mean of a sample of size 5 is 13.3, is this sufficient to reject $H_o : \mu = 12$ in favor of the alternative $H_a : \mu > 12$? Use $\alpha = 0.05$.
b. The manufacturer wants to be fairly certain that H_o is rejected if $\mu = 13$. How large a sample should be taken to make this probability 0.92?

4.15 ▶ Hypothesis Tests on Two Samples

A basic scientific problem is that of comparing two samples, possibly one from a control group and the other an experimental group. The investigator may want to decide whether or not the two populations from which the samples are drawn have the same mean value, or interest may center on the equality of the true variances of the populations. We begin with a comparison of population means.

Test on Two Means

Example 4.15.1

Suppose an investigator is comparing two methods of teaching students to use a popular computer algebra program. One group (X) is taught by the conventional lecture-demonstration method while the second (Y) group is divided into small groups and uses cooperative learning. After some time of instruction, the groups are given the same examination with the following results:

$$n_x = 13, \ \bar{x} = 77, \ s_x^2 = 193.7,$$
$$\text{and}$$
$$n_y = 9, \ \bar{y} = 84, \ s_y^2 = 309.4.$$

We wish to test the hypothesis

$$H_0 : \mu_x = \mu_y$$

against

$$H_a : \mu_x < \mu_y.$$

Assume that the sampling is from normal distributions. We know that

$$E(\overline{X} - \overline{Y}) = \mu_x - \mu_y$$

and that

$$\mathrm{Var}(\overline{X} - \overline{Y}) = \frac{\sigma_x^2}{n_x} + \frac{\sigma_y^2}{n_y},$$

so, from the central limit theorem,

$$z = \frac{(\overline{X} - \overline{Y}) - (\mu_x - \mu_y)}{\sqrt{\dfrac{\sigma_x^2}{n_x} + \dfrac{\sigma_y^2}{n_y}}}$$

is a $N(0, 1)$ variable.

Now z can be used to test hypotheses or to construct confidence intervals if the variances are known. Consider for the moment that we know that the populations have equal variances, say $\sigma^2 = 289$. Then

$$z = \frac{(77 - 84) - 0}{\sqrt{\dfrac{289}{13} + \dfrac{289}{9}}} = -0.9496.$$

If the test had been at the 5% level, then the null hypothesis would be accepted because $z > -1.645$.

We could also use z to construct a confidence interval. Here a one-sided interval is appropriate because of H_a. We have

$$P\left[(\overline{X} - \overline{Y}) - 1.645 \sqrt{\frac{\sigma_x^2}{n_x} + \frac{\sigma_y^2}{n_y}} \leq \mu_x - \mu_y \right] = 0.95,$$

which in this case becomes the interval greater than -19.126. Since 0 is in this interval, the hypothesis of equal means is accepted. ◄

Example 4.15.2

A situation more common than that in Example 4.15.1 occurs when the population variances are unknown. There are then two possibilities: they are equal or they are not. We consider first the case where the variances are unknown, but they are known to be equal. Denote the common value for the variances by σ^2. The variable

$$z = \frac{(\overline{X} - \overline{Y}) - (\mu_x - \mu_y)}{\sqrt{\dfrac{\sigma_x^2}{n_x} + \dfrac{\sigma_y^2}{n_y}}}$$

is a $N(0, 1)$ variable. Now $\frac{(n_x-1)s_x^2}{\sigma^2} + \frac{(n_y-1)s_y^2}{\sigma^2}$ is a χ^2 variable with $(n_x - 1) + (n_y - 1) = n_x + n_y - 2$ degrees of freedom because each of the summands is a chi-squared variable. Since a t variable is the ratio of a $N(0, 1)$ variable to the square root of a chi-squared variable divided by its number of degrees of freedom, it follows that

$$t_{n_x+n_y-2} = \frac{\dfrac{(\overline{X} - \overline{Y}) - (\mu_x - \mu_y)}{\sqrt{\dfrac{\sigma^2}{n_x} + \dfrac{\sigma^2}{n_y}}}}{\sqrt{\dfrac{\dfrac{(n_x - 1)s_x^2}{\sigma^2} + \dfrac{(n_y - 1)s_y^2}{\sigma^2}}{n_x + n_y - 2}}}.$$

This can be simplified to

$$t_{n_x+n_y-2} = \frac{(\overline{X} - \overline{Y}) - (\mu_x - \mu_y)}{s_p\sqrt{\dfrac{1}{n_x} + \dfrac{1}{n_y}}}$$

where

$$s_p^2 = \frac{(n_x - 1)s_x^2 + (n_y - 1)s_y^2}{n_x + n_y - 2}.$$

s_p^2 is called the *pooled variance*.

Using the data in Example 4.15.1, we find that

$$s_p^2 = \frac{12(193.7) + 8(309.4)}{13 + 9 - 2} = 239.98$$

and

$$t_{20} = \frac{(77 - 84) - 0}{\sqrt{239.98(\frac{1}{13} + \frac{1}{9})}} = -1.04.$$

Since the one-sided test rejects H_0 if $t_{20} < -1.725$, the hypothesis is accepted if $\alpha = 0.05$. ◀

Example 4.15.3

Now we consider the case where the population variances are unknown and cannot be assumed to be equal. (Later in this chapter we will show how that hypothesis may be tested also.) Unfortunately, there is no exact solution to this problem, known in the statistical literature as the Behrens-Fisher Problem. Several approximate solutions are known; we give one here due to Welch [27].

Welch's approximation is as follows: The variable

$$T = \frac{(\overline{X} - \overline{Y}) - (\mu_x - \mu_y)}{\sqrt{\frac{s_x^2}{n_x} + \frac{s_y^2}{n_y}}}$$

is approximately a t variable with ν degrees of freedom where

$$\nu = \frac{\left(\frac{s_x^2}{n_x} + \frac{s_y^2}{n_y}\right)^2}{\frac{\left(\frac{s_x^2}{n_x}\right)^2}{n_x - 1} + \frac{\left(\frac{s_y^2}{n_y}\right)^2}{n_y - 1}}.$$

Using the data in the previous examples, we find that $\nu = 14.6081$, so we must use a t variable with 14 degrees of freedom. This gives

$$T_{14} = -0.997178,$$

a result quite comparable to previous results. The critical t value is -1.761. The Welch approximation will make a very significant difference if the population variances are quite disparate. ◀

4.15.1   Tests on Two Variances

It is essential to determine whether or not the population variances are equal before testing the equality of population means. It is possible to test this using two samples from the populations.

If χ_a^2 and χ_b^2 are independent chi-squared variables, then the random variable

$$\frac{\dfrac{\chi_a^2}{a}}{\dfrac{\chi_b^2}{b}} = F(a, b)$$

where $F(a, b)$ denotes the F random variable with a and b degrees of freedom respectively. A proof of this fact will not be given here. The reader is referred to Hogg and Craig [14] for a proof. A table of some critical values of the F distribution can be found in Appendix 3.

The probability density function for $F(a, b)$ is

$$f(x) = \frac{\Gamma\left(\dfrac{a+b}{2}\right)}{\Gamma\left(\dfrac{a}{2}\right) \cdot \Gamma\left(\dfrac{b}{2}\right)} \cdot a^{a/2} \cdot b^{b/2} \cdot (ax + b)^{-(a+b)/2} \cdot x^{(a/2)-1}, \ x \geq 0.$$

The F variable has two numbers of degrees of freedom; one is associated with the numerator and the other with the denominator. Due to the definition of F, it is clear that

$$\frac{1}{F(a, b)} = F(b, a).$$

So the reciprocal of an F variable is an F variable with the numbers of degrees of freedom interchanged.

Several F curves are shown in Figure 4.17.

The F distribution can be used in testing the equality of variances in the following way. If the sampling is from normal populations, then

$$\frac{(n_x - 1)s_x^2}{\sigma_x^2} \text{ and } \frac{(n_y - 1)s_y^2}{\sigma_y^2}$$

are independent chi-squared variables. It follows that

$$\frac{\dfrac{(n_x - 1)s_x^2}{\sigma_x^2(n_x - 1)}}{\dfrac{(n_y - 1)s_y^2}{\sigma_y^2(n_y - 1)}}$$

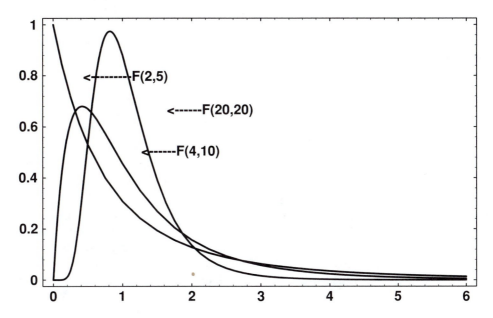

Figure 4.17   Some F distributions.

is the ratio of two independent chi-squared random variables each divided by its number of degrees of freedom. It follows that this variable, simplified as

$$\frac{\dfrac{s_x^2}{\sigma_x^2}}{\dfrac{s_y^2}{\sigma_y^2}} = F(n_x - 1, n_y - 1).$$

Now consider the hypotheses

$$H_0 : \sigma_x^2 = \sigma_y^2$$
$$H_a : \sigma_x^2 \neq \sigma_y^2.$$

If the null hypothesis is true, then the F variable becomes

$$F(n_x - 1, n_y - 1) = \frac{s_x^2}{s_y^2}.$$

This is used as the test statistic with a two-tailed critical region.

Example 4.15.4

As an example, consider the data used in the previous section, where

$$n_x = 13, \quad s_x^2 = 193.7$$

and

$$n_y = 9, \quad s_y^2 = 309.4.$$

Here $F(12, 8) = \frac{193.7}{309.4} = 0.626$. The critical values, choosing $\alpha = 0.05$, are 4.1995 and 0.2002, so the null hypothesis is accepted.

One-sided tests can also be performed with the F statistic; it is common not to worry if a variance is too small, but in many instances, care must be taken that the variance has not become too large. In that case, a large variation may result in a manufacturing process producing too great a percentage of product that does not meet specifications. We will discuss this further in section 4.17. ◄

Exercises 4.15

1. To test the hypothesis that the resistance of wire can be reduced by at least 0.050 ohms by alloying, samples of 12 for each type of wire gave the following results:

	Mean	Standard Deviation
Alloyed wire	0.083	0.003
Standard wire	0.136	0.002

   a. Test $H_o : \sigma_1^2 = \sigma_2^2$ using $\alpha = 0.05$.
   b. Do the data substantiate the claim?

2. Two analysts took repeated readings on the hardness of city water with the following results:

Analyst A	Analyst B
x	y
0.46	0.82
0.62	0.61
0.37	0.89
0.40	0.51
0.44	0.33
0.58	0.48
0.48	0.23
0.53	0.25
	0.67
	0.88

a. Test $H_o : \mu_x = 0.55$ against the alternative $H_1 : \mu_x \neq 0.55$ using $\alpha = 0.05$.

b. Test the hypothesis in part **a** again, assuming now that $\sigma_x^2 = 0.0081$.

c. Test $H_o : \mu_x = \mu_y$ against the alternative $H_a : \mu_x < \mu_y$ with $\alpha = 5\%$.

3. Over a long period of time, 10 patients selected at random are each given 2 different treatments for a disease. The results of standard tests are as follows:

Patient	Treatment 1	Treatment 2
1	47	52
2	38	35
3	50	52
4	33	35
5	47	46
6	23	27
7	40	45
8	42	41
9	15	17
10	36	41

Test $H_o : \mu_1 = \mu_2$ against $H_a : \mu_1 < \mu_2$ at the 1% level of significance, assuming that the population variances are equal.

4. An experiment compared two different processes for producing steel. The measurements represent the thickness of the steel. The samples gave

$$n_x = 8 \quad \bar{x} = 6.701 \quad s_x = 0.108$$
$$n_y = 6 \quad \bar{y} = 6.841 \quad s_y = 0.155$$

a. Using $\alpha = 0.02$, test $H_o : \sigma_x^2 = \sigma_y^2$ against the alternative $H_a : \sigma_x^2 \neq \sigma_y^2$.

b. Now test $H_o : \mu_x = \mu_y$ against the alternative $H_o : \mu_x \neq \mu_y$ using a 5% test.

5. Samples from normal populations gave

$$n_x = 6 \quad \bar{x} = 22.6 \quad s_x^2 = 102.4$$
$$n_y = 8 \quad \bar{y} = 31.9 \quad s_y^2 = 89.6$$

Find a 98% confidence interval for σ_x^2 / σ_y^2.

6. In comparing times to failure (in hours) of 2 different types of light bulbs, 2 samples gave

$$n_x = 13 \quad \bar{x} = 984 \quad s_x^2 = 8742$$
$$n_y = 15 \quad \bar{y} = 1121 \quad s_y^2 = 9411$$

Find a 95% confidence interval for the difference of the true population means, $\mu_x - \mu_y$,

a. assuming $\sigma_x^2 = \sigma_y^2$.

   b. assuming $\sigma_x^2 = 9000$ and $\sigma_y^2 = 9500$.

7. In a batch chemical process, 2 catalysts are being compared for their effect on the output of the process reaction. A sample of 11 batches was prepared using catalyst 1, and a sample of 9 batches was prepared using catalyst 2. The sample results are

$$n_1 = 11 \quad \overline{x_1} = 85 \quad s_1^2 = 16$$
$$n_2 = 9 \quad \overline{x_2} = 81 \quad s_2^2 = 25$$

Assuming that the true variances are equal, find a 95% confidence interval for the differences between the means, $\mu_1 - \mu_2$.

8. To determine yield strengths, a study of 10 pieces of cold-rolled steel (X) gave a sample mean of 29.8 kilograms per square inch (kg/in.^{2}) and a sample variance of 4.2 kg/in.^{2}. A second sample of 13 pieces of galvanized steel (Y) gave a sample mean of 34.7 kg/in.^{2} and a sample variance of 4.9 kg/in.^{2}.

   a. Assuming the true variances are equal, find a 95% confidence interval for the difference between the true strengths, $\mu_x - \mu_y$.

   b. Repeat part **a** assuming the true variances are $\sigma_x^2 = 4$ and $\sigma_y^2 = 5$.

9. An experiment is conducted to compare the crash resistance of two different types of automobile bumpers. Type A bumpers were mounted on 12 cars and type B bumpers were mounted on 9 cars. The cars were driven into a concrete wall at 10 mph and the resulting damage (in $ to repair) was assessed. The results were

$$\overline{A} = 235 \quad s_A^2 = 421$$
$$\overline{B} = 286 \quad s_B^2 = 511$$

   a. Would the hypothesis $\sigma_A^2 = \sigma_B^2$, when tested against $\sigma_A^2 \neq \sigma_B^2$ with $\alpha = 2\%$, be accepted or rejected?

   b. Find a 98% confidence interval for σ_A^2/σ_B^2.

   c. Test $H_o : \mu_A = \mu_B$ against the alternative $H_a : \mu_A < \mu_B$ in a test of size 1%.

10. The vending machines in the student lounge and in the cafeteria should dispense the same amount of coffee. However, some students believe that the mean amount of coffee dispensed in the lounge (L) is less than that dispensed in the cafeteria (C). The following summary statistics were obtained from samples from each machine.

$$n_L = 12 \quad \overline{x_L} = 10.1 \quad s_L = 0.8$$
$$n_C = 10 \quad \overline{x_C} = 9.8 \quad s_C = 1.4$$

Is there statistical evidence to support the students' claim? Assume that the amounts dispensed are approximately normal and use a test of size 5%.

11. A recent study of accident victims in a hospital gave the following results:

	n	Mean	Standard Deviation
Seat Belts	15	565	220
No Seat Belts	12	1200	540

The data indicate the cost of hospitalization.

a. Assuming that σ_{SB} (the true standard deviation for seat belt wearers) is 220 and σ_{NSB} (the true standard deviation for non-seat belt wearers) is 540, find a 95% confidence interval for $\mu_{SB} - \mu_{NSB}$.

b. Answer part **a** assuming that the true standard deviations are unknown but equal.

c. Is it tenable to believe that the population variances are equal? State the smallest p value at which the data would reject the hypothesis of equal variances (against the alternative of unequal variances).

12. The following data represent the running times of films produced by two motion picture companies:

Company	Time (minutes)						
A	102	86	98	109	92		
B	81	165	97	134	92	87	114.

a. Test $H_o : \sigma_A^2 = \sigma_B^2$ against the alternative $H_a : \sigma_A^2 \neq \sigma_B^2$ with $\alpha = 2\%$.

b. Test $H_o : \mu_A = \mu_B - 10$ against the alternative $\mu_A < \mu_B - 10$ with $\alpha = 1\%$.

13. Wire cable is manufactured by 2 processes. It is desired to determine if the process affects the mean breaking strength of the cable. Laboratory tests are performed by putting samples under tension and recording the load required to break the cable. Following are the sample data:

Sample	Size	Mean	Variance
X	6	8.2	2.0
Y	7	11.2	4.0

a. Test $H_o : \sigma_x^2 = \sigma_y^2$ against the alternative $H_a : \sigma_x^2 \neq \sigma_y^2$ with $\alpha = 2\%$.

b. Now test $H_o : \mu_x = \mu_y$ against the alternative $H_a : \mu_x \neq \mu_y$ using a 5% test.

14. Five samples of a ferrous-type substance are to be used to determine if there is a difference between a laboratory chemical analysis and an X-ray fluorescence analysis of iron content. Each sample was split into 2 sub-samples and the 2 types of analysis were applied with the following data, which represent percent yield:

	Sample				
Analysis	1	2	3	4	5
X-Ray	11.0	2.0	8.3	3.1	2.4
Chemical	11.2	1.9	8.5	3.3	2.4.

Assuming the population of measurements to be normal, test whether or not the 2 methods of analysis give, on average, the same result. Use $\alpha = 5\%$.

15. An automobile designer suggests that painting a racing car reduces its top speed. He selects 6 cars and tests them with and without paint. The results are:

	Top Speed (mph)	
Car	Unpainted	Painted
1	189	186
2	186	185
3	183	179
4	188	184
5	185	183
6	188	186

Use the data to decide whether or not painting the cars reduces the top speed. Use $\alpha = 5\%$.

16. The golf scores of 2 competitors, A and B, are recorded over a period of 10 different days on which weather conditions varied widely. Golfer A claims that his game is better than golfer B's. Do the data support this claim?

Day	A	B
1	87	89
2	86	85
3	79	83
4	82	87
5	78	76
6	87	90
7	84	85
8	81	78
9	83	85
10	81	84

 17. A wire manufacturer alters the production process hoping to increase the resistance of the wire. Below are the results of samples taken from the old process and the new process. Has the resistance of the wire increased?

New Process	Old Process
0.140	0.135
0.138	0.140
0.143	0.136
0.142	0.142
0.144	0.138
0.137	0.140

18. Two randomly selected groups of industrial trainees are taught a new assembly line operation by 2 different methods. Measurements were made on the time to complete the operation, with the following results:

Group	Size	Mean	Standard Deviation
I	10	60.43	20.2
II	10	31.23	26.8

a. Assuming normality, test $H_o : \sigma_I^2 = \sigma_{II}^2$ against the alternative $H_a : \sigma_I^2 \neq \sigma_{II}^2$ at the 5% level of significance.

b. Test $H_o : \mu_I = \mu_{II}$ against the alternative $H_a : \mu_I \neq \mu_{II}$ at the 5% level of significance.

c. Determine the values of k for which $H_o : \mu_I = \mu_{II} + k$ would be accepted at the 5% level of significance when tested against the alternative $H_a : \mu_I \neq \mu_{II} + k$.

19. Two samples are drawn from normal populations, each with variance 100. Due to sampling costs, it is possible to select $2n$ items from population A, but only n items from population B. In testing $H_o : \mu_A = \mu_B$ against the alternative $H_a : \mu_A = \mu_B + 3$ it is desired to have $\alpha = \beta = 0.10$. Find n so that this is approximately so and then discuss the implications of rounding n to an integer.

20. Suppose samples of sizes n_X and n_Y are available from normal populations whose variances and means are unknown. It is suspected that $\sigma_X^2 = 2\sigma_Y^2$.

a. Explain, by establishing the distribution of $\frac{s_X^2}{2s_Y^2}$, how $\frac{s_X^2}{2s_Y^2}$ can be used to test the hypothesis that $\sigma_X^2 = 2\sigma_Y^2$.

b. Assuming that the hypothesis in part **a** is accepted, show that a test of $H_o : \mu_X = \mu_Y$ can be based on

$$r = \frac{(\overline{X} - \overline{Y}) - (\mu_X - \mu_Y)}{s_w \sqrt{\dfrac{2}{n_X} + \dfrac{1}{n_Y}}}$$

where

$$s_w^2 = \frac{(n_X - 1)s_X^2 + 2(n_Y - 1)s_Y^2}{2(n_X + n_Y - 2)}$$

by establishing the distributions of r and s_w^2.

4.16 ▶ Least Squares Linear Regression

The estimation of unknown parameters was considered earlier in this chapter; we return to that problem here and introduce a new principle of estimation, that of *least squares*. This is a commonly used principle when a straight line or other curve must be fitted to a set of data, and when one wants the "best"-fitting straight line or curve to the experimental data.

Example 4.16.1

Suppose we are given the data set $\{x_1, x_2, \ldots, x_n\}$. We want to find a number a such that

$$S = \sum_{i=1}^{n}(x_i - a)^2$$

is as small as possible.

That is, we want to minimize the sum of the squared deviations from a. Such estimates, when they exist, are known as *least squares estimates*. In this case letting the derivative of S with respect to a equal 0 gives

$$\frac{dS}{da} = -2\sum_{i=1}^{n}(x_i - a) = 0$$

with the solution

$$\widehat{a} = \frac{\sum_{i=1}^{n} x_i}{n} = \overline{x},$$

the mean of the data set.

So the sum of the squared deviations, $\sum_{i=1}^{n}(x_i - a)^2$, is minimized when $\widehat{a} = \overline{x}$.   ◀

Example 4.16.2

A researcher suspects that the achievement score on a standard mathematics examination, Y, for a group of students is a linear function of the student's IQ score, X. In order to investigate this hypothesis, data are collected and a *model* of the situation is presumed. In this case the model is composed of a *linear* part, $a + bx_i$, reflecting the researcher's hypothesis, and a *random* part, ε_i, reflecting the fact that the relationship between Y and X may be subject to other factors that are not accounted for in the experiment. The random part ε_i is in fact an observation of a random variable. Often this is a normal random variable, as we will see. The model chosen is

$$y_i = a + bx_i + \varepsilon_i, i = 1, 2, \ldots, n.$$

Here y_i is the ith observation of Y, and x_i is the ith observation of X.

There are two fundamental problems here. One is the estimation from the data of the unknowns a and b; the other is, given a and b, does the line fit the data well or not?

To answer the first question, we will use the principle of least squares to estimate the parameters a and b. This principle chooses values of a and b that minimize a sum of squares, namely,

$$S = \sum_{i=1}^{n} \varepsilon_i^2 = \sum_{i=1}^{n}(y_i - a - bx_i)^2.$$

We take the partial derivatives of S with respect to a and b and equate each to 0:

$$\frac{\partial S}{\partial a} = 2\sum_{i=1}^{n}(y_i - a - bx_i)(-1) = 0$$

$$\frac{\partial S}{\partial b} = 2\sum_{i=1}^{n}(y_i - a - bx_i)(-x_i) = 0.$$

Summing and simplifying gives

$$\sum_{i=1}^{n} y_i = n\widehat{a} + \widehat{b}\sum_{i=1}^{n} x_i$$

and

$$\sum_{i=1}^{n} x_i y_i = \widehat{a}\sum_{i=1}^{n} x_i + \widehat{b}\sum_{i=1}^{n} x_i^2 \qquad (4.2)$$

Equations 4.2 are called *least squares equations*. Their simultaneous solution is

$$\overline{y} = \widehat{a} + \widehat{b}\overline{x}$$

and

$$\widehat{b} = \frac{n \sum\limits_{i=1}^{n} x_i y_i - \sum\limits_{i=1}^{n} x_i \sum\limits_{i=1}^{n} y_i}{n \sum\limits_{i=1}^{n} x_i^2 - (\sum\limits_{i=1}^{n} x_i)^2} = \frac{\sum\limits_{i=1}^{n} (x_i - \overline{x})(y_i - \overline{y})}{\sum\limits_{i=1}^{n} (x_i - \overline{x})^2}.$$

Usually $\widehat{b}$ is found and then $\widehat{a}$ is found from the equation $\overline{y} = \widehat{a} + \widehat{b}\overline{x}$. Suppose now that the data collected are as follows:

Math. score	92	86	104	109	75	100	91	110	128
IQ	104	91	123	102	86	99	92	114	99

A scatter plot of the data points is shown in Figure 4.18. The data appear to be somewhat linear, with considerable variation.

Substituting in Formula 4.2 we find that the least squares estimates for a and b are

$$\widehat{a} = 63.0792 \text{ and } \widehat{b} = 0.38244$$

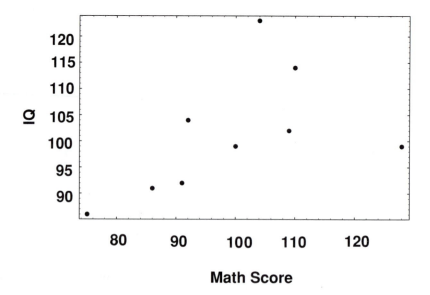

Figure 4.18   Scatter plot of data.

so that the least squares line, which is called the *regression of Y on X,* is

$$\widehat{y}_i = 63.0792 + 0.38244x_i.$$

Figure 4.19 shows this line plotted with the data points.

The line does not appear to predict the Y values very well, so we consider whether or not the line fits the data satisfactorily. We begin with the values the line predicts for the X values in the data set. Following is a table of the data points, the predicted values, and the *residuals* (the observed Y values minus the predicted Y values).

Math. Score (X)	92	86	104	109	75	100	91	110	128
IQ (Y)	104	91	123	102	86	99	92	114	99
Prediction	98.26	95.97	102.85	104.77	91.77	101.32	97.88	105.15	112.03
Residual (error)	5.74	−4.97	20.15	−2.77	−5.76	−2.32	−5.88	8.85	−13.03

The absolute size of these residuals does not tell much except, as we previously noted, some of the residuals are quite large. To show a specific test of the hypothesis that the straight line fits the data well, consider the *total sum of squares of the residuals:*

$$\sum_{i=1}^{n}(y_i - \widehat{y}_i)^2 \text{ where } \widehat{y}_i = \widehat{a} + \widehat{b}x_i.$$

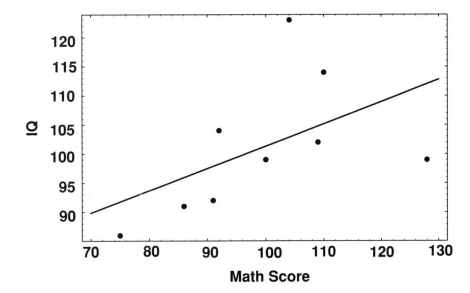

Figure 4.19   Regression line and data points.

Now we show a remarkable identity by adding and subtracting $\overline{y}$:

$$\sum_{i=1}^{n}(y_i - \hat{y}_i)^2 = \sum_{i=1}^{n}[(y_i - \overline{y}) - (\hat{y}_i - \overline{y})]^2.$$

$$\sum_{i=1}^{n}(y_i - \hat{y}_i)^2 = \sum_{i=1}^{n}(y_i - \overline{y})^2 - 2\sum_{i=1}^{n}(y_i - \overline{y})(\hat{y}_i - \overline{y}) + \sum_{i=1}^{n}(\hat{y}_i - \overline{y})^2.$$

Using Formula 4.2, the last two terms can be combined and we have

$$\sum_{i=1}^{n}(y_i - \hat{y}_i)^2 = \sum_{i=1}^{n}(y_i - \overline{y})^2 - \sum_{i=1}^{n}(\hat{y}_i - \overline{y})^2$$

or   (4.3)

$$\sum_{i=1}^{n}(y_i - \overline{y})^2 = \sum_{i=1}^{n}(\hat{y}_i - \overline{y})^2 + \sum_{i=1}^{n}(y_i - \hat{y}_i)^2.$$

This identity is an example of an *analysis of variance* identity. Such identities arise frequently in the analysis of experimental data. The terms have individual interpretations with respect to the regression problem:

$$\sum_{i=1}^{n}(y_i - \overline{y})^2 \text{ is the } total\ sum\ of\ squares,$$

$$\sum_{i=1}^{n}(\hat{y}_i - \overline{y})^2 \text{ is the } sum\ of\ squares\ due\ to\ regression,$$

and

$$\sum_{i=1}^{n}(y_i - \hat{y}_i)^2 \text{ is the } residual\ or\ error\ sum\ of\ squares.$$

The identity 4.3 then partitions the total sum of squares into two parts: the sum of squares due to regression, and the residual sum of squares. It is beyond the scope of this book, but, presuming the error term ε_i in the original model to be normally distributed with mean 0 and variance σ^2, the sum of squares due to regression can be shown to have a chi-squared distribution, it can also be shown that the error sum of squares, divided by $n - 2$, is a chi-squared variable. Moreover, the two chi-squared variables are independent. It follows that the ratio of the chi-squared variables is an *F* variable. This information is usually exhibited in an *analysis of variance* table:

Analysis of variance table

Source	Sum of Squares	df	Mean Square	$F(1, n-2)$
Regression	$\sum_{i=1}^{n}(\widehat{y}_i - \overline{y})^2$	1	$\sum_{i=1}^{n}(\widehat{y}_i - \overline{y})^2/1$	$\dfrac{\sum_{i=1}^{n}(\widehat{y}_i - \overline{y})^2/1}{\sum_{i=1}^{n}(y_i - \widehat{y}_i)^2/(n-2)}$
Error	$\sum_{i=1}^{n}(y_i - \widehat{y}_i)^2$	$n-2$	$\sum_{i=1}^{n}(y_i - \widehat{y}_i)^2/(n-2)$	
Total	$\sum_{i=1}^{n}(y_i - \overline{y})^2$	$n-1$		

If the data points are linearly related, then we would expect some of the predicted values to differ significantly from the average of the y values; so we would expect $\sum_{i=1}^{n}(\widehat{y}_i - \overline{y})^2$ to be large; and we would expect that the error sum of squares $\sum_{i=1}^{n}(y_i - \widehat{y}_i)^2$ to be small because the predicted values and the observed values should be close together. So, if the regression is truly linear, we would expect the F ratio to be large. This leads to a one-tailed test of the hypothesis that the data follow a linear relationship.

The analysis of variance table for the data in this example is as follows:

Analysis of variance table

Source	Sum of Squares	Degrees of Freedom	Mean Square	F(1,7)
Regression	284.368	1	284.368	2.5117
Error	792.521	7	113.217	
Total	1076.889	8		

The probability that $F(1, 7) \geq 2.5117 = 0.157022$. The test in this case can be shown to be one-tailed, with the rejection region in the right tail. We conclude that the regression is not significant in this case. The analysis shows that there is more random scatter in the data than there is a linear relationship; this might be suspected from the preceding sums of squares because the sum of squares for regression is 284.368 while the error sum of squares, 792.521, is much larger.

The ratio of the sum of squares due to regression to the total sum of squares is called the *coefficient of determination*, or the square of the *correlation coefficient, r*:

$$r^2 = \frac{\sum_{i=1}^{n}(\widehat{y}_i - \overline{y})^2}{\sum_{i=1}^{n}(y_i - \overline{y})^2}.$$

Since the numerator in r^2 is, at most, equal to the denominator, it follows that

$$0 \le r^2 \le 1$$

and so

$$-1 \le r \le 1.$$

In this example, $r^2 = 0.264065$ or $r = 0.5139$. So only about 26% of the total variation in the y values is due to a linear relationship; 74% is due to randomness.

Computer algebra systems and statistical analysis packages make calculation of the analysis of variance table easy. This should always be done because a least squares fit without a test for linearity is quite meaningless. The interpretation of a possible linear relationship based on the correlation coefficient alone is not advised.　　　◄

Exercises 4.16

1. The following data represent the weight x in units of 1000 pounds, and the fuel consumption y in gallons per 100 miles, for six different brands of automobiles:

x	3.4	4.1	2.6	2.0	1.9	3.4
y	5.5	6.5	3.6	2.9	3.1	4.9

   a. Make a scatter plot of the data.
   b. Fit a least squares regression line to the data.
   c. Show the analysis of variance table and state the conclusions that can be drawn from it.

2. Ohm's law can be written in the form of a regression as $I = \beta_o + \beta_1 V$ where $\beta_o = 0$ and $\beta_1 = 1/R$. Since V is set by the experimenter, it can be thought of as the independent variable whereas I is viewed as the dependent variable. Data from an experiment on one wire are as follows:

V	0.5	1.0	1.5	1.8	2.0
I	0.52	1.19	1.62	2.00	2.40

   a. Plot the data. Do there appear to be any unusual points?
   b. Find the least squares fit for the data.
   c. Interpret the analysis of variance table and state conclusions this has for the experiment.

3. A recent paper reported a study on the relationship between applied stress (the independent variable X in kg/mm) and the time to fracture (the dependent variable Y in hr) for a type of stainless steel under uniaxial stress in a solution at a constant temperature. Ten different settings of applied stress were used and the following data resulted:

	X	Y
1	2.5	63
2	5.0	58
3	10.0	55
4	15.0	61
5	17.5	62
6	20.0	37
7	25.0	38
8	30.0	45
9	35.0	46
10	45.0	19

a. Plot the data. Do there appear to be any unusual points?

b. Find the least squares fit for the data.

c. Interpret the analysis of variance table and state any conclusions this has for the experiment.

4. A small study on productivity in a factory compared hours worked (x) with parts assembled (y). The data are:

x	y
1	2
2	3
3	5
4	6
5	4

a. Find the equation of the least squares regression line.

b. Show the analysis of variance table and state any conclusions that can be drawn from it.

c. Find the correlation coefficient.

5. The following data represent the total number of items produced by a manufacturing process (Y) and the total cost involved in production (X).

X	10	12	20	21	22	20	19
Y	10	15	20	20	25	30	30

a. Show a scatter plot of the data.

b. Find the equation of the least squares regression line predicting Y from X.

c. Find the equation of the least squares regression line predicting X from Y. Explain why the answer here is not equivalent to the answer in part **b**.

6. A study was done to compare engine size X (measured by cubic inches of displacement) and miles per gallon estimates Y for 8 compact automobiles. The data are:

CID (X)	121	120	97	98	122	97	85	122
MPG (Y)	30	31	34	27	29	34	38	32

a. Find the equation of the least squares regression line predicting Y from X.

b. Show the analysis of variance table and discuss any conclusions that can be drawn from it.

c. Find the correlation coefficient.

7. Raw material used in the production of a synthetic fiber is stored in a place that has no humidity control. Measurements of the relative humidity X in the storage place and the moisture content of a sample of the raw material Y (both in percentages) on 12 days were as follows:

Humidity (X)	46	53	37	42	34	29	60	44	41	48	33	40
Moisture (Y)	12	14	11	13	10	8	17	12	10	15	9	13

a. Find the equation of the least squares regression line predicting Y from X.

b. Find the correlation coefficient. What is the interpretation of this number?

8. The yield Y of a chemical process is thought to be a function of the amount of catalyst X added to the reaction. An experiment gave the following data:

Yield (Y)	60.54	63.86	63.76	60.15	66.66	71.66	70.81	65.72
Catalyst (X)	0.9	1.4	1.6	1.7	1.8	2.0	2.1	2.3

a. Find the least squares regression line predicting Y from X.

b. Find the correlation coefficient.

9. An experimenter has a data set $\{(x_1, y_1), (x_2, y_2), ..., (x_n, y_n)\}$ and wishes to fit an equation of the form $y_i = \beta x_i^2$ to the data.

a. Use the principle of least squares to find a formula for $\widehat{\beta}$, the least squares estimator for β.

b. Use the result in part **a** to calculate $\widehat{\beta}$ for the data:

x	1	2	3	4	5	6	7
y	2	3	10	15	22	31	50

10. An experimenter wishes to fit an equation of the form $y_i = \alpha + \frac{\beta}{x_i}$ to the data set $\{(x_1, y_1), (x_2, y_2), ..., (x_n, y_n)\}$. Show the least squares equations and show how to use these to estimate α and β.

11. In a regression situation it is known that the regression line passes through the origin. The model is:

$$y_i = \beta x_i + \varepsilon_i, \ i = 1, 2, ..., n$$

where the ε_i's are distributed independently and normally with mean 0 and variance σ^2.

a. Find the least squares estimator of β, $\widehat{\beta}$.

b. Let $a_i = \dfrac{x_i}{\sum_{i=1}^{n} x_i^2}$, a constant for each value of i. Find $E[\widehat{\beta}]$ and $Var[\widehat{\beta}]$.

c. In the usual regression situation, $\sum_{i=1}^{n}(y_i - \widehat{y}_i) = 0$. Show that this is not necessarily true in this case.

d. Show that

$$1]\ \sum_{i=1}^{n}(y_i - \widehat{y}_i)^2 = \sum_{i=1}^{n} y_i^2 - \widehat{\beta}^2 \sum_{i=1}^{n} x_i^2.$$

$$2]\ E[\sum_{i=1}^{n}(y_i - \widehat{y}_i)^2] = (n-1)\sigma^2.$$

12. Suppose that an experimenter wishes to fit a straight line of the form $y_i = \alpha + m x_i$ to the data set $\{(x_1, y_1), (x_2, y_2), \ldots, (x_n, y_n)\}$ where m is a known constant.

a. Find the least squares estimate of α, $\widehat{\alpha}$.

b. Show that $E(\widehat{\alpha}) = \alpha$.

4.17 ► Quality Control Chart for $\overline{X}$

Manufacturers often monitor product quality through periodic sampling during a production process. The samples taken are usually small in size and are frequently reduced to simple statistics, such as the sample mean or range, for each sample. These statistics are then plotted in a graph indicating the time series of the measurements so that monitoring of the process can be done as time progresses. Such charts are called *quality control charts*. We will consider one type of quality control chart in this section and some of the mathematics behind the analysis of the data collected. We consider a specific example.

Example 4.17.1

A manufacturer of ball bearings takes periodic random samples of size 4 from the production line and measures the mean diameter $\overline{X}$, of the ball bearings, for each sample. The sample data (which have been coded for convenience) are shown together with $\overline{X}$ and the sample standard deviation s for each sample.

Data	$\overline{X}$	s
$3, -1, -6, 4$	0	4.54606
$9, 0, 3, -2$	2.5	4.79583
$-12, 4, -9, -6$	-5.75	6.94622
$11, 9, 4, 1$	6.25	4.57347
$-1, -2, 4, -1$	0	2.70801
$8, 1, -2, 3$	2.5	4.20317
$-1, -4, -9, -3$	-4.25	3.40343
$-8, -3, -6, -4$	-5.25	2.21736
$1, -2, -2, 1$	-0.5	1.73205
$-2, -2, -3, -2$	-2.25	0.50000
$0, 4, 1, -2$	0.75	2.50000

The sample means are plotted in time-order sequence in Figure 4.20. What does the chart tell us about the process? First, we seek a central value for the means. The sample size is small, and generally nothing is known about the mean of the population from which the samples were drawn (in fact that population may be changing, resulting in a change in the population mean, which is one reason the control chart is being used). Since the central limit theorem indicates that the means are approximately normal, this central value is taken as the mean of the sample means, $\overline{\overline{X}}$.

Our second problem concerns the variation in the measurements. If σ^2 were known, then we could find a confidence interval for the true mean μ, but we don't know this variance. Usually, an estimate of the confidence interval is found as

$$\overline{\overline{X}} \pm 3 \frac{\widehat{\sigma}}{\sqrt{n}}$$

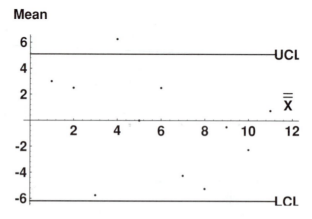

Figure 4.20   A control chart for sample means.

where $\widehat{\sigma}$ is an estimate of the unknown standard deviation. The multiplier 3 is commonly used in industry and produces about 0.0027 of the data outside the limits. It is fairly safe to assume that an observation outside the limits did not arise by chance, but is due to some alteration in the production process. The limits are called *upper and lower control limits*.

How is σ estimated? There are many ways to do this; we show here a method based on the sample standard deviations. We know that

$$\frac{(n-1)s^2}{\sigma^2} \sim \chi^2_{n-1}$$

and that

$$E(s^2) = \sigma^2.$$

with large samples, $E(s) \approx \sigma$, but this, unfortunately, is not true with small samples, so we will find an expression for $E(s)$ that will hold for any sample size. We find first the distribution of χ_{n-1}.

Suppose that X is a χ_r^2 random variable so that

$$f(x) = \frac{1}{\Gamma\left[\frac{r}{2}\right] \cdot 2^{r/2}} x^{(r/2)-1} \cdot e^{-x/2}, \; x > 0$$

and let $Y = \sqrt{X}$. Then $G(y) = P(Y \leq y) = P(\sqrt{X} \leq y) = P(X \leq y^2) = F(y^2)$, so

$$g(y) = 2 \cdot y \cdot f(y^2)$$

or

$$g(y) = \frac{2}{\Gamma\left[\frac{r}{2}\right] \cdot 2^{r/2}} \cdot y^{r-1} \cdot e^{-y^2/2}, \; y > 0$$

and

$$E(Y) = \frac{2}{\Gamma\left[\frac{r}{2}\right] \cdot 2^{r/2}} \int_0^\infty y^r e^{-y^2/2} \, dy.$$

Letting $z = \frac{y^2}{2}$ in the integral,

$$E(Y) = \frac{\sqrt{2}}{\Gamma\left[\frac{r}{2}\right]} \int_0^\infty z^{(r-1)/2} e^{-z} \, dz = \frac{\sqrt{2}}{\Gamma\left[\frac{r}{2}\right]} \Gamma\left[\frac{r+1}{2}\right].$$

Now, letting $r = n - 1$,

$$E(Y) = \frac{\sqrt{2}}{\Gamma\left[\frac{n-1}{2}\right]} \Gamma\left[\frac{n}{2}\right].$$

But $Y = \dfrac{\sqrt{n-1}\,s}{\sigma}$, so

$$E(s) = \frac{\sqrt{2}}{\Gamma\left[\frac{n-1}{2}\right]} \Gamma\left[\frac{n}{2}\right] \cdot \frac{\sigma}{\sqrt{n-1}}.$$

The factor $\dfrac{\Gamma\left[\frac{n}{2}\right]}{\Gamma\left[\frac{n-1}{2}\right]} \cdot \dfrac{\sqrt{2}}{\sqrt{n-1}}$ is denoted by c_4 in quality control literature.

Values of c_4 are shown in Table 1. While practical interest centers on small values for n, note that c_4 approaches 1 quite rapidly.

n	c_4
2	0.797885
3	0.886227
4	0.921318
5	0.939986
6	0.951533
7	0.959369
8	0.965030
9	0.969311
10	0.972659
11	0.975350
12	0.977559
13	0.979406
14	0.980971
15	0.982316
16	0.983484
17	0.984506

Table 1.

A graph of these values is shown in Figure 4.21.

For the data in this example, $\overline{\overline{X}} = -0.54545$, and the average of the sample standard deviations is $\overline{s} = 3.46596$, so

$$\widehat{\sigma} = \frac{3.46596}{0.921318} = 3.76196$$

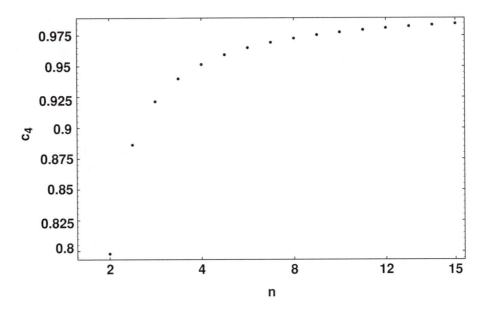

Figure 4.21   Factors c_4 for a quality control chart.

giving control limits

$$UCL = -.5455 + 3 \cdot \frac{3.76196}{\sqrt{4}} = 5.0974$$

and

$$LCL = -.5455 - 3 \cdot \frac{3.76196}{\sqrt{4}} = -6.188.$$

These are the limits shown on the control chart in Figure 4.20. We see that the fourth sample has mean 6.25, which exceeds the upper control limit. Many manufacturers would investigate the production process at that point.

Another common method for estimating the process standard deviation is based on the mean range of the samples. One reason for using the range is that it is easily calculated on the production floor. Since the range ignores all of the sample except for two values, it is not surprising to find that it is not as efficient as the sample standard deviations for estimating σ.

We see that one value of the control chart is the glimpse it gives of the production process through the use of a sample, which usually is small. Many other types of control charts are used in industry. Interested readers are referred to Duncan [5] and Grant and Leavenworth [10] for more information.   ◀

Exercises 4.17

1. Ten samples, shown below, were taken in order to establish control limits in an industrial process.

Sample	Values			
1	10.6	10.1	11.3	9.1
2	10.2	11.6	10.5	10.5
3	10.1	9.8	8.8	9.3
4	10.1	9.5	10.3	10.6
5	8.7	11.6	9.7	9.3
6	10.1	9.8	10.8	8.9
7	11.2	11.5	10.9	11.6
8	10.6	9.6	10.3	9.9
9	9.8	7.7	9.4	9.9
10	10.0	8.4	10.6	8.8

   a. Calculate the sample mean and the sample standard deviation for each sample.

   b. Calculate upper and lower control limits using the results in part **a**.

   c. Plot the control chart for the sample means. Are any of the data points unusual?

2. A new machine fills cereal boxes by weight. It is desired to start a control chart on the average weight of the boxes. Ten samples, each of size 5, are taken with the following results:

Sample	Values				
1	16.1	16.2	15.9	16.0	16.1
2	16.2	16.4	15.8	16.1	16.2
3	16.0	16.1	15.7	16.3	16.1
4	16.1	16.2	15.9	16.4	16.6
5	16.5	16.1	16.4	16.4	16.2
6	16.8	15.9	16.1	16.3	16.4
7	16.1	16.9	16.2	16.5	16.5
8	15.9	16.2	16.8	16.1	16.4
9	15.7	16.7	16.1	16.4	16.8
10	16.2	16.9	16.1	17.0	16.4

   a. Calculate the sample mean and the sample standard deviation for each sample.

   b. Calculate upper and lower control limits using the results in part **a**.

   c. Plot the control chart for the sample means. Are any of the data points unusual?

Chapter Review

Drawing conclusions from samples – statistical inference – has been the subject of this chapter. Statistical inference is a central part of the scientific method because it involves the analysis of sample data gathered in the course of a scientific investigation.

Any quantity calculated from a sample is called a *statistic*. Statistics are thus random variables with probability distributions, means, and variances of their own. From the central limit theorem, we know that the mean of a sample, $\overline{X}$, has, approximately, a normal distribution with mean μ and variance σ^2/n, where n is the sample size and where μ and σ^2 are the true population values. This theorem is a primary tool in establishing tests of hypotheses on single means and on the difference between means from two populations.

The purpose of this chapter was to establish tests of hypotheses and confidence intervals for some common parameters of statistical distributions, as well as to introduce the principle of least squares in fitting a linear function to a data set. In order to establish the probability distributions of some common statistics determined from samples, we first establish the probability distributions of some common *functions of random variables*. We begin with a procedure for finding the probability distribution function for a function of a random variable, say $Y = H(X)$. We use the fact that $g(y)$, the probability density function for Y, is

$$g(y) = \frac{dG(y)}{dy}$$

where $G(y) = P(Y \le y) = P[H(X) \le y]$. We then solve for X and express $G(y)$ in terms of $F(x)$, the distribution function for X.

We established the fact that if $X \sim N(0, 1)$, then X^2 has a χ^2_1 distribution. We also established the distribution of the maximum of a set of uniformly distributed random variables.

The most important function of a random variable involves sums of independent random variables. In section 4.4 we used the fact that

$$P(X + Y = z) = \sum_k P(X = x) \cdot P(Y = z - k).$$

This sum can be evaluated for a variety of probability distributions. Important facts here are that the sum of independent binomials with common value for p is binomial, and that the sum of independent Poissons is Poisson, where the parameter for the sum is the sum of the parameters of the individual Poissons. Binomial and Poisson variables are called *reproductive* because their sum has the same kind of distribution as the individual summands.

Variables are not commonly reproductive, however, and we calculated the probability distribution for the sum of independent uniform variables from some special cases. The sums appeared to become normal; that is indeed the case, a fact established in subsequent parts of the chapter.

The probability distribution of a random variable can be summarized by the *probability generating function* for the random variable. If it exists, this function, (denoted by $P_X(t)$), is the expected value of a special function of X, t^X, so

$$P_X(t) = E(t^X).$$

It follows that

$$P_X(t) = \sum_x t^X \cdot P(X = x)$$

since we used this function only for discrete random variables.

Sections 4.5 and 4.6 established some properties of probability generating functions. These were, supposing $A(t)$ and $B(t)$ to be probability generating functions for variables X and Y, respectively, that

1. the coefficient of t^k in a power series expansion of $P_X(t)$ is $P(X = k)$. It is in this sense that $P_X(t)$ summarizes or characterizes the random variable since it is possible to find *all* the probabilities from $P_X(t)$.
2. $E(X) = P_X'(1)$.
3. $\text{Var}(X) = P_X''(1) + P_X'(1) - [P_X'(1)]^2$.
4. The coefficient of t^k in $A(t) \cdot B(t)$ is $P(X + Y = k)$.

These facts are of great importance in establishing probability distributions for sums.

Probability generating functions for some specific random variables were derived in section 4.5. There it was found that

$$P_X(t) = (q + pt)^n \text{ for a binomial variable with parameters } n \text{ and } p, \text{ and}$$

$$P_X(t) = \frac{pt}{1 - qt} \text{ for a geometric random variable with parameter } p.$$

We then discussed a series of dependent binomial trials called *Poisson's trials* in which the probabilities in binomial trials vary.

The probability generating function is not often used for continuous random variables. Instead, a function we call the *moment generating function* is used. It too characterizes probability distributions. Its definition is

$$M[X; t] = E(e^{tX}) = \int_{-\infty}^{\infty} e^{tx} \cdot f(x)\, dx,$$

provided of course that the integral exists. The moment generating function generates moments (although it is uncommon to use it for this purpose) in either of two ways:

1. The coefficient of $t^k/k!$ in the power series expansion of $M[X; t]$ is $E(X^k)$, or

2. $\dfrac{d^k M[X; t]}{dt^k}\bigg|_{t=0} = E(X^k)$.

We calculated some specific moment generating functions:

1. If $f(x) = 1$, $0 \le x \le 1$, then $M[X; t] = \frac{1}{t}(e^t - 1)$.
2. If $f(x) = \lambda e^{-\lambda x}$, $x \ge 0$, then $M[X; t] = \frac{\lambda}{\lambda - t}$.
3. If $X \sim N(0, 1)$, then $M[X; t] = e^{t^2/2}$.

This last fact takes on enormous importance because if we can show that a moment generating function approaches that of the standard normal, we can conclude that the variable approaches the standard normal distribution. In this sense the moment generating function is of great importance since it can be used to establish the distributions of functions of random variables and, in particular, their sums.

These properties of moment generating functions are important:

1. $M[c + X; t] = e^{ct} \cdot M[X; t]$, and
2. $M[cX; t] = M[X; ct]$.

These facts help establish the fact that if $X \sim N(\mu, \sigma)$ then $M[X; t] = e^{\mu t + t^2/(2\sigma^2)}$.

Sums of random variables were considered in section 4.10; it was found that under quite general conditions these approach normality as the number of summands increases. Examples were shown that demonstrate that

1. Sums of normals are normal.
2. Sums of exponentials are normal.
3. Sums of binomials are normal.

These facts help explain the persistence of normality in many examples earlier in the book.

In Section 4.11 we stated and demonstrated the *central limit theorem*:

THEOREM: If X has mean μ and variance σ^2, and if X has a moment generating function, then

$$\overline{X} \rightarrow N(\mu, \sigma/\sqrt{n}) \qquad \blacksquare$$

Before examining confidence intervals and tests of hypotheses, we showed that the probability distribution of the sample variance can be determined from the fact that

$$\frac{(n-1)s^2}{\sigma^2} \sim \chi^2_{n-1}$$

when the sample of size n is chosen from a normal distribution with known variance σ^2.

The central limit theorem allows us to *test hypotheses* on single means. We also considered the distribution of the sample mean when the population variance is unknown, and finally the distribution of the difference between sample means. The tests of various hypotheses considered are summarized here. Critical regions are presumed to be constructed so that the size of the test is α.

1. To test $H_o : \mu = \mu_o$ against $H_a : \mu \neq \mu_o$, the best critical region is $\overline{X} > a$ or $\overline{X} < b$. One-sided tests are used in testing one-sided alternatives. If the population standard deviation σ is known, then a and b can be determined using the fact that

$$\frac{\overline{X} - \mu}{\frac{\sigma}{\sqrt{n}}}$$

   is a normal variable.

2. To test $H_o : \mu = \mu_o$ against $H_a : \mu \neq \mu_o$ and the population standard deviation is unknown, then the best critical region is $\overline{X} > a$ or $\overline{X} < b$ where values for a and b can be determined using the fact that

$$\frac{\overline{X} - \mu}{\frac{s}{\sqrt{n}}}$$

   follows a t_{n-1} distribution.

3. If two samples are drawn and the population variances are both known, and the hypothesis $H_o : \mu_x = \mu_y$ is to be tested against $H_a : \mu_x \neq \mu_y$ then the test statistic is

$$z = \frac{(\overline{X} - \overline{Y}) - (\mu_x - \mu_y)}{\sqrt{\frac{\sigma_x^2}{n_x} + \frac{\sigma_y^2}{n_y}}}$$

   where z is a normal random variable.

   If the population variances are unknown but can be presumed to be equal, and if the samples are chosen from normal populations, then

$$t_v = \frac{(\overline{X} - \overline{Y}) - (\mu_x - \mu_y)}{s_p \sqrt{\frac{1}{n_x} + \frac{1}{n_y}}}$$

   where

$$s_p^2 = \frac{(n_x - 1)s_x^2 + (n_y - 1)s_y^2}{n_x + n_y - 2}$$

   and

$$v = n_x + n_y - 2.$$

If the population variances are unknown and known to be unequal, then no exact test is known for the hypothesis that the population means are equal. An approximate test, due to Welch, is

$$T_v = \frac{(\overline{X} - \overline{Y}) - (\mu_x - \mu_y)}{\sqrt{\dfrac{s_x^2}{n_x} + \dfrac{s_y^2}{n_y}}}$$

where

$$v = \frac{\left(\dfrac{s_x^2}{n_x} + \dfrac{s_y^2}{n_y}\right)^2}{\dfrac{\left(\dfrac{s_x^2}{n_x}\right)^2}{n_x - 1} + \dfrac{\left(\dfrac{s_y^2}{n_y}\right)^2}{n_y - 1}}.$$

4. A test of $H_o : \sigma_x^2 = \sigma_y^2$ and the alternative $H_a : \sigma_x^2 \neq \sigma_y^2$ is based on the fact that

$$\frac{\dfrac{s_x^2}{\sigma_x^2}}{\dfrac{s_y^2}{\sigma_y^2}} = F(n_x - 1, n_y - 1).$$

We then considered *simple linear regression*, or the fitting of data to a straight line of the form $y_i = a + bx_i$, $i = 1, 2, \ldots, n$. The principle of least squares chooses those estimates that minimize

$$S = \sum_{i=1}^{n} (y_i - a - bx_i)^2.$$

The result is a set of *least squares equations*:

$$\sum_{i=1}^{n} y_i = n\widehat{a} + \widehat{b} \sum_{i=1}^{n} x_i$$

and

$$\sum_{i=1}^{n} x_i y_i = \widehat{a} \sum_{i=1}^{n} x_i + \widehat{b} \sum_{i=1}^{n} x_i^2.$$

Their simultaneous solution is

$$\widehat{b} = \frac{n\sum\limits_{i=1}^{n} x_i y_i - \sum\limits_{i=1}^{n} x_i \sum\limits_{i=1}^{n} y_i}{n\sum\limits_{i=1}^{n} x_i^2 - (\sum\limits_{i=1}^{n} x_i)^2} = \frac{\sum\limits_{i=1}^{n}(x_i - \overline{x})(y_i - \overline{y})}{\sum\limits_{i=1}^{n}(x_i - \overline{x})^2}$$

and

$$\overline{y} = \widehat{a} + \widehat{b}\overline{x}.$$

Finally in this chapter we considered *quality control charts* for sample means. These charts plot means calculated from periodic samples and establish *upper and lower control limits*, indicating that the process may be out of control. These limits are $\overline{\overline{X}} \pm 3\dfrac{\widehat{\sigma}}{\sqrt{n}}$ where $\widehat{\sigma}$ is an estimate of the unknown population standard deviation σ. Table 1 gives divisors of the average sample standard deviations that are used to find $\widehat{\sigma}$.

Problems for Review

Exercises 4.3 - # 1, 3, 4, 8
Exercises 4.4 - # 1, 2, 4
Exercises 4.5 - # 1, 4
Exercises 4.7 - # 1, 2, 4, 6, 7, 10
Exercises 4.8 - # 2, 3, 4, 5, 6, 9
Exercises 4.10 - # 1, 2, 5, 9
Exercises 4.11 - # 1, 2, 4, 5
Exercises 4.13- # 1, 2, 5, 7
Exercises 4.14- # 1, 2, 5, 6, 8, 10, 17
Exercises 4.15- # 1, 2, 3, 6, 8, 9, 11, 14, 15, 19
Exercises 4.16 - # 1, 3, 5, 9
Exercises 4.17- # 1

Supplementary Exercises for Chapter 4

1. For the triangular distribution $f(x) = \frac{2}{a}\left(1 - \frac{x}{a}\right)$, $0 < x < a$:

  a. Find the moment generating function.
  b. Use the moment generating function to find the mean and variance of X and check these results by direct calculation.

2. Find the mean and variance of X where X has the Pareto distribution, $f(x) = a \cdot b^a \cdot x^{-(a+1)}$, $a > 0$, $b > 0$, $x > b$.

3. Consider the truncated exponential distribution, $f(x) = e^x$, $0 \le x \le \ln 2$.

    a. Find the moment generating function for X and expand it in a power series.

    b. From the series in part **a**, find the mean and variance of X.

4. Random variable X denotes the number of green marbles drawn when a sample of 2 is selected without replacement from a box containing 3 green and 7 yellow marbles.

    a. Find the moment generating function for X.

    b. Verify that $E(X^3) = 1$.

5. Find $E(X^k)$ if X is a Weibull random variable with parameters α and β.

6. Let $S = \sum_{i=1}^{n} X_i$ where X_i is a uniform random variable on the interval $(0, 1)$. Find the moment generating function for $Z = \frac{S-\mu}{\sigma}$ where μ and σ are the mean and standard deviation, respectively, of S. Then show that this moment generating function approaches the moment generating function for a standard normal random variable as $n \to \infty$.

7. A fair quarter is tossed until it comes up heads; suppose X is the number of tosses necessary. If $X = x$, then x fair pennies are tossed; let Y denote the number of heads on the pennies. Find $P(Y = 3)$, simplifying the result as much as you can.

8. A coin loaded so as to come up heads $\frac{1}{3}$ of the time is tossed until a head appears. This is followed by the toss of a coin loaded so as to come up heads with a probability of 1/4 until that coin comes up heads.

    a. Find the probability distribution of Z, the total number of tosses necessary.

    b. Find the mean and variance of Z.

9. Customers at a gasoline station buy regular or premium unleaded gasoline with probabilities p and $q = 1 - p$, respectively. The number of customers in a daily period is Poisson with mean μ. Find the probability distribution for the number of customers buying regular unleaded gasoline.

10. A company claims that the actual resistance of resistors is normally distributed with mean 200 ohms and variance $4 \cdot 10^{-4} \text{ohms}^2$.

    a. What is the probability that a resistor drawn at random from this set of resistors will have resistance greater than 200.025 ohms?

    b. A sample of 25 resistors drawn at random from this set has an average resistance of 200.01 ohms. Would you conclude that the true population mean is still 200 ohms?

11. A sample of size n is drawn from a population about which nothing is known except that the variance is 4. How large a sample must be drawn so that the probability is at least 0.95 that the sample average $\overline{X}$ is within 1 unit of the true population mean, μ?

12. Suppose 12 fair dice are thrown. Let X denote the total number of spots showing on the 12 uppermost faces. Use the central limit theorem to estimate $P(25 \leq X \leq 40)$.

13. Mathematical and verbal SAT scores are, individually, $N(500, 100)$.

   a. Find the probability that the total mathematical plus verbal SAT score for an individual is at least 1100, assuming that the scores are independent.

   b. What is the probability that the average of five individual total scores is at least 1100?

14. A student makes 100 check transactions in a period covering her bank statement. Rather than subtract the amount she spends exactly, she rounds each checkbook entry off to the nearest dollar. Assume that the errors are uniformly distributed on $[-\frac{1}{2}, \frac{1}{2}]$. What is the probability that the total error is more than \$5?

15. The time a construction crew takes to construct a building is normally distributed with mean 90 days and standard deviation 10 days. After construction, it takes additional time to install utilities and finish the interior. Assume the additional time is independent of the construction time, and is normally distributed with mean 30 days and standard deviation 5 days.

   a. Find the probability that it takes at least 101 days for the construction only of a building.

   b. Find the probability that it takes an average of 101 days for the construction only of 4 buildings.

   c. What is the probability that the total completion time for one building is, at most, 130 days?

   d. What is the probability that the average additional completion time for 5 buildings is at least 35 days?

16. A random variable X has the probability distribution function

$$f(x) = \frac{1}{3} \text{ for } x = -1, 0, \text{ or } 1.$$

   a. Find $M[X; t]$, the moment generating function for X.

   b. If X_1 and X_2 are independent observations of X, find $M[X_1 + X_2; t]$ without first finding the probability distribution of $X_1 + X_2$.

   c. Verify the result in part **b** by finding the probability distribution of $X_1 + X_2$.

17. A random variable X has the probability density function $f(x) = 2(1 - x)$, $0 < x < 1$.

   a. Find the moment generating function for X.

   b. Use the moment generating function to find a formula for $E(X^k)$.

   c. Let $Y = \frac{1}{2}(X + 1)$. Find $M[Y; t]$.

18. A random variable X has $M[X; t] = e^{-6t + 32t^2}$. Find $P(-4 \leq X \leq 16)$.

19. A discrete random variable X has the probability distribution function

$$f(x) = \begin{cases} \frac{1}{2}, & x = 1 \\ \frac{1}{3}, & x = 2 \\ \frac{1}{6}, & x = 3. \end{cases}$$

   a. Find μ_x and σ_x^2.

   b. Find $M[X; t]$.

   c. Verify the results in part **a** using the moment generating function.

20. Find the moment generating function for a random variable X with probability density function $f(x) = e - e^x$ if $0 \le x \le 1$.

21. A random variable has $M[X; t] = \frac{2}{5}e^t + \frac{1}{5}e^{2t} + \frac{2}{5}e^{3t}$.

   a. What is the probability distribution function for X?

   b. Expand $M[X; t]$ in a power series and find μ_x and σ_x^2.

22. Suppose that X is the number of 6's in n_1 tosses of a fair die and that Y is the number of 3's in n_2 tosses of another fair die. Use moment generating functions to show that $S = X + Y$ has a binomial distribution with parameters $n = n_1 + n_2$ and $p = 1/6$.

23. A rectifier following a square law has the characteristic $Y = kX^2$, $x > 0$, where X and Y are the input and output voltages, respectively. If the input to the rectifier is noise with the probability density function

$$f(x) = \frac{2x}{\beta} e^{-x^2/\beta}, \ x \ge 0, \ \beta > 0,$$

find the probability density function of the output.

24. If $X \sim N(0, 1)$,

   a. find $P(X^2 \ge 5)$.

   b. Suppose $X_1, X_2, \ldots, X_8$ are independent observations of X. Find $P(X_1^2 + X_2^2 + \cdots + X_8^2 > 10)$.

25. Suppose X has the probability density function

$$f(x) = \begin{cases} x + 1, & -1 \le x \le 0 \\ 1 - x, & 0 \le x \le 1. \end{cases}$$

   a. Find the probability density function of $Y = X^2$.

   b. Show that the result in part **a** is a probability density function.

26. The resistance R of a resistor has the probability density function

$$f(r) = \frac{r}{200} - 1, \; 200 < r < 220.$$

A fixed voltage of 5 v is placed across the resistor.

  a. Using the fact that $V = I \cdot R$, find the probability density function of the current I through the resistor.

  b. What is the expected value of the current?

27. If X is uniformly distributed on $[-1, 1]$, find the probability density function of $Y = \sqrt{1 - X^2}$.

28. Let X be uniformly distributed on $[0, \; 1]$.

  a. Find the probability density function for $Y = \frac{1}{X+1}$ and prove that your result is a probability density function.

  b. Explain how values of X could be used to sample from the distribution $f(x) = 2x, \; 0 < x < 1$.

29. Random variable X has the probability density function $f(x) = 2x$, $0 < x < 1$. Let $Y = \frac{1}{X}$. Find $E(Y)$ by

  a. first finding $g(y)$.

  b. not using $g(y)$.

30. Given: $f(x) = 2e^{-2x}$, for $x > 0$. Find the probability density function for $Y = e^{-X}$ and, from it, find $E[e^{-X}]$.

31. A random variable X has the probability density function $f(x) = \frac{2}{9}x(3 - x)$, for $0 \leq x \leq 3$. Find the probability density function for $Y = X^2 - 1$.

32. The moment generating function for a random variable Y is $M[Y; t] = \frac{e^{4t} - e^{2t}}{2t}$. Expand $M[Y; t]$ in a power series in t and find μ_y and σ_y^2.

33. Suppose that X is uniform on $[-1, 2]$. Find the probability density function for $Y = |X|$.

34. The following data represent radiation readings, in milliroentgens per hour, taken from television display areas in different department stores: 0.40, 0.48, 0.60, 0.15, 0.50, 0.80, 0.50, 0.36, 0.16, and 0.89.

  a. Find a 95% confidence interval for μ if it is known that $\sigma^2 = 1$.

  b. Find a 95% confidence interval for μ if σ^2 is unknown.

35. The variance of a normally distributed industrial measurement is known to be 225. If a random sample of 14 measurements is taken and the sample variance computed, what is the probability that the sample variance is twice the true variance?

36. A random sample of 21 observations is taken from a normal distribution with variance 100. What is the probability that the sample variance exceeds 140?

37. A machine that produces ball bearings is sampled periodically. The mean diameter of the ball bearings produced is known to be under control, but the variability of these diameters is of concern. If the machine is working properly, the variance is 0.50 mm^2. If a sample of 31 measurements shows a sample variance of 0.94 mm^2, should the operator of the machine be concerned that something is wrong with the machine? Use $\alpha = 0.05$.

38. A manufacturer of piston rings for automobile engines assumes that piston ring diameter is approximately normally distributed. If a random sample of 15 rings has mean diameter 74.036 mm and sample standard deviation 0.008 mm, construct a 98% confidence interval for the true mean piston ring diameter.

39. A commonly used method for determining the specific heat of iron has a standard deviation 0.0100. A new method of determination yielded a standard deviation of 0.0086 based on 9 test runs. Assuming a normal distribution, is there evidence at the 10% level that the new method reduces the standard deviation?

40. **a.** A random sample of 10 electric light bulbs is selected from a normal population. The standard deviation of the lifetimes of these bulbs is 120 hours. Find 95% confidence limits for the variance of all such bulbs manufactured by the company.
    b. Find 95% confidence limits for the standard deviation if the sample size is 100.

41. A city draws a random sample of employees from its labor force of 5000 people. The number of years each employee has worked for the city is 8.2, 5.6, 4.7, 9.6, 7.8, 9.1, 6.4, 4.2, 9.1, and 5.6. Assume that the time employees have been employed is approximately normal. Calculate a 90% confidence interval for the average number of years an employee has worked for the city.

42. The number of ounces of liquid a soft drink machine dispenses into a bottle is a normal random variable with unknown mean μ but known variance 0.25 oz^2. A random sample of 75 bottles filled by this machine has mean 12.2 oz.

    a. Determine a 95% two-sided confidence interval for μ.
    b. It is desired to be 99% confident that the error in estimating the mean is less than 0.1 oz. What should the sample size be?

43. The maximum acceptable level for exposure to microwave radiation in the United States is an average of 10 microwatts per square centimeter. It is feared that a large television transmitter may be polluting the air by exceeding a safe level of microwave radiation.

a. Test $H_o : \mu = 10$ against the alternative hypothesis $H_a : \mu > 10$ with $\alpha = 0.05$ if a sample of 36 readings gives a sample mean of 10.3 microwatts and a sample standard deviation of 2.1 microwatts.

b. Find a 98% confidence interval for μ.

44. A machine producing washers is found to produce washers whose variance is 30 in^2.

    a. A sample of 36 washers is taken and the mean diameter $\overline{X}$ is found. Find the probability that $\overline{X}$ is within 0.1 units of μ, the true mean diameter.

    b. How large a sample is necessary so that the probability that $\overline{X}$ is within 0.2 units of μ is 0.90?

45. **a.** To determine with 94% confidence the average hardness of a large number of selenium-alloy ball bearings, how many would have to be tested to obtain an estimate within 0.009 units of the true mean hardness if σ^2 is known to be 0.0016?

    b. A small study of 5 bearings in part **a** gives $\overline{X} = 2.057$. What is the probability that this differs from the true mean hardness by at least 0.009 units?

46. Heat transfer coefficients of 65, 63, 60, 68, and 72 were observed in a sample of heat exchangers made by a company. Find a 95% confidence interval for the true average heat transfer coefficient μ if

    a. σ^2 is known to be 17.64.

    b. σ^2 is unknown.

47. Find the probability that a random sample of 25 observations from a normal population with variance 6 will have a sample variance between 3.100 and 10.750.

48. The hardness (in degrees) of a certain rubber is claimed to be 65. A sample of 14 specimens gave $\overline{X} = 63.1$.

    a. If σ^2 is known to be 12.25 degrees^2 for this rubber, can $H_o : \mu = 65$ be accepted against the alternative hypothesis $H_a : \mu \neq 65$ if $\alpha = 5\%$?

    b. Answer part **a** if the sample variance is 10.18 degrees^2 .

49. A manufacturer of steel rods considers that the process is working properly if the mean length of the rods is 8.6 in. The standard deviation of these rods is approximately 0.3 in. Suppose that when 36 rods were tested, the sample mean was 8.45.

    a. Test the hypothesis that the average length is 8.6 in. against the alternative that it is less than 8.6 in., using a 5% level of significance.

    b. Since short rods must be scrapped, it is extremely important to know when the process began to produce rods of mean length less than 8.6. Find the probability of a Type II error when the alternative hypothesis is $H_a : \mu = 8.4$ in.

50. A coffee vending machine is supposed to dispense 6 oz per cup. The machine is tested 9 times yielding an average fill $\overline{X} = 6.1$ oz, with standard deviation 0.15 oz.

    a. Find a 90% confidence interval for μ, the true mean fill per cup.

    b. Find a 90% confidence interval for σ^2, the true variance of the fill per cup.

51. A random sample of 22 freshman mathematics SAT scores at a large university gives a sample mean of 680 and a standard deviation of 35. Find a 99% confidence interval for μ, the true population mean.

52. A population has unknown mean μ but known standard deviation of 5. How large a sample is necessary so that we can be 95% confident that $\overline{X}$ is within 1.5 units of the true mean?

53. A fuel oil company claims that 20% of the homes in a city are heated by oil. Do we have reason to doubt this claim if 236 homes in a sample of 1000 homes are heated by oil? Use $\alpha = 1\%$.

54. A brand of car battery claims that the standard deviation of the battery's lifetime is 0.9 years. If a random sample of 10 of these batteries has $s = 1.2$, test $H_o : \sigma^2 = 0.81$ against the alternative hypothesis, $H_a : \sigma^2 > 0.81$, if $\alpha = 0.05$.

55. A researcher is studying the weights of male college students. She wishes to test $H_o : \mu = 68$ kg against the alternative hypothesis $H_a : \mu \neq 68$ kg. A sample of 64 students has $\overline{X} = 68.90$ kg and $s = 4$ kg.

    a. Is the hypothesis accepted or rejected?

    b. Find β for the alternative $\mu = 69.3$ kg.

56. Fractures in metals have been studied and it is thought that the rate at which fractures expand is normally distributed. A sample of 14 pieces of a particular steel gives $\overline{X} = 3205$ ft/sec .

    a. Find a 95% confidence interval for μ, the true average rate of expansion, if σ is assumed to be 53 ft/sec.

    b. Now suppose σ is unknown. The sample variance is 6686.53 $(\text{ft/sec})^2$. Find a 95% confidence interval for μ.

57. Engineers think that a design change will improve the gasoline mileage of a certain brand of automobile. Previously such cars averaged 18 mpg under test conditions. A sample of 15 cars has $\overline{X} = 19.5$ mpg.

    a. Test $H_o : \mu = 18$ against the alternative hypothesis $H_a : \mu > 18$ assuming $\sigma^2 = 9$ and $\alpha = 5\%$.

    b. Test the hypothesis in part **a** at the 5% level if the sample variance is 7.4.

58. One-hour carbon monoxide concentrations in 10 air samples from a city had mean 11.5 ppm and variance 40 $(\text{ppm})^2$. After imposing smog control measures on a local industry, 12 air samples had mean 10 ppm and variance 43 $(ppm)^2$. Estimate the true difference in average carbon monoxide concentrations in a 98% confidence interval. What assumptions are necessary for your answer to be valid?

59. Specifications for a certain type of ribbon call for a mean breaking strength of 185 lb. In order to monitor the process, a random sample of 30 pieces selected from different rolls is taken each hour, and the sample mean is used to decide if the mean breaking strength has shifted. The test is of the hypothesis $H_o : \mu = 185$ against the alternative hypothesis $H_a : \mu < 185$ with $\alpha = 0.05$. Assuming $\sigma = 10$ lb,

   a. find the critical region in terms of $\overline{X}$.
   b. find β for the alternative $\mu = 179.5$.

60. To test $H_o : \mu = 46$ against the alternative hypothesis $H_a : \mu > 46$, a random sample of 24 is taken. The critical region is $\overline{X} > 51.7$. Suppose $\sigma^2 = 100$.

   a. Find α.
   b. Find β for the alternative $\mu = 48$.

61. In 16 test runs the gasoline consumption of an experimental engine had sample standard deviation 2.2 gallons. Construct a 95% confidence interval for σ, the true standard deviation of gasoline consumption of the engine. What assumptions are necessary for your analysis to be valid?

62. A production supervisor wants to determine if changes in a production process reduce the amount of time necessary to complete a subassembly. Specifically, she wishes to test $H_o : \mu = 30$ against the alternative hypothesis, $H_a : \mu < 30$, with $\alpha = 5\%$. The measurements are in minutes.

   a. Find the critical region for the test (in terms of $\overline{X}$) if a sample of 4 times is taken and the true variance is assumed to be 1.2.
   b. Now suppose a sample gave $\overline{X} = 29.06$ and $s^2 = 1.44$. Is the hypothesis accepted or not?

► *Chapter 5*

Bivariate Probability Distributions

▌5.1 ▶ Introduction

So far, we have studied a single random variable defined on the points of a sample space. Scientific investigations, however, most commonly involve several random variables arising in the course of an investigation. A physicist, for example, may be interested in studying the effects of transmissions in a fiber optic cable when transmission rates and the composition of the cable are varied. Sample surveys usually ask several questions of the respondents, creating separate random variables for each question. Educators studying grade point averages for college students find that these averages are dependent upon aptitude test scores, entrance examinations, rank in high school class, as well as many other factors that could be considered. Each of these examples suggests a sample space on which more than one random variable is defined.

While these variables could be considered individually as the *univariate* variables studied in the previous chapters, studies of the individual random variables will provide no information at all on how the variables behave together. Separate studies offer no information on how the variables *interact*, or are *correlated* with each other; this is often crucial information in scientific investigations, since the manner in which the variables act together may indicate the most important factors in explaining the outcome. Because of this, investigations involving only one factor at a time are becoming increasingly rare. The interactions revealed in studies are often of greater importance than the effects of the individual variables alone, but measuring them requires that we consider combinations of the variables. In this chapter we will study *jointly distributed random variables* and some of their characteristics. This is an essential prelude to the actual measurement of the influence of separate variables and interactions. Inferences from these measurements are statistical problems that are normally discussed in texts on statistics.

▌5.2 ▶ Joint and Marginal Distributions

Example 5.2.1

In Example 2.9.1 we examined tossing two fair coins and recording X, the number of heads that occur. The coins that come up heads are put aside and only those that come up tails the first time are tossed again. Let Y denote the number of heads obtained in the second set of tosses. The variable Y is of primary interest here, but to investigate it we must consider X as well. While this might appear to be a purely theoretical exercise, the result is applicable when a number of components in a system fail according to a binomial model; interest centers on when all the components will fail, so our example is a generalization of this situation. We use five coins here and only two group tosses (so that it may be that not all the coins will turn up heads), but the extension to any number of coins is very similar to this special case.

Y is clearly dependent upon X. In fact, since $5 - X$ coins came up tails the first time and then were tossed again by a binomial process, it follows that if $X = x$, then the conditional probability that $Y = y$ is given by a binomial probability:

$$P(Y = y | X = x) = \binom{5-x}{y} \left(\frac{1}{2}\right)^y \left(\frac{1}{2}\right)^{5-x-y}, \quad y = 0, 1, \ldots, 5 - x.$$

X itself is also a random variable, and so the unconditional probability that $X = x$ is

$$P(X = x) = \binom{5}{x} \left(\frac{1}{2}\right)^5, \quad x = 0, 1, \ldots, 5.$$

If we call

$$f(x, y) = P(X = x \text{ and } Y = y),$$

which we also denote as

$$f(x, y) = P(X = x, Y = y),$$

the *joint probability distribution* of X and Y, then

$$f(x, y) = P(X = x, Y = y) = P(X = x) \cdot P(Y = y | X = x)$$

where $P(Y = y | X = x)$ is the conditional probability that $Y = y$ if $X = x$.

In this example, the conditional probability $P(Y = y \mid X = x)$ is also binomial with $5 - x$ trials and probability of success at any trial $\frac{1}{2}$, as we have seen, so

$$f(x, y) = \binom{5}{x} \left(\frac{1}{2}\right)^x \left(\frac{1}{2}\right)^{5-x} \cdot \binom{5-x}{y} \left(\frac{1}{2}\right)^y \left(\frac{1}{2}\right)^{5-x-y},$$

$$x = 0, 1, \ldots, 5; \quad y = 0, 1, \ldots, 5 - x,$$

which can be simplified to

$$f(x, y) = \binom{5}{x} \binom{5-x}{y} \left(\frac{1}{2}\right)^{10-x}, \quad x = 0, 1, \ldots, 5; \quad y = 0, 1, \ldots, 5 - x.$$

These probabilities are exhibited in the following table:

TABLE 5.2.1   Joint distribution for the coin tossing example.

		Y						
		0	**1**	**2**	**3**	**4**	**5**	**f(x)**
	0	$\frac{1}{1024}$	$\frac{5}{1024}$	$\frac{10}{1024}$	$\frac{10}{1024}$	$\frac{5}{1024}$	$\frac{1}{1024}$	$\frac{32}{1024}$
	1	$\frac{10}{1024}$	$\frac{40}{1024}$	$\frac{60}{1024}$	$\frac{40}{1024}$	$\frac{10}{1024}$	0	$\frac{160}{1024}$
	2	$\frac{40}{1024}$	$\frac{120}{1024}$	$\frac{120}{1024}$	$\frac{40}{1024}$	0	0	$\frac{320}{1024}$
X	**3**	$\frac{80}{1024}$	$\frac{160}{1024}$	$\frac{80}{1024}$	0	0	0	$\frac{320}{1024}$
	4	$\frac{80}{1024}$	$\frac{80}{1024}$	0	0	0	0	$\frac{160}{1024}$
	5	$\frac{32}{1024}$	0	0	0	0	0	$\frac{32}{1024}$
	g(y)	$\frac{243}{1024}$	$\frac{405}{1024}$	$\frac{270}{1024}$	$\frac{90}{1024}$	$\frac{15}{1024}$	$\frac{1}{1024}$	1

Notice that the entries in the table must all be non-negative (since they represent probabilities) and that the sum of these entries is 1.

Probabilities can be found from the table. For example,

$$P(X \geq 2, Y \geq 2) = \frac{120}{1024} + \frac{80}{1024} + \frac{40}{1024} = \frac{15}{64}.$$

A scatter plot of the joint probability distribution is also useful (see Figure 5.1).

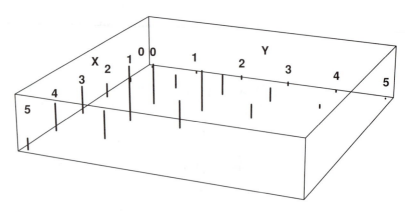

Figure 5.1   Scatter plot for the coin-tossing example.

Now suppose we want to recover information on the variables X and Y separately. These, individually, are random variables on their own. What are their probability distributions?

To find $P(X = 3)$, for example, because the three events $X = 3$ and $Y = 0$, $X = 3$ and $Y = 1$, and $X = 3$ and $Y = 2$ are mutually exclusive, we see that

$$P(X = 3) = P(X = 3, Y = 0) + P(X = 3, Y = 1) + P(X = 3, Y = 2)$$

$$= \binom{5}{3}\binom{2}{0}\left(\frac{1}{2}\right)^7 + \binom{5}{3}\binom{2}{1}\left(\frac{1}{2}\right)^7 + \binom{5}{3}\binom{2}{2}\left(\frac{1}{2}\right)^7$$

$$= \binom{5}{3}\left(\frac{1}{2}\right)^7 \left\{\binom{2}{0} + \binom{2}{1} + \binom{2}{2}\right\}$$

$$= \binom{5}{3}\left(\frac{1}{2}\right)^7 2^2 = \binom{5}{3}\left(\frac{1}{2}\right)^5 = \frac{10}{32} = \frac{5}{16}$$

We find this probability entered in the side margin of the table at $X = 3$. It is found by adding the probabilities across the row, thus considering all the possible values of Y when $X = 3$.

Other values of $P(X = x)$ could be found in a similar manner and so we make the following definition.

Definition:

The *marginal distribution* of X is given by

$$P(X = x) = \sum_y P(X = x, Y = y)$$

where the sum is over all possible values of y.

We denote $P(X = x)$ by $f(x)$. The term *marginal distribution of X* arises because the distribution occurs in the margin of the table.

So,

$$f(x) = P(X = x) = \sum_y P(X = x, Y = y)$$

where the sum is over all possible values of y.

To find $f(x)$ then in this example, we must calculate

$$f(x) = \sum_{y=0}^{5-x} f(x, y) = \sum_{y=0}^{5-x} \binom{5}{x}\binom{5-x}{y}\left(\frac{1}{2}\right)^{10-x}$$

$$= \binom{5}{x}\left(\frac{1}{2}\right)^{10-x} \cdot \sum_{y=0}^{5-x}\binom{5-x}{y} = \binom{5}{x}\left(\frac{1}{2}\right)^{10-x} \cdot 2^{5-x}$$

$$\text{so } f(x) = \binom{5}{x}\left(\frac{1}{2}\right)^{5}, x = 0, 1, 2, \ldots, 5.$$

This verifies that X is binomial with $n = 5$ and $p = \frac{1}{2}$.

If we denote the marginal distribution of Y by $g(y)$ then, reasoning in the same way as we did for $f(x)$, we conclude that

$$g(y) = P(Y = y) = \sum_x P(X = x, Y = y)$$

where the sum is over all values of x.

The functions $f(x)$ and $g(y)$ are given in the margins in Table 5.1.

In this case it is not so easy to see the pattern in the distribution of Y, but there is one. First, by the definition of $g(y)$,

$$P(Y = y) = g(y) = \sum_{x=0}^{5-y} f(x, y) = \sum_{x=0}^{5-y} \binom{5}{x}\binom{5-x}{y}\left(\frac{1}{2}\right)^{10-x},$$

and this can be written as

$$g(y) = \sum_{x=0}^{5-y} \binom{5}{y}\binom{5-y}{x}\left(\frac{1}{2}\right)^{10-x}.$$

Now we remove common factors, rearrange, and insert the factor 1^{5-y-x}, to find that we can write $g(y)$ as

$$g(y) = \binom{5}{y}\left(\frac{1}{2}\right)^{10} \sum_{x=0}^{5-y}\binom{5-y}{x}2^x \cdot 1^{5-y-x}$$

$$= \binom{5}{y}\left(\frac{1}{2}\right)^{10}(2+1)^{5-y}$$

by the binomial theorem. It follows that

$$g(y) = \binom{5}{y} \cdot \left(\frac{1}{2}\right)^{10} \cdot 3^{5-y}, \quad y = 0, 1, \ldots, 5.$$

Some characteristics of the distribution of Y may be of interest. A graph of its values is shown in Figure 5.2.

Finally we find $E(Y)$, the expected number of heads as a result of this experiment. A computer algebra system will evaluate $E(Y)$ $= \sum_y y \cdot g(y) = \frac{5}{4}$. This also has an intuitive interpretation. One can argue that as a result of the first set of tosses, $\frac{5}{2}$ coins are expected to be tails, and of these, $\frac{1}{2}$ can be expected to result in heads on the second set of tosses, producing $\frac{5}{4}$ as $E(Y)$. Note that by this argument $E(Y)$ was found without using the probability distribution for Y. It is often possible, and on occasion desirable, to do this. We will give this process more validity later in this chapter. Now we consider a continuous example. ◄

Example 5.2.2

An investigator, intending to make a certain type of steel stronger, is examining the content of the steel. He considers adding carbon (X) and molybdenum (Y) to the steel and measuring the resulting strength. However, the carbon and molybdenum interact in a complex way in the steel being considered, so the investigator takes some data by varying the values

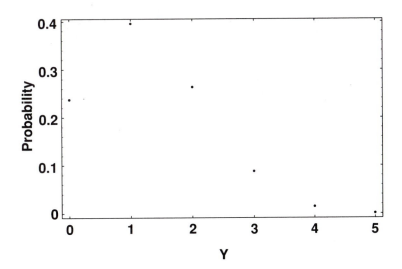

Figure 5.2 Marginal distribution for Y in Example 5.2.1.

of X and Y (whose values have been coded here for convenience). He finds that the resulting strength of the steel can be approximated by the function

$$f(x, y) = x^2 + \frac{8}{3}xy \text{ for } 0 < x < 1 \text{ and } 0 < y < 1.$$

A graph of this surface is shown in Figure 5.3.
We find that

$$\int_0^1 \int_0^1 f(x, y)\, dy\, dx = 1 \text{ and that } f(x, y) \geq 0.$$

Because of these two facts, and in analogy with univariate probability densities, we call $f(x, y)$ a *continuous bivariate probability density function.* Note again the distinction between discrete probability *distributions* and continuous probability *densities.*

Rather than sum, as we did in the discrete example, we integrate to find the marginal densities. We let

$$f(x) = \int_y f(x, y)\, dy$$

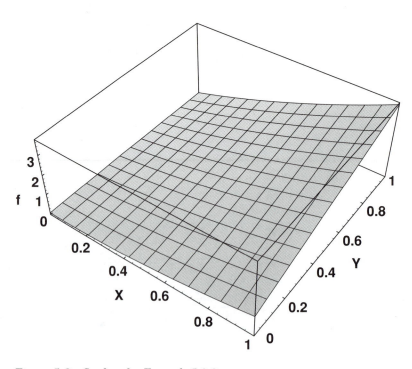

Figure 5.3   Surface for Example 5.2.2.

and

$$g(y) = \int_x f(x, y) \, dx,$$

provided, of course, that the integrals exist. In this case,

$$f(x) = \int_0^1 \left(x^2 + \frac{8}{3}xy \right) dy = x^2 + \frac{4x}{3}, \quad 0 < x < 1,$$

and

$$g(y) = \int_0^1 \left(x^2 + \frac{8}{3}xy \right) dx = \frac{4y + 1}{3}, \quad 0 < y < 1.$$

Graphs of these probability densities are shown in Figures 5.4 and 5.5.

We can verify that each of these is a univariate probability density. In Example 5.2.1 we found $E(Y)$ rather simply and without making use of the probability density for Y alone. In this case it is easy to verify that $E(Y) = \frac{11}{18}$, but it is not so easy to see how we would find this without finding $g(y)$ first. We will show how this can be done in section 5.4. ◀

Example 5.2.3

Since the volume under the bivariate probability density function is 1, parts of that volume represent probabilities. In this example we find that

$$P\left(X > \frac{1}{2}, \ Y < \frac{2}{3}\right) = \int_{\frac{1}{2}}^1 \int_0^{\frac{2}{3}} \left(x^2 + \frac{8}{3}xy \right) dy \, dx = \frac{5}{12}.$$

We can also compute more complex probabilities, such as $P(X > Y)$. To calculate this we must integrate over the triangular region in the sample space where $X > Y$. This gives

$$P(X > Y) = \int_0^1 \int_0^x \left(x^2 + \frac{8}{3}xy \right) dy \, dx$$

$$= \int_0^1 x^2 y + \frac{4xy^2}{3} \bigg|_0^x \, dx$$

$$= \int_0^1 \frac{7}{3}x^3 \, dx = \frac{7}{12}. \qquad ◀$$

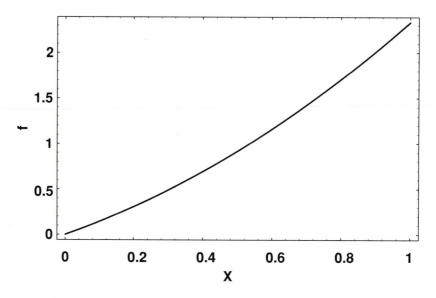

Figure 5.4   Marginal distribution for X, Example 5.2.2.

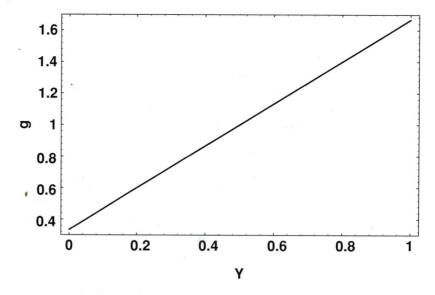

Figure 5.5   Marginal distribution for Y, Example 5.2.2.

Exercises 5.2

1. An engineering college has made a study of the grade-point averages of graduating engineers, denoted by the random variable Y. It is desired to study these as a function of high school grade-point averages, denoted by the random variable X.

The joint probability distribution is shown, where the grade point averages have been combined into five categories for each variable.

		X				
		2.0	2.5	3.0	3.5	4.0
	2.0	0.05	0	0	0	0
	2.5	0.10	0.04	0	0.01	0
Y	3.0	0.02	0.10	0.05	0.10	0.01
	3.5	0	0	0.10	0.20	0.10
	4.0	0	0	0.05	0.02	0.05

a. Find the marginal distributions for X and Y.

b. Find $E(X)$ and $E(Y)$.

c. Find $P(X \geq 3, Y \geq 3)$.

2. A random sample of 6 items is drawn from a factory's daily production. Let the random variables X and Y denote the number of good items and the number of defective items chosen, respectively. If the production contains 40 good and 10 defective items, find

a. the joint probability distribution function for X and Y.

b. the marginal distributions of X and Y.

3. Two cards are chosen without replacement from a deck of 52 cards. Let X denote the number of threes and Y denote the number of kings that are drawn.

a. Find the joint probability distribution of X and Y.

b. Find the marginal distributions of X and Y.

4. Suppose that X and Y are continuous random variables with joint probability density function

$$f(x, y) = k, \ 1 < x < 2, \ 2 < y < 4,$$

where k is a constant. (The random variables X and Y are said to have a *joint uniform probability density*.)

a. Find k.

b. Find the marginal densities for X and Y.

5. Suppose the joint (discrete) probability distribution function for discrete random variables X and Y is

$$P(X = x, Y = y) = k, \ x = 1, 2, \ldots, 10; \ y = 10 - x, 11 - x, \ldots, 10.$$

a. Find k.

b. Find the marginal distributions for X and Y.

6. A researcher finds that 2 random variables of interest, X and Y, have joint probability density function

$$f(x, y) = 24xy, \ 0 < x < 1, \ 0 < y < 1 - x.$$

 a. Show a graph of the joint probability density function.
 b. Calculate the marginal densities.
 c. Find $P(X > \frac{1}{2}, Y < \frac{1}{4})$.

7. A researcher is conducting a sample survey and is interested in a particular question to which respondents answer "yes" or "no." Suppose the probability that a respondent answers "yes" is p, and that the respondents' answers are independent. Let X denote the number of yeses in the first n_1 trials and Y denote the number of yeses in the next n_2 trials.

 a. Show the joint probability distribution function.
 b. Find the probability distribution of the random variable $X + Y$.

8. Refer to the previous exercise. If, in the second set of trials, the probability of a "yes" response has become $p_1 \neq p$, find the joint probability distribution function and the marginal distributions. Explain why the variable $X + Y$ is *not* binomial.

9. The number of telephone calls X that come into an office during a certain period of the day is distributed as a Poisson random variable with $\lambda = 6$ per hour. The calls are answered according to a binomial process with $p = \frac{3}{4}$. Let Y denote the number of calls answered.

 a. Find the joint probability distribution of X and Y.
 b. Express $P(Y = y)$ in simple form.
 c. Find $E(Y)$ without using the result in part **b** above.

10. Suppose that random variables X and Y have joint probability density

$$f(x, y) = \frac{1}{2\pi} e^{-(x^2+y^2)/2}, \ -\infty < x < \infty, \ -\infty < y < \infty.$$

 a. Show that the marginal densities are normal.
 b. Find $P(X > Y)$.

11. Three students are randomly selected from a group of 3 freshmen, 2 sophomores, and 2 juniors. Let X denote the number of freshmen selected and Y denote the number of sophomores selected. Find the joint probability distribution of X and Y.

12. Random variables X and Y are jointly distributed random variables with $f(x, y) = k$, $x = 0, 1, 2, \ldots$, and $y = 0, 1, 2, \ldots, 3 - x$.

 a. Find k.
 b. Find the marginal densities for X and Y.

13. Suppose that random variables X and Y have joint probability density $f(x, y) = kxy$ on the region bounded by the curves $y = x^2$ and $y = x$ in the first quadrant.

    a. Show that $k = 24$.

    b. Find the marginal densities $f(x)$ and $g(y)$.

14. Let X and Y be random variables with joint probability density function $f(x, y) = \frac{k}{x}$, $0 < y < x$, $0 < x < 1$.

    a. Show that $k = 1$.

    b. Find $P(X > \frac{1}{2}, Y < \frac{1}{4})$.

15. Random variables X and Y have joint probability density

$$f(x, y) = kx, \ x - 1 < y < 1 - x, \ 0 < x < 1.$$

    a. Find k.

    b. Find $g(y)$, the marginal density for Y.

    c. Find $\text{Var}(X)$.

16. Suppose that random variables X and Y have joint probability distribution function

$$f(x, y) = \frac{1}{21}(x + y), \ x = 1, 2, 3; \ y = 1, 2.$$

    a. Find the marginal densities for X and Y.

    b. Find $P(X + Y \le 3)$.

17. A fair coin is flipped 3 times. Let Y be the total number of heads on the first 2 tosses and let W be the total number of heads on the last 2 tosses.

    a. Determine the joint probability distribution of W and Y.

    b. Find the marginal distributions.

18. An environmental engineer measures the amount (by weight) of particulate pollution in air samples of a given volume collected over the smokestack of a coal-operated power plant. X denotes the amount of pollutant per sample collected when a cleaning device on the stack is not in operation, and Y denotes the same amount when the cleaning device is operating. It is known that the joint probability density function for X and Y is

$$f(x, y) = k, \ 0 \le x \le 2, \ 0 \le y \le 1, \ x > 2y.$$

    a. Find k.

    b. Find the marginal densities for X and Y.

    c. Find the probability that the amount of pollutant with the cleaning device in operation is at most $\frac{1}{3}$ of the amount without the cleaning device in operation.

19. Random variables X and Y have joint probability density function

$$f(x, y) = k, \ x \geq 0, \ y \geq 0, \ \frac{x}{4} + y \leq 1.$$

a. Show that $k = \frac{1}{2}$.
b. Find $P(X \geq 2, \ Y \geq \frac{1}{4})$.
c. Find $P(X \leq 1)$.

20. A fair die is thrown once; let X denote the result. Then X fair coins are thrown; let Y denote the number of heads that occur.

a. Find an expression for $P(Y = y)$.
b. Explain why $E(Y) = \frac{7}{4}$.

21. Suppose that X and Y are random variables whose joint probability density function is

$$f(x, y) = 3y, \ 0 < x < y < 1.$$

a. Show that $f(x, y)$ is a joint probability density function.
b. Find the marginal densities.
c. Show that $E\left[\dfrac{X}{Y}\right] = \dfrac{E[X]}{E[Y]}$. Does $E\left[\dfrac{Y}{X}\right] = \dfrac{E[Y]}{E[X]}$?

22. A fair coin is tossed until a head appears for the first time; denote the number of trials necessary by X. Then X fair coins are tossed; let Y denote the number of heads that appear.

a. Find the joint distribution of X and Y assuming that the coins are fair.
b. Find the marginal distributions of X and Y. Note: X is geometric; to simplify the distribution of Y, consider the binomial expansion $(1 - x)^{-n}$.
c. Show that $E(Y) = 1$ whether the coins are fair or not.
d. Find the marginal distribution for Y assuming that the coins are loaded to come up heads with probability p.

5.3 ▶ Conditional Distributions and Densities

In Example 5.2.1 we tossed 5 coins and recorded X, the number of heads that appeared. We then tossed the $5 - X$ coins that came up tails again and recorded Y, the number of heads in the second set of tosses. The joint probability distribution function is shown in Table 5.2.1.

We might be interested in some conditional probabilities such as the probability that the second set of tosses showed at least 2 heads, given that one head appeared on the first toss, or $P(Y \geq 2|X = 1)$.

We cannot look at the row for $X = 1$ and add the probabilities for $Y \geq 2$ because the probabilities in the row for $X = 1$ do not add up to 1; that is, the row for $X = 1$ is not a probability distribution. However, we know that

$$P(Y \geq 2|X = 1) = \frac{P(Y \geq 2, X = 1)}{P(X = 1)},$$

so a probability distribution can be created from the entries in the row for $X = 1$ by dividing each of them by $P(X = 1)$. If we do this, we find

$$P(Y \geq 2|X = 1) = \frac{\dfrac{60 + 40 + 10}{1024}}{\dfrac{160}{1024}} = \frac{11}{16}.$$

We conclude generally that

$$P(Y = y|X = x) = \frac{P(Y = y, X = x)}{P(X = x)} \text{ if } P(X = x) \neq 0.$$

This clearly holds for the case of discrete random variables. We proceed in the same way for continuous random variables, leading to the following definition.

Definition:
The *conditional probability distributions* $f(y \mid X = x)$ and $f(x \mid Y = y)$, which we denote by $f(y|x)$ and $f(x|y)$, are defined as

$$f(y \mid X = x) = f(y|x) = \frac{f(x, y)}{f(x)}$$

$$f(x \mid Y = y) = f(x|y) = \frac{f(x, y)}{g(y)}$$

where $f(x, y)$, $f(x)$, and $g(y)$ are the joint and marginal distributions for X and Y respectively.

Example 5.3.1

In Example 5.2.2 we considered the joint probability density function of the continuous variables X and Y where

$$f(x, y) = x^2 + \frac{8}{3}xy, \ 0 < x < 1, \ 0 < y < 1.$$

The conditional densities can be seen geometrically as the intersections of the joint probability density surface and vertical or horizontal planes. These curves of intersection are generally not probability densities because they do not have area 1, so they must be divided by the marginal densities to achieve this.

Since the marginal densities are $f(x) = x^2 + \frac{4x}{3}$, $0 < x < 1$, and $g(y) = \frac{4y+1}{3}$, $0 < y < 1$, it follows that

$$f(x|y) = \frac{3x^2 + 8xy}{4y + 1}, \quad 0 < x < 1$$

and

$$f(y|x) = \frac{x^2 + \frac{8}{3}xy}{x^2 + \frac{4}{3}x} = \frac{3x^2 + 8xy}{3x^2 + 4x}, \quad 0 < y < 1.$$

The domain of each variable is denoted above; the remaining symbol is understood to be fixed.

That each of these is a probability density can be verified by calculating that $\int_0^1 f(x|y)\, dx = 1$, and $\int_0^1 f(y|x)\, dy = 1$ and noting that $f(x \mid y) \geq 0$ and $f(y \mid x) \geq 0$. Areas under these conditional densities are probabilities. For example, if $Y = \frac{3}{4}$ then $f(x \mid y = \frac{3}{4}) = \frac{3}{4}(x^2 + 2x)$, so

$$P(X < \frac{1}{2}|Y = \frac{3}{4}) = \frac{3}{4}\int_0^{\frac{1}{2}} x^2 + 2x\, dx = \frac{7}{32}.$$

The mean values of the conditional densities are also of interest. These are denoted as $E(Y|X = x)$ and $E(X|Y = y)$ respectively. We see that

$$E(Y|X = x) = \int_y y \cdot f(y|x)\, dy$$

and

$$E(X|Y = y) = \int_x x \cdot f(x|y)\, dx.$$

In this example,

$$E(Y \mid X = x) = \int_0^1 y \cdot \frac{x^2 + \frac{8}{3}x \cdot y}{x^2 + \frac{4}{3}x}\, dy = \frac{9x + 16}{18x + 24}$$

and

$$E(X|Y = y) = \int_0^1 x \cdot \frac{x^2 + \frac{8}{3}x \cdot y}{\frac{4}{3}y + \frac{1}{3}}\, dx = \frac{9 + 32y}{12 + 48y}.$$

We note that $E(X|Y = y)$ and $E(Y|X = x)$ are functions of y and x respectively.   ◀

Example 5.3.2

Now we consider a slightly more complex continuous example. Let

$$f(x, y) = 2e^{-x-y}, \ x \geq 0, \ y \geq x.$$

Here we must be cautious in determining the limits of integration. A picture of the sample space is shown in Figure 5.6.
  The bivariate function itself, shown in Figure 5.7, is also interesting. The marginal densities are

$$f(x) = \int_{x}^{\infty} 2e^{-x-y} \, dy = 2e^{-2x}, \ x \geq 0$$

and

$$g(y) = \int_{0}^{y} 2e^{-x-y} dx = 2e^{-y}\left(1 - e^{-y}\right), \ y \geq 0.$$

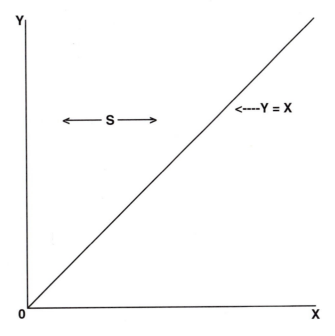

Figure 5.6   Sample space for $f(x, y) = 2e^{-x-y}$, $x \geq 0$, $y \geq x$.

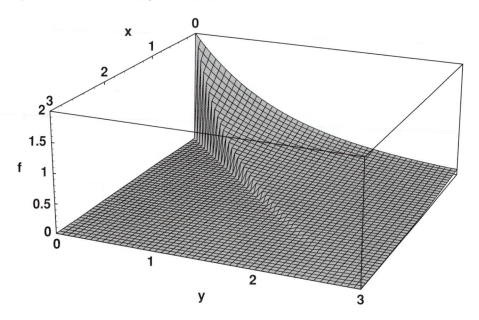

Figure 5.7   Probability surface for Example 5.3.2.

The conditional densities are then

$$f(x|y) = \frac{e^{-x}}{1 - e^{-y}}, \ 0 \le x \le y$$

and

$$f(y|x) = e^{x-y}, \ y \ge x.$$

The reader might verify that each of these is a probability density. The conditional expectations are then

$$E(X|Y = y) = \frac{e^y - y - 1}{e^y - 1} \text{ and } E(Y|X = x) = 1 + x.$$ ◀

Exercises 5.3

1. Suppose the joint probability density function of random variables X and Y is given by

$$f(x, y) = c \cdot (x + y), \ 0 < x < 1, \ 0 < y < 1.$$

   a. Show that $c = 1$.
   b. Find the marginal densities and the conditional densities, verifying in each case that these are probability densities.

c. Find $E(Y|X = x)$ and $E(X|Y = y)$.

2. Suppose X and Y are discrete random variables with

$$f(x, y) = \frac{x + y}{21}, \quad x = 1, 2, 3; \ y = 1, 2.$$

   a. Show that $f(x, y)$ is a probability distribution function.
   b. Find the conditional distributions.
   c. Find $E(Y|X = x)$ and $E(X|Y = y)$.

3. Let $f(x, y) = kxy, \ 0 < x < 1, \ 0 < y < x$ for random variables X and Y.

   a. Show that $k = 8$.
   b. Find the marginal densities and the conditional densities.

4. Random variables X and Y have joint probability density function

$$f(x, y) = c \cdot x \cdot (2 - x - y), \ 0 < x < 1, \ x < y < 1.$$

   a. Show that $c = 8$.
   b. Find the marginal densities.
   c. Verify that $f(y|x)$ is a probability density.

5. Suppose that random variables X and Y have joint probability density
   $f(x, y) = k \cdot x \cdot y$ for $x \geq 0$ and $y \geq 0$ on the disk $x^2 + y^2 \leq 1$.

   a. Find k.
   b. Find $E(Y|X = x)$ and then find $E(Y)$.

6. Let X and Y have joint probability distribution $f(0, 0) = \frac{1}{3}, \ f(0, 1) = \frac{1}{2},$
   $f(1, 1) = \frac{1}{6}$.

   a. Show that $f(x, y)$ is a joint probability distribution.
   b. Find the marginal and conditional distributions.

7. Random variables X and Y have $f(x, y) = k \cdot x \cdot y$, where the sample space is
   the finite area between $y = x^2$ and $y = x$.

   a. Find $E(Y|X = x)$ and $E(X|Y = y)$.
   b. Find $P(X > \frac{1}{2}|Y = \frac{1}{3})$.

8. Suppose X and Y have joint probability density function

$$f(x, y) = \frac{3}{8}(x + y^2), \ 0 < x < 2, \ 0 < y < 1.$$

   a. Find $P(X > 1, \ Y < \frac{1}{2})$.
   b. Find $f(y \mid x)$.
   c. Evaluate $P(Y > \frac{1}{2} \mid X = 1)$.

9. Let X and Y be random variables with joint probability density function

$$f(x, y) = 3x + 1, \ 0 < x < 1, \ 0 < y < 1 - x.$$

   a. Find the marginal densities.

   b. Find $f(y \mid x)$.

 10. Random variables X and Y have joint probability density function

$$f(x, y) = k \cdot x \cdot y^2, \ 0 < x < 1, \ x < y < 1.$$

   a. Find k.

   b. Find the marginal densities $f(x)$ and $g(y)$.

   c. Find the conditional density $f(y \mid x)$.

5.4 ▶ Expected Values and the Correlation Coefficient

If random variables X and Y have a joint probability density $f(x, y)$ then, as we have seen, X and Y are univariate random variables and hence have means and variances. It is true, of course, that

$$E(X) = \int_x x \cdot f(x) \, dx$$

and

$$E(Y) = \int_y y \cdot g(y) \, dy,$$

but these values can also be found from the joint probability density as

$$E(X) = \int_x \int_y x \cdot f(x, y) \, dy \, dx$$

and   (5.1)

$$E(Y) = \int_x \int_y y \cdot f(x, y) \, dy \, dx.$$

That these relationships are true can be easily established. Consider the above expression for $E(X)$ and factor out x from the inner integral. This gives

$$E(X) = \int_x x \cdot \left[\int_y f(x, y) dy \right] dx = \int_x x \cdot f(x) \, dx,$$

so the formulas are equivalent. Formulas 5.1 show that the marginals are not needed, however, since the order of the integration can often be reversed and so the expectations can be found without finding the marginal densities.

Now we turn to measuring the degree of dependence of one random variable with another. The idea of *independence* is easy to anticipate.

Definition:
Random variables X and Y are *independent* if, and only if,

$$f(x, y) = f(x) \cdot g(y) \text{ for all values of } x \text{ and } y,$$

where $f(x)$ and $g(y)$ are the marginal distributions or densities.

Usually it is not necessary to consider the joint density of independent variables because probabilities can be calculated from the marginal densities. If X and Y are independent, then

$$P(a < X < b, \ c < Y < d) = \int_a^b \int_c^d f(x, y) \, dy \, dx$$

$$= \int_a^b \int_c^d f(x) \cdot g(y) \, dy \, dx$$

$$= \int_a^b f(x) \, dx \cdot \int_c^d g(y) \, dy$$

so

$$P(a < X < b, \ c < Y < d) = P(a < X < b) \cdot P(c < Y < d),$$

showing that the joint density is not needed.

Referring to Example 5.3.2,

$$f(x, y) = 2e^{-x-y} \neq 2e^{-2x} \cdot 2e^{-y}(1 - e^{-y}) = f(x) \cdot g(y),$$

so X and Y are not independent. This raises the idea of measuring the extent of their dependence. In order to do this we first define the *covariance* of random variables X and Y as

Definition:
The *covariance* of random variables X and Y is

$$Covariance(X, Y) = \text{Cov}(X, Y) = E\left[[X - E(X)][Y - E(Y)]\right]. \quad (5.2)$$

As a special case, if $X = Y$, then $\text{Cov}(X, Y) = \text{Cov}(X, X)$ and the formula becomes the variance of X, $\text{Var}(X)$. But, unlike the variance, the covariance can be negative. Before calculating that, however, consider Formula 5.2. By expanding it we find that

$$\text{Cov}(X, Y) = E\,[X \cdot Y - X \cdot E(Y) - Y \cdot E(X) + E(X) \cdot E(Y)]$$

$$= E(X \cdot Y) - E(X) \cdot E(Y) - E(X) \cdot E(Y) + E(X) \cdot E(Y)$$

$$\text{Cov}(X, Y) = E(X \cdot Y) - E(X) \cdot E(Y),$$

a result that is often very useful.

In the example we are considering, $E(X \cdot Y) = 1$, $E(X) = \frac{1}{2}$, and $E(Y) = \frac{3}{2}$, so $\text{Cov}(X, Y) = \frac{1}{4}$.

The covariance is also used to define the *correlation coefficient*, $\rho(x, y)$, as we do now.

Definition:

The *correlation coefficient* of the random variables X and Y is

$$\rho(x, y) = \frac{\text{Cov}(x, y)}{\sigma_x \sigma_y},$$

where σ_x and σ_y are the standard deviations of X and Y respectively.

In this example we find that $\sigma_x = \frac{1}{2}$ and $\sigma_y = \frac{\sqrt{5}}{2}$, so $\text{Cov}(X, Y) = \frac{1}{\sqrt{5}} = 0.447214$. Now consider jointly distributed random variables X and Y.

$$E(X + Y) = \int_x \int_y (x + y) \cdot f(x, y)\, dy\, dx$$

$$= \int_x \int_y x \cdot f(x, y) dy dx + \int_x \int_y y \cdot f(x, y)\, dy\, dx$$

so

$$E(X + Y) = E(X) + E(Y),$$

or

The expectation of a sum is the sum of the expectations.

It is also easy to see that, if a and b are constants,

$$E(aX + bY) = aE(X) + bE(Y)$$

because the constants can be factored out of the integrals. The result can be easily generalized.

$$E(aX + bY + cZ + \cdots) = aE(X) + bE(Y) + cE(Z) + \cdots$$

As might be expected, variances of sums are a bit more complicated than expectations of sums. We begin with $\text{Var}(aX + bY)$. By definition this is

$$\text{Var}(aX + bY) = E\,[aX + bY - E\,[aX + bY]]^2$$

$$= E\,[a\,[X - E(X)] + b\,[Y - E(Y)]]^2\,.$$

Squaring, factoring out the constants, and taking the expectation term by term, we find

$$\text{Var}(aX + bY) = a^2 E\,[X - E(X)]^2$$

$$+ 2abE\,[[X - E(X)][Y - E(Y)]] + b^2 E[Y - E(Y)]^2,$$

and we recognize the terms in this expression as

$$\text{Var}(aX + bY) = a^2 \text{Var}(X) + 2ab\,\text{Cov}(X, Y) + b^2 \text{Var}(Y). \qquad (5.3)$$

So we can't say, as we did with expectations, that the variance of a sum is the sum of the variances, but this would be true if the covariance were zero. When does this occur? If X and Y are independent, then

$$E(X \cdot Y) = \int_x \int_y xy f(x, y)\, dy\, dx$$

$$= \int_x \int_y x \cdot y \cdot f(x) \cdot g(y)\, dy\, dx = \int_x x \cdot \left[\int_y y \cdot g(y)dy\right] \cdot f(x)\, dx$$

$$E(X \cdot Y) = E(Y) \cdot \int_x xf(x)dx = E(X) \cdot E(Y).$$

Since $E(X \cdot Y) = E(X) \cdot E(Y)$, $\text{Cov}(X, Y) = 0$. So we can say that, if X and Y are independent, then

$$\text{Var}(aX + bY) = a^2 \text{Var}(X) + b^2 \text{Var}(Y).$$

But the converse of this assertion is false: That is, if $\text{Cov}(X, Y) = 0$, then X and Y are not necessarily independent. An example will establish this. Consider the joint distribution of X and Y as given in the following table.

		Y		
		-1	0	1
	-1	a	b	a
X	0	b	0	b
	1	a	b	a

We select a and b so that $4a + 4b = 1$. Take $a = \frac{1}{6}$ and $b = \frac{1}{12}$ as an example among many choices that could be made. The symmetry in the table shows that $E(X) = E(Y) = 0$ and that $E(X \cdot Y) = 0$. So X and Y have $\text{Cov}(X, Y) = 0$. But $P(X = -1, Y = -1) = \frac{1}{6} \neq (\frac{5}{12}) \cdot (\frac{5}{12}) = \frac{25}{144}$, so X and Y are not independent. To take the more general case,

$$P(X = -1, Y = -1) = a \neq (2a + b)^2$$

so X and Y are not independent.

If $\text{Cov}(X, Y) = 0$, we call X and Y *uncorrelated*. We conclude that if X and Y are independent, then they are uncorrelated, but uncorrelated variables are not necessarily independent.

Finally in this section we establish a useful fact; namely, that the correlation coefficient ρ is always in the interval from -1 to 1:

$$-1 \leq \rho(x, y) \leq 1.$$

As a proof of this, consider variables X and Y that each have mean 0 and variance 1. (If this is not the case, transform the variables by subtracting their means and dividing by their standard deviations, producing X and Y.)

Since the variance of any variable is non-negative,

$$\text{Var}(X - Y) \geq 0$$

or, by Formula 5.3,

$$\text{Var}(X - Y) = \text{Var}(X) - 2\text{Cov}(X, Y) + \text{Var}(Y).$$

But $\text{Var}(X) = \text{Var}(Y) = 1$ and $\text{Cov}(X, Y) = \rho$, so

$$1 - 2\rho + 1 \geq 0,$$

which implies that $\rho \leq 1$.

The other half of the inequality can be established in a similar way by noting that $\text{Var}(X + Y) \geq 0$.

The reader will be asked in Exercise 5 of Section 5.5 to show that the transformations done to insure that the variables have mean 0 and variance 1 do not affect the correlation coefficient.

Example 5.4.1

The fact that the expectation of a sum is the sum of the expectations, and the fact that, if the summands are independent, the variance of the sum is the sum of the variances can be used to provide an elegant derivation of the mean and variance of a binomial random variable.

Suppose the random variable X represents the number of successes of n independent trials of a binomial random variable with probability p of success at any trial. Now define the variables

$$X_1 = \begin{cases} 1 & \text{if the first trial is a success} \\ 0 & \text{otherwise} \end{cases}$$

$$X_2 = \begin{cases} 1 & \text{if the second trial is a success} \\ 0 & \text{otherwise} \end{cases}$$

$$\vdots$$

$$X_n = \begin{cases} 1 & \text{if the } n\text{th trial is a success} \\ 0 & \text{otherwise.} \end{cases}$$

The $X_i's$ are often called *indicator random variables.*

Since X_i is 1 only when a success occurs and is 0 when a failure occurs,

$$X = X_1 + X_2 + \cdots + X_n = \sum_{i=1}^{n} X_i.$$

Now

$$E(X_i) = 1 \cdot p + 0 \cdot (1 - p) = p$$

and

$$E(X_i^2) = 1^2 \cdot p + 0^2 \cdot (1 - p) = p$$

so

$$E(X) = E\left(\sum_{i=1}^{n} X_i\right) = \sum_{i=1}^{n} E(X_i) = \sum_{i=1}^{n} p = np.$$

Also,

$$\text{Var}(X_i) = E(X_i^2) - \left[E(X_i)\right]^2 = p - p^2 = p(1 - p) = pq,$$

so

$$\text{Var}(X) = \text{Var}\left(\sum_{i=1}^{n} X_i\right) = \sum_{i=1}^{n} \text{Var}(X_i) = \sum_{i=1}^{n} pq = npq.$$

We have again established the formulas for the mean and variance of a binomial random variable. ◄

▌5.5 ► **Conditional Expectations**

Recall Example 5.2.1 where five fair coins were tossed, the coins showing heads being put aside and those showing tails tossed again. We called the random variable X the number of coins showing heads on the first toss and the random variable Y the number of coins showing heads on the second set of tosses. $E(Y)$ is of interest.

First note that $E(Y|X = x)$ and $E(X|Y = y)$ are functions of x and y respectively. If we let

$$E(Y|X = x) = k(x),$$

then we could consider the function $k(X)$ of the random variable X. This itself is a random variable. We denote this random variable by $E(Y|X)$ so

$$E(Y|X) = k(X).$$

We now return to our example. Since there are $5 - x$ coins to be tossed the second time and the probability of success is $\frac{1}{2}$, it follows that $E(Y|X = x) = (5 - x)/2$. This conditional expectation $E(Y \mid X)$ is a function of X. It also has an expectation that is

$$E\left[E(Y|X)\right] = E\left[\frac{5 - X}{2}\right] = \frac{5}{2} - \frac{E(X)}{2} = \frac{5}{2} - \frac{5/2}{2} = \frac{5}{4}$$

which we found as $E(Y)$. So we conjecture that $E[E(Y|X)] = E(Y)$ and that $E[E(X|Y)] = E(X)$.

We now justify this process of establishing *unconditional* expectations based on *conditional* expectations.

It is essential to note here that $E(Y|X)$ is a function of X; its expectation is found using the marginal distribution of X. Similarly, $E(X|Y)$ is a function of Y, and its expectation is found using the marginal distribution of Y.

We will give a proof that $E[E(Y|X)] = E(Y)$ using a continuous bivariate distribution, say $f(x, y)$. First we note that

$$E[E(Y|X)] = \int_x E(Y|X) \cdot f(x)\, dx$$

$$= \int_x \left[\int_y y \cdot \frac{f(x, y)}{f(x)}\, dy\right] \cdot f(x)\, dx$$

$$= \int_x \int_y y \cdot f(x, y)\, dy\, dx = E(Y).$$

The proof that $E[E(X|Y)] = E(X)$ is similar.

Example 5.5.1

We apply these results to Example 5.3.2. In that example, $f(x, y) = 2e^{-x-y}$, $x \geq 0$, $y \geq x$. We found that $f(y|x) = e^{x-y}$, $y \geq x$, and that $E(Y|X) = 1 + X$. Now we calculate, using $g(y)$,

$$E(Y) = \int_0^\infty 2 \cdot y \cdot e^{-y} \cdot (1 - e^{-y}) \, dy = \frac{3}{2}.$$

Also,

$$E[E(Y|X)] = E(1 + X) = \int_0^\infty (1 + x) \cdot 2 \cdot e^{-2x} dx.$$

This integral, as the reader can easily check, is also $\frac{3}{2}$. ◀

Example 5.5.2

An observation X is taken from a uniform density on $(0, 1)$, then observations are taken until the result exceeds X. Call this observation Y. What is the expected value of Y?

It appears obvious that, on average, the first observation is $\frac{1}{2}$. Then Y can be considered to be a uniform variable on the interval $(\frac{1}{2}, 1)$, so its expectation is $\frac{3}{4}$. Let us make these calculations more formal so that the technique could be applied in a less obvious situation.

We have that X is uniform on $(0, 1)$ so that $f(x) = 1$, $0 < x < 1$, and Y is uniform on $(x, 1)$ so that

$$f(y \mid x) = \frac{1}{1 - x}, \quad x < y < 1.$$

The joint density is then

$$f(x, y) = f(x) \cdot f(y \mid x) = \frac{1}{1 - x}, \quad x < y < 1, \ 0 < x < 1.$$

The marginal density for Y is then

$$g(y) = \int_0^y \frac{1}{1 - x} dx = -\ln(1 - y), \quad 0 < y < 1,$$

and

$$E(Y) = \int_0^1 -y \ln(1 - y) \, dy = \frac{3}{4},$$

verifying our earlier informal result. ◀

Exercises 5.5

1. Show that $\rho(X, Y)$ in Example 5.2.1 is $-1/\sqrt{3}$.

2. Find the correlation coefficient between X and Y for the probability density
$$f(x, y) = x + y, \ 0 < x < 1, \ 0 < y < 1.$$

3. Show that
$$\text{Cov}(aX + bY, cX + dY) = ac\text{Var}(X) + (ad + bc)\text{Cov}(X, Y) + bd\text{Var}(Y).$$

4. Show that $\text{Cov}(X - Y, X + Y) = \sigma_x^2 - \sigma_y^2$.

5. Show that $\rho(aX + b, cY + d) = \rho(X, Y)$ provided that $a > 0$ and $b > 0$.

6. Let $f(x, y) = k, 0 \le x \le 2, 0 \le y \le 1, x \ge 2y$.
   a. Are X and Y independent?
   b. Find $P(Y < \frac{X}{3})$.

7. Let $f(x, y) = \frac{3}{40}(x + \frac{1}{2}xy^2), \ 0 < x < 2; -2 < y < 2$.
   a. Show that $f(x, y)$ is a joint probability density function.
   b. Show that X and Y are independent.

8. Let $f(x, y) = (\frac{3}{8})(x^2 + y^2), \ -1 < x < 1; -1 < y < 1$.
   a. Verify that $f(x, y)$ is a joint probability density function.
   b. Find the marginal densities $f(x)$ and $g(y)$.
   c. Find the conditional densities $f(x|y)$ and $f(y|x)$.
   d. Verify that $E[E(X \mid Y)] = E(X)$.
   e. Find $P(X > \frac{1}{2} \mid Y = \frac{1}{2})$.

9. Let $f(x, y) = 1 + \frac{x}{2} - \frac{1}{4}xy^2, \ -\frac{1}{2} < x < \frac{1}{2}; -\frac{1}{2} < y < \frac{1}{2}$.
   a. Show that $f(x, y)$ is a joint probability density function.
   b. Show that $f(x|y) = f(x, y)$.

10. Let $f(x, y) = k \cdot x^2 \cdot (8 - y), \ x < y < 2x, \ 0 < x < 2$.
    a. Find the marginal and conditional densities.
    b. Find $E(Y)$ and then verify that $E[E(Y|X)] = E(Y)$.

11. Random variables X and Y have joint probability density function
$$f(x, y) = kx, \ 0 < x < 2, \ x^2 < y < 4.$$
    a. Show that $k = \frac{1}{4}$.

    b. Find $f(x|y)$.

    c. Find $E(X|Y = y)$.

12. Suppose that the joint probability density function for random variables X and Y is

$$f(x, y) = 2x + 2y - 4xy, \ 0 \le x \le 1, \ 0 \le y \le 1.$$

Are X and Y independent? Why or why not?

13. The joint probability density function for random variables X and Y is

$$f(x, y) = k, \ -1 < x < 1, \ x^2 < y < 1.$$

Find the conditional densities $f(x|y)$ and $f(y|x)$ and show that each of these is a probability density function.

14. Suppose that random variables X and Y have joint probability density function $f(x, y) = kx(2 - x - y), \ 0 < y < 1, \ 0 < x < y$. Find $f(x|y)$ and show that this is a probability density function.

15. Random variables X and Y have joint probability distribution function

$$f(x, y) = \frac{1}{6}, \ x = 1, 2, 3 \text{ and } y = 1, 2, \dots, 4 - x.$$

    a. Find formulas for $f(x)$ and $g(y)$, the marginal densities.

    b. Find a formula for $f(y|x)$ and explain why the result is a probability distribution function.

16. Find the correlation coefficient between X and Y if their joint probability density function is

$$f(x, y) = k, \ 0 \le x \le y, \ 0 \le y \le 1.$$

17. Random variables X and Y are discrete with joint distribution given by

		Y	
		0	1
X	0	$\frac{1}{6}$	$\frac{1}{3}$
	1	$\frac{1}{3}$	$\frac{1}{6}$

Find the correlation coefficient between X and Y.

18. Random variables X and Y have joint probability density function $f(x, y) = \frac{1}{\pi}, \ x^2 + y^2 \le 1$. Are X and Y independent?

19. Let random variables X and Y have joint probability density function $f(x, y) = kxy$ on the region bounded by $0 \le x \le 2, \ 0 < y < x$.

a. Show that $k = \frac{1}{2}$.

b. Find the marginal densities $f(x)$ and $g(y)$.

c. Are X and Y independent?

d. Find $P(X > \frac{3}{2}|Y = 1)$.

20. Random variables X and Y are uniformly distributed on the region $x^2 \le y \le 1, \ 0 \le x \le 1$. Verify that $E(Y \mid X) = E(Y)$.

21. Suppose that X and Y are random variables with $\sigma_x^2 = 10$, $\sigma_y^2 = 20$, and $\rho = \frac{1}{2}$. Find $\mathrm{Var}(2X - 3Y)$.

22. Let X_1, X_2, and X_3 be random variables with $E(X_i) = 0$ and $\mathrm{Var}(X_i) = 1$, $i = 1, 2, 3$. Also, $\mathrm{Cov}(X_i, X_j) = -\dfrac{1}{2}$ if $i \ne j$. Find $\mathrm{Var}(\sum_{i=1}^3 i X_i)$.

23. Let $X_1, X_2, X_3 \ldots, X_n$ be uncorrelated random variables with common variance σ^2. Show that $\overline{X} = \frac{1}{n} \sum_{i=1}^n X_i$ and $X_j - \overline{X}$ are uncorrelated, $j = 1, 2, \ldots, n$.

24. A box contains 5 red and 2 green marbles. Two marbles are drawn out, the first not being replaced before the second is drawn. Let X be 1 if the first marble is red and 0 otherwise; Y is 1 if the second marble is red and is 0 otherwise. Find the correlation coefficient between X and Y.

25. In 4 tosses of a fair coin, let X be the number of heads and Y be the length of the longest *run* of heads (a sequence of tosses of heads). For example, in the sequence $HTHH$, $X = 3$ and $Y = 2$. Find the correlation coefficient between X and Y.

26. An observation X is taken from the exponential distribution $f(x) = e^{-x}$, $x \ge 0$. Sampling then continues until an observation Y is, at most, X. Show that $E(Y) = 2 - \frac{\pi^2}{6}$. (Hint: Expand the integral in a power series to show that

$$\int_0^{\infty} \frac{xe^{-2x}}{1 - e^{-x}} dx = \frac{1}{2^2} + \frac{1}{3^2} + \frac{1}{4^2} + \ldots .)$$

27. Variances as well as expected values can be found by conditioning. Consider the formula

$$\mathrm{Var}(X) = E[\mathrm{Var}(X|Y)] + Var[E(X|Y)].$$

a. Verify that the formula gives the correct result for the joint probability distribution

$$f(x, y) = 8xy, 0 < y < x, \ 0 < x < 1.$$

b. Prove that the formula is correct in general.

28. A fair wheel is spun once and the result X is recorded. Then the wheel is spun again until the result is less than X. Call the second result Y.

a. Find the joint probability density for X and Y.
b. Find $E(X \mid Y)$ and $E(Y \mid X)$.
c. Verify that $E[E(Y \mid X)] = \frac{1}{4}$.

■ 5.6 ▶ Bivariate Normal Densities

The bivariate extension of the normal density is a very important example of a bivariate density. We study this density and some of its applications in this section.

We say that X and Y have a *bivariate normal density* if

$$f(x, y) = \frac{1}{2\pi\sigma_x\sigma_y\sqrt{1 - \rho^2}}\, e^{-\frac{1}{2(1-\rho^2)}\left\{\left(\frac{x-\mu_x}{\sigma_x}\right)^2 - 2\rho\left(\frac{x-\mu_x}{\sigma_x}\right)\left(\frac{y-\mu_y}{\sigma_y}\right) + \left(\frac{y-\mu_y}{\sigma_y}\right)^2\right\}}$$

for $-\infty < x < \infty$ and $-\infty < y < \infty$. Note that there are five parameters – the means and variances of each variable as well as ρ, the correlation coefficient between X and Y.

A graph of a typical bivariate normal surface is shown in Figure 5.8. The surface is a *standard bivariate normal surface* because X and Y each have mean 0 and variance 1, and ρ has been taken to be 0.

As one might expect, the marginal and conditional densities (and indeed the intersection of the surface with any plane perpendicular to the X, Y plane) are normal

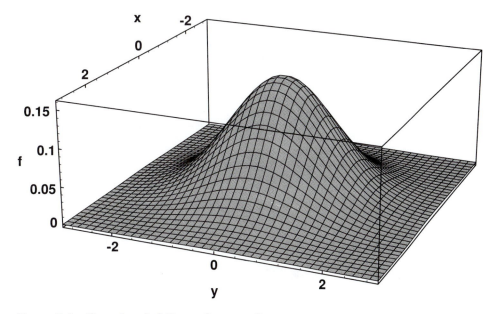

Figure 5.8   Normal probability surface, $\rho = 0$.

densities. We now establish some of these facts. For convenience, and without any loss of generality, we consider a bivariate normal surface with $\mu_x = \mu_y = 0$ and $\sigma_x = \sigma_y = 1$. We begin with a proof that the volume under the surface is 1.
The function we are considering is now

$$f(x, y) = \frac{1}{2\pi\sqrt{1-\rho^2}} e^{-(x^2-2\rho xy+y^2)\big/[2(1-\rho^2)]},$$

$$-\infty < x < \infty \text{ and } -\infty < y < \infty.$$

Completing the square in the exponent gives

$$f(x, y) = \frac{1}{2\pi\sqrt{1-\rho^2}} e^{-(x-\rho y)^2\big/[2(1-\rho^2)]-(1/2y^2)}.$$

So $\int_{-\infty}^{\infty}\int_{-\infty}^{\infty} f(x, y)\, dy\, dx$

$$= \int_{-\infty}^{\infty} \left[\int_{-\infty}^{\infty} \frac{1}{\sqrt{2\pi}\sqrt{1-\rho^2}} e^{-(x-\rho y)^2\big/[2(1-\rho^2)]}\, dx \right] \frac{1}{\sqrt{2\pi}} e^{-1/2y^2}\, dy.$$

The inner integral represents the area under a normal curve with mean ρy and variance $1 - \rho^2$ and so is 1. The outer integral is the area under a standard normal curve and so is 1 also, showing that $f(x, y)$ has volume 1.
This also shows that the marginal density for Y is

$$g(y) = \frac{1}{\sqrt{2\pi}} e^{-1/2y^2}, \quad -\infty < y < \infty,$$

with a similar result for $f(x)$.
Finding $\dfrac{f(x, y)}{g(y)}$ above gives

$$f(x|y) = \frac{1}{\sqrt{2\pi}\sqrt{1-\rho^2}} e^{-(x-\rho y)^2\big/[2(1-\rho^2)]},$$

which is a normal curve with mean ρy and standard deviation $\sqrt{1-\rho^2}$.

Now let's return to the general case where the density is not a standard bivariate normal density. It is easy to show that the marginals are now $N(\mu_x, \sigma_x)$ and $N(\mu_y, \sigma_y)$ and the conditional densities are

$$f(y|x) = N\left[\mu_y + \rho\frac{\sigma_y}{\sigma_x}(x - \mu_x), \sigma_y\sqrt{1 - \rho^2}\right]$$

and

$$f(x|y) = N\left[\mu_x + \rho\frac{\sigma_x}{\sigma_y}(y - \mu_y), \sigma_x\sqrt{1 - \rho^2}\right].$$

The expected value of Y given $X = x$, $E(Y \mid X = x)$, is called the *regression of Y on X*. Here we find that

$$E(Y \mid X = x) = \mu_y + \rho\frac{\sigma_y}{\sigma_x}(x - \mu_x),$$

a straight line.

If $\rho = 0$, note that $f(x, y) = f(x) \cdot g(y)$, so that X and Y are independent. In this case it is probably easiest to use the individual marginal densities in finding probabilities. If $\rho \neq 0$, it is probably best to standardize the variables before calculating probabilities. Some computer algebra systems can calculate bivariate probabilities.

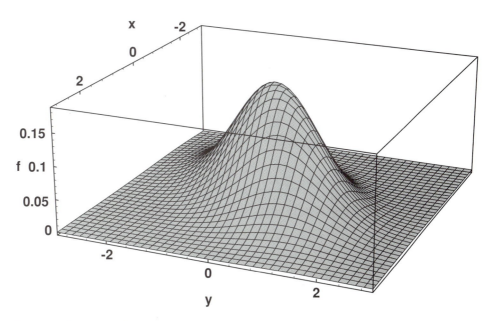

Figure 5.9   Normal bivariate surface, $\rho = 0.5$.

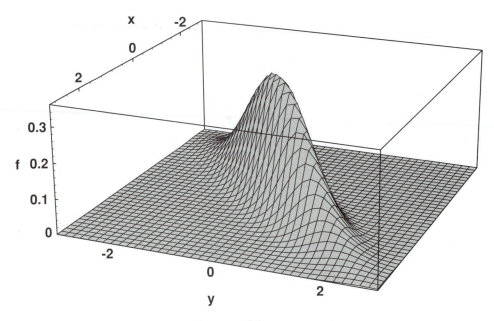

Figure 5.10   Normal bivariate surface, $\rho = 0.9$.

In Figures 5.9 and 5.10 we show two graphs to indicate the changes that result in the shape of the surface when the correlation coefficient varies.

5.6.1   Contour Plots

Contours or *level curves* of a surface show the locations of points for which the function, or height of the surface, takes on constant values. The contours are then slices of the surface with planes parallel to the X, Y plane.

If $\rho = 0$, we expect the contours to be circles, as the graph in Figure 5.11 shows.

If $\rho = 0.9$, however, the contours become ellipses, as the graph in Figure 5.12 shows.

Exercises 5.6

1. Let X and Y have a standard bivariate normal density with $\rho = 0.6$.

  a. Show that the marginal densities are normal.

  b. Show that the conditional densities are normal.

  c. Calculate $P(-2 < X < 1,\ 0 < Y < 2)$.

2. Height (X) and intelligence (Y) are presumed to be random variables that have a slight positive correlation coefficient. Suppose that these characteristics for a group

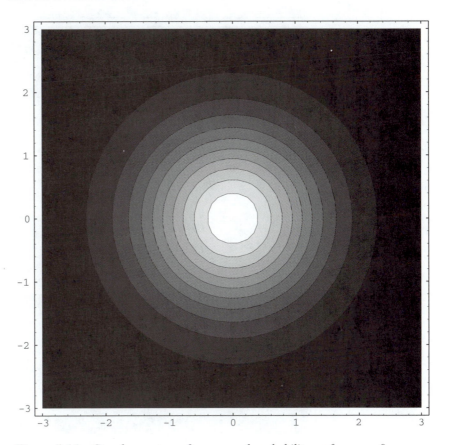

Figure 5.11   Circular contours for a normal probability surface, $\rho = 0$.

of people are distributed according to a bivariate normal curve with $\mu_x = 67$ in, $\sigma_x = 4$ in, $\mu_y = 114$, $\sigma_y = 10$, and $\rho = 0.20$.

a. Find $P(66 < X < 70,\ 107 < Y < 123)$.

b. Find the probability that a person whose height is 5 ft 7 in has an intelligence quotient of at least 121.

c. Find the regression of Y on X.

3. Show that ρ is the correlation coefficient between X and Y when these variables have a bivariate normal density.

4. The guidance system for a missile is being tested. The aiming point of the missile (X, Y) is presumed to be a bivariate normal density with $\mu_x = 0$, $\sigma_x = 1$, $\mu_y = 0$, $\sigma_y = 6$, and $\rho = 0.42$. Find the probability that the missile lands within 2 units of the origin.

5. Show that uncorrelated bivariate normal random variables are independent.

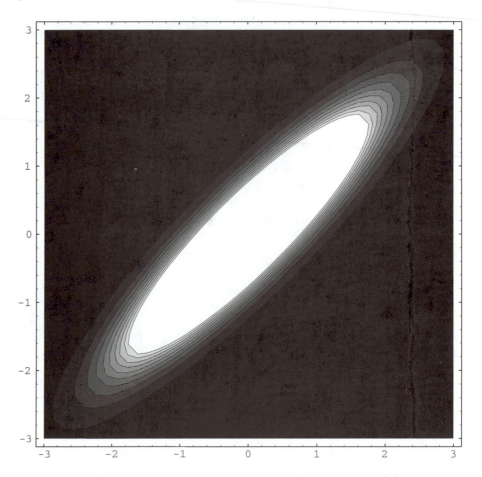

Figure 5.12   Elliptical contours for a normal probability surface, $\rho = 0.9$.

5.7 ▶ Functions of Random Variables

Joint distributions can be used to establish the probability densities of sums of random variables and can also be utilized to find the probability densities of products and quotients. We show how this is done through examples.

Example 5.7.1

Consider again two independent variables, X and Y, each uniformly distributed on $(0, 1)$. The joint density is then $f(x, y) = 1$, $0 < x < 1$, $0 < y < 1$.

   If $Z = X + Y$, the distribution function of Z, $G(z)$, can be found by calculating volumes beneath the joint density. The diagram in Figure 5.13 will help in doing this.

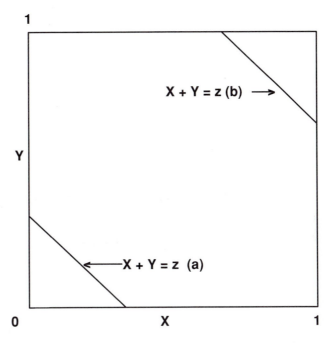

Figure 5.13   Jointly distributed uniform variables.

Computing volumes under the joint density function we find that

(a) $G(z) = P(X + Y \le z) = \frac{1}{2}z^2$ if $0 < z < 1$

and

(b) $G(z) = 1 - \frac{1}{2}[1 - (z - 1)]^2$ if $1 < z < 2$.

It follows from this that

$$g(z) = \begin{cases} z, & 0 < z < 1 \\ 2 - z, & 1 < z < 2. \end{cases}$$

This gives the triangular density we have seen previously.

The technique is feasible for more difficult densities as the next example shows.   ◀

Example 5.7.2

Suppose X and Y are each independently exponentially distributed with $f(x) = \lambda e^{-\lambda x}$, $x \ge 0$, with a similar distribution for Y. The distribution function for $Z = X + Y$ can be found by considering first the sample space shown in Figure 5.14.

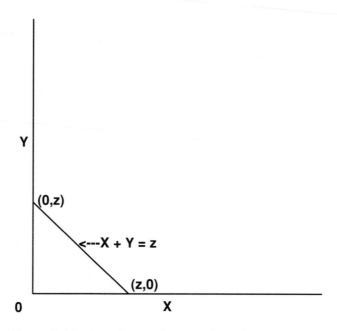

Figure 5.14   Sample space for two independent exponential variables.

Now,

$$G(z) = \int_0^z \int_0^{z-x} \lambda^2 e^{-\lambda(x+y)} \, dy \, dx,$$

which can be found to be

$$G(z) = 1 - e^{-\lambda z} - \lambda z e^{-\lambda z},$$

so that

$$g(z) = \lambda^2 z e^{-\lambda z}, \; z \geq 0.$$

This is the gamma density we have seen before.

This technique will work nicely on other random variables such as sums or quotients. An example of each follows.   ◄

Example 5.7.3

Let X and Y be independently distributed uniform random variables on $(0, 1)$ and consider the product, $Z = X \cdot Y$. Figure 5.15 will help in seeing that the appropriate volume under the joint density gives

$$G(z) = z + \int_z^1 \int_0^{z/x} 1 \, dy \, dx,$$

which is found to be

$$G(z) = z - z \ln z$$

so that

$$g(z) = -\ln z, \ 0 < z < 1. \qquad \blacktriangleleft$$

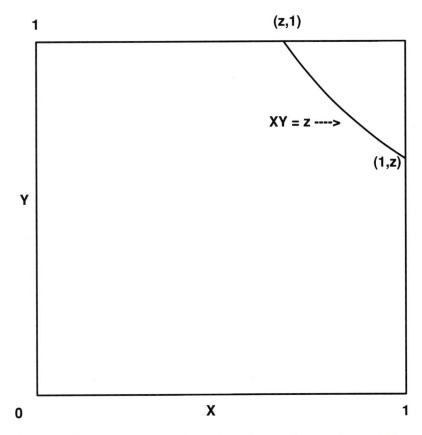

Figure 5.15 Sample space for the product of two uniform random variables.

Example 5.7.4

As a final example in this section, we consider the quotient of two independently distributed uniform variables, $Z = \frac{X}{Y}$. Figure 5.16 shows the relevant geometry.

Separating the two cases here and noting that $Z = \frac{X}{Y} \geq z$ above the line $Y = \frac{X}{z}$, we have

$$(a) \qquad G(z) = 1 - \int_0^1 \int_0^{x/z} 1 \, dy \, dx = 1 - \frac{1}{2z}, \quad 1 < z < \infty$$

and

$$(b) \qquad G(z) = \int_0^z \int_{x/z}^1 1 \, dy \, dx = \frac{z}{2}, \quad 0 \leq z \leq 1.$$

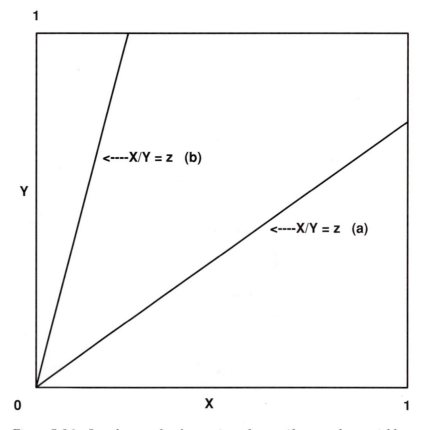

Figure 5.16 Sample space for the quotient of two uniform random variables.

This gives the density of the quotient as

$$g(z) = \begin{cases} \frac{1}{2}, & 0 < z < 1 \\ \frac{1}{2z^2}, & 1 < z < \infty. \end{cases}$$

◀

Exercises 5.7

1. Find the probability density for $Z = \frac{X}{Y}$ if X has an exponential density with mean 1 and Y has an independent exponential density with mean $\frac{1}{2}$.

2. Suppose X and Y are independent observations from $f(x) = 2x$, $0 < x < 1$. Find the probability density for $Z = X \cdot Y$.

3. Find the probability density for $Z = X + Y$ where X is an observation from $f(x) = 2x$, $0 < x < 1$, and Y is an independent observation from $g(y) = 2(1 - y)$, $0 < y < 1$.

4. Find the probability density for $Z = X \cdot Y$ where X and Y are independent observations from $f(x) = 3x^2$, $0 < x < 1$.

Chapter Review

In this chapter we studied random variables that are defined on the same sample space and considered their *joint probability distributions* or *joint probability densities*

$$f(x, y) = P(X = x, Y = y) \text{ if } X \text{ and } Y \text{ are discrete}$$

and

$$P(a < X < b, \ c < Y < d) = \int_c^d \int_a^b f(x, y) \, dy \, dx \text{ if } X \text{ and } Y \text{ are continuous.}$$

Our study has been confined to two random variables producing *bivariate* probability distributions or densities, although the techniques used in this chapter can often be extended to three or more random variables.

X and Y are random variables and their individual densities are called *marginal densities*. In the continuous case, denoting these by $f(x)$ and $g(y)$ respectively,

$$f(x) = \int_{-\infty}^{\infty} f(x, y) \, dy$$

and

$$g(y) = \int_{-\infty}^{\infty} f(x, y)\, dx.$$

Appropriate sums are used in the discrete case.

It is best to think of $f(x, y)$ geometrically and to plot the surface. Slices of the surface for which X or Y are constant produce curves that, because their areas are not usually one, are rarely probability densities. *Conditional densities* can be produced, however, by dividing the conditional density by the area under the curve:

$$f(x \mid Y = y) = f(x \mid y) = \frac{f(x, y)}{f(x)}$$

and

$$f(y \mid X = x) = f(y \mid x) = \frac{f(x, y)}{g(y)}.$$

Bivariate distributions or densities, like univariate distributions or densities, are summarized by means and variances. However, unlike univariate distributions or densities, means and variances for bivariate distributions or densities can be calculated in more than one way. For example,

$$E(X) = \int_{-\infty}^{\infty} x \cdot f(x)\, dx = \int_{-\infty}^{\infty}\int_{-\infty}^{\infty} x \cdot f(x, y)\, dy\, dx$$

with a similar result for $E(Y)$.

Functions of random variables are of interest, sums being of particular importance. We noted that

$$E(X + Y) = E(X) + E(Y)$$

for any random variables X and Y. However, a direct calculation shows that

$$\mathrm{Var}(X + Y) = \mathrm{Var}(X) + 2\mathrm{Cov}(X, Y) + \mathrm{Var}(Y)$$

where

$\mathrm{Cov}(X, Y) = E\big[[X - E(X)][Y - E(Y)]\big]$ denotes the *covariance* of X and Y.

If X and Y are independent, then $\mathrm{Var}(X + Y) = \mathrm{Var}(X) + \mathrm{Var}(Y)$. If X and Y are not independent, then they can be partially or totally dependent on one another. The degree of dependence can be measured by the *correlation coefficient, ρ*:

$$\rho = \frac{\mathrm{Cov}(X, Y)}{\sigma_x \sigma_y}.$$

It is known that $-1 \le \rho \le 1$.

We showed that for the random variables $E(X \mid Y)$ and $E(Y \mid X)$,

$$E[E(X \mid Y)] = E(X)$$

and

$$E[E(Y \mid X)] = E(Y).$$

Normal bivariate densities comprise an important family of bivariate densities. Their probability densities in standard form can be written as

$$f(x, y) = \frac{1}{2\pi\sqrt{1-\rho^2}}\, e^{-(x^2-2\rho xy+y^2)/[2(1-\rho^2)]}, \quad -\infty < x < \infty, \ -\infty < y < \infty.$$

Finally, we considered products and quotients of random variables, finding the distribution functions of these functions of two random variables. By differentiating the result, we found their respective probability density functions.

Problems for Review

Exercises 5.2 - # 2, 4, 5, 7, 8
Exercises 5.3 - # 1, 2, 5, 7
Exercises 5.5 - # 2, 6, 9, 10, 13
Exercises 5.6 - # 1, 2
Exercises 5.7 - # 2, 3

Supplementary Exercises for Chapter 5

1. For the joint probability density function $f(x, y) = k(x^2 + y^2)$, $-2 < x < 2$, $-2 < y < 2$,

   a. find k.
   b. Are X and Y independent?
   c. Find the marginal and conditional densities.

2. Given $f(x, y) = k \sin(x) \cos(y) e^{-x}$, $0 < x < \pi/2, -\pi/2 < y < \pi/2$,

   a. find k.
   b. find the marginal and conditional densities.

3. Two fair dice are thrown. Let X denote the largest face that appears and Y denote the smallest face that appears.

   a. Find the joint probability distribution.

b. Find the conditional distribution $f(x|y)$ and show that

$$E(X \mid Y = y) = \frac{42 - y^2}{13 - 2y}, \quad y = 1, 2, \ldots, 6.$$

Then use this result to find $E(X)$.

c. Show that $E(Y|X = x) = \frac{x^2}{2x-1}$, $x = 1, 2, \ldots, 6$ and use this result to find $E(Y)$.

d. Explain why $E(X + Y) = 7$.

4. Given $g(x, y) = (\frac{1}{12})(3x - 2y + 1)$, $1 < x < 3$, $0 < y < 1$,

a. find the marginal densities.

b. find the conditional densities.

c. verify that $E[E[X|Y]] = E[X]$ and that $E[E[Y|X]] = E[Y]$

5. Suppose that $f(x) = 1$, $0 < x < 1$ and $g(y) = 2y$, $0 < y < 1$. Find $P(Y < X)$ if X and Y are independent.

6. Consider the joint probability density function $f(x, y) = kx$, $x - 1 < y < 1 - x$, $0 < x < 1$, for random variables X and Y.

a. Show that $k = 3$.

b. Find the marginal density of Y.

c. Find the variance of X.

7. Suppose $f(x, y) = k$ is a joint probability density function on the area in the first quadrant between the curves $y = 4$ and $y = x^2$.

a. Find k.

b. Find $E(X|Y = y)$.

8. Random variables X and Y have $g(y) = \lambda^2 y e^{-\lambda y}$, $y \geq 0$ and $f(x|y) = 1/y$, $0 < x < y$.

a. Find the marginal density for X.

b. Find $E[Y|X = x]$.

9. X is an observation from $f(x) = 2x$, $0 < x < 1$, and Y is an observation from the same density but $Y \geq X$. Find $E(Y)$.

10. Show that random variables X and Y are independent for the joint probability density function

$$f(x, y) = k(4x - x^2 - 2xy^2 + \frac{1}{2}x^2 y^2), \quad 0 < x < 2, \ 0 < y < 1.$$

11. For what value of k is $f(x, y) = ke^{-\frac{x+y}{3}}$ a joint probability density function for $0 \le x \le 1$ and $0 \le y \le 1$?

12. X and Y have joint probability density function $f(x, y) = k$, $0 < x < 2$, $x^2 < y < 4$.

    a. Show that $k = \frac{3}{16}$.
    b. Find the marginal densities $f(x)$ and $g(y)$.
    c. Find $f(x \mid y)$.
    d. Find $E(X \mid Y = y)$.

13. Random variables X and Y have $g(y \mid X = x) = \frac{1}{2x}$, $0 < y < 2x$ and $f(x) = 24x^2$, $0 < x < \frac{1}{2}$.

    a. Find the joint probability density function $f(x, y)$.
    b. Find $g(y)$, the marginal density for Y.
    c. Are X and Y independent? Justify your answer.

14. Random variables X and Y have $E(X) = -5$, $E(Y) = 8$, $E(X^2) = 100$, $E(Y^2) = 364$, and $\text{Cov}(X, Y) = 100$. Find ρ.

15. The joint probability density function of X and Y is given by

$$f(x, y) = kx(x - y), \quad 0 < x < 1, \quad -x < y < x.$$

    a. Find k.
    b. Are X and Y independent? Support your answer.
    c. Find $P\left(X > \frac{1}{2} \mid Y = \frac{1}{4}\right)$.

16. A lot of television sets contains, unknown to the manufacturer, 3 with defective picture tubes, 4 with defective sound systems, and 5 that have no defective parts. Three of the sets are selected, without replacement. Let X denote the number in the sample with defective picture tubes and Y denote the number in the sample with defective sound systems.

    a. Find the joint probability distribution of X and Y.
    b. Find the marginal distributions.
    c. Find the mean and variance of X.
    d. Find $\text{Cov}(X, Y)$.
    e. Find ρ.
    f. Find $\text{Var}(X - Y)$.

17. Random variables X and Y have the following characteristics: $E(X) = 3$, $\text{Var}(X) = 10$, $E(Y) = 2$, $\text{Var}(Y) = 30$, $E(X \cdot Y) = 4$.

    a. Find $E(X^2)$.
    b. Find $\text{Cov}(X, Y)$.

    c. Find $\rho_{X,Y}$.

    d. Let $Z = 3X - 2$. Find $\rho_{X,Z}$.

    e. Find Var$(5X + 1)$.

    f. Find Var$(X - Y)$.

18. Suppose that a fair coin is tossed 3 times. Let X denote the total number of heads and let Y be the number of heads on just the first toss.

    a. Find the joint probability distribution of X and Y.

    b. Calculate $P(X \geq 1 \mid Y = 0)$.

    c. Find the marginal distributions of X and Y.

    d. Calculate the correlation coefficient.

19. Let R be the region bounded by $0 \leq x \leq 1$ and $0 \leq y \leq x^2$, and let the joint probability density function of X and Y be given by

$$f(x, y) = \begin{cases} kxy & \text{if } (x, y) \text{ is in } R \\ 0 & \text{otherwise.} \end{cases}$$

    a. Determine k.

    b. Find the marginal densities, $f(x)$ and $g(y)$.

    c. Are X and Y independent?

    d. Calculate $P\left(X > \frac{1}{2}, Y > \frac{1}{4}\right)$.

20. Consider a box that has 5 white marbles and 2 yellow marbles. A marble is drawn out and not replaced, and then a second marble is chosen. Random variable X is 1 if the first marble is yellow and is 0 otherwise. Random variable Y is 1 if the second marble is yellow and is 0 otherwise.

    a. What is the joint probability distribution of X and Y?

    b. Find the correlation coefficient.

    c. Are X and Y independent? Explain.

    d. Find Var$(X + Y)$.

21. In 4 tosses of a fair coin let X denote the number of heads and Y denote the longest run of heads. (A run of heads is a sequence of successive heads.)

    a. Show the joint probability distribution of X and Y in a table and determine the marginal distribution of Y.

    b. Find $P(X = 2 \mid Y < 2)$.

    c. Find the expected value of X.

22. Suppose X and Y are independent random variables with marginal densities $f(x) = \lambda e^{-\lambda x}$, $x \geq 0$ and $g(y) = \lambda e^{-\lambda y}$, $y \geq 0$.

    a. Find the joint probability density function of X and Y.

    b. Find $E(X \cdot Y)$.

c. Find $P(Y > 2X)$.

23. Let X and Y be random variables with probability density function

$$f(x, y) = \frac{k}{\sqrt{xy}}, \quad 0 < x < 1, \ 0 < y < 1.$$

a. Find k.
b. Find the marginal densities.
c. Find $P\left(Y > \frac{X}{2}\right)$.

24. Let random variables X and Y denote, respectively, the temperature and the time in minutes that it takes a diesel engine to start. The joint density for X and Y is

$$f(x, y) = c(4x + 2y + 1), \quad 0 \le x \le 4, \ 0 \le y \le 2.$$

a. Find c.
b. Find the marginal densities and decide whether or not X and Y are independent.
c. Find $f(x \mid Y = 1)$.

Chapter 6

Recursions and Markov Chains

▌6.1 ▶ Introduction

In this chapter we study two important parts of the theory of probability: recursions and Markov chains.

It happens that many interesting probability problems can be posed in a recursive manner; indeed, it is often most natural to consider some problems in this way. While we have seen several recursive functions in our previous work, we have not established their solution. Some of our examples will be recalled here, and we will show the solution of some recursions. We will also show some new problems and solve them using recursive functions. In particular we will devote some time to a class of probability problems known as *waiting time* problems.

Finally, we will consider some of the theory of *Markov chains,* which arises in a number of practical situations. This theory is quite extensive, so we are able to give only a brief introduction in this book.

▌6.2 ▶ Some Recursions and Their Solutions

We return now to problems involving recursions, considerably expanding our work in this area and considering more complex problems than we have previously.

Consider again the simple problem of counting the number of permutations of n distinct objects, letting P_n denote the number of permutations of these n distinct objects. P_n is, of course, a function of n. If we were to permute $n-1$ of these objects, a new object, the nth, could be placed in any one of the $n-2$ positions between the objects or in one of the two end positions, a total of n possible positions for the nth object. For a given permutation of the $n-1$ objects, each one of the n choices for the nth object gives a distinct permutation. This reasoning shows that

$$P_n = n \cdot P_{n-1}, \quad n \geq 1 \tag{6.1}$$

Formula 6.1 expresses one value of a function, P_n, in terms of another value of the same function, P_{n-1}. For this reason, Formula 6.1 is called a *recursion* or *recurrence relation* or *difference equation.*

We have encountered recursions several times previously in this book. Recall that in Chapter 1 we observed that the number of combinations of n distinct objects taken r at a time, $\binom{n}{r}$, could be characterized by the recursion

$$\binom{n}{r} = \frac{n-r+1}{r} \cdot \binom{n}{r-1}, \quad r = 1, 2, \ldots, n. \tag{6.2}$$

We saw in Chapter 2 that values of the binomial probability distribution, where

$$P(X = x) = \binom{n}{x} p^x q^{n-x}, \quad x = 0, 1, \ldots, n,$$

are related by the recursion

$$P(X = x) = \frac{n - x}{x + 1} \cdot P(X = x - 1), \quad x = 1, 2, \ldots, n.$$

This recursion was used to find the mean and the variance of a binomial random variable.

One of the primary values of a recursion is that, given a starting point, any value of the function can be calculated. In the example regarding the permutations of n distinct objects, it would be natural to let $P_1 = 1$. Then we find, by repeatedly applying the recursion, that

$$P_2 = 2P_1 = 2 \cdot 1 = 2$$
$$P_3 = 3P_2 = 3 \cdot 2 \cdot 1 = 6$$
$$P_4 = 4P_3 = 4 \cdot 3 \cdot 2 \cdot 1 = 24$$

and so on. It is easy to conjecture that the general *solution* of the recursion 6.1 is $P_n = n!$. The conjecture can be proved to be correct by showing that it satisfies the original recursion. To do this we check that

$$P_n = n \cdot P_{n-1}$$

giving

$$n! = n \cdot (n - 1)!,$$

which is true. So we have found a solution for the recursion.

In this example it is easy to see a pattern arising from some specific cases, and this led us to the general solution. Soon we will require a specific procedure for determining solutions for recursions where specific cases do not provide a hint of the general solution.

As another example, we saw in Chapter 1 that the solution for Equation 6.2 is

$$\binom{n}{r} = \frac{n!}{r!(n - r)!}$$

where a starting point is $\binom{n}{0} = 1$. Solutions for recursions are not, however, always so simple.

We want to abandon purely combinatorial examples now and turn our attention to recursions that arise in connection with problems in probability. It happens that many interesting problems can be described by recursions; we will show several of these, together with the solutions of these difference equations.

We begin with an example.

Example 6.2.1

A quality control inspector, thinking that he might make his work a bit easier, decides on the following inspection plan as either good or nonconforming items come off his assembly line: If an item is inspected, the next item is inspected with probability p; if an item is not inspected, the next item is inspected with probability $1 - p$. The inspector decides to make p small, hoping that he will inspect only a few items this way. Will his plan work?

A model of the situation can be constructed by letting a_n denote the probability that the nth item is inspected. So,

$$a_n = P(n\text{th item is inspected}).$$

The nth item will be subject to inspection in two mutually exclusive ways: either the $n - 1$ item is inspected, or it is not. Therefore,

$$a_n = pa_{n-1} + (1 - p) \cdot (1 - a_{n-1}) \text{ for } n \geq 2. \tag{6.3}$$

Letting $a_1 = 0$ (so the first item from the production line is not inspected), Recursion 6.3 gives the following values for some small values of n:

$$a_1 = 0$$
$$a_2 = 1 - p$$
$$a_3 = p(1 - p) + (1 - p)p = 2p(1 - p)$$
$$a_4 = 2p^2(1 - p) + (1 - p)[1 - 2p(1 - p)]$$
$$= 1 - 3p + 6p^2 - 4p^3.$$

If further values are required, it is probably most sensible to proceed using a computer algebra system that will calculate any number of these values. This is, in fact, one of the most useful features of a recursion—within reason, any of the values can be calculated from the problem itself. In the short term then, the question may be as valuable as the answer! In some cases, the question is *more* valuable than the answer because the answer may be very complex. In many cases the answer cannot be found at all, so we must be satisfied with a number of special cases.

To return to the problem, how do the values of a_n behave as n increases? First we note that if $p = \frac{1}{2}$, then $a_2 = a_3 = a_4 = \frac{1}{2}$, prompting us to look at Recursion 6.3 when $p = \frac{1}{2}$. It is easy to see that in that case, $a_n = \frac{1}{2}$ for all values of n, so if the inspector takes $p = \frac{1}{2}$, he will inspect $\frac{1}{2}$ of the items.

The inspector now searches for other values of p, hoping to find some that lead to less work.

A graph of a_{10} as a function of p, found using a computer algebra system, is shown in Figure 6.1.

This shows the inspector that, alas, any reasonable value for p leads to inspection about half the time! The graph indicates that a_{10} is very close to $\frac{1}{2}$ for $\frac{1}{4} \le p \le \frac{3}{4}$, but even values of p outside this range affect the value

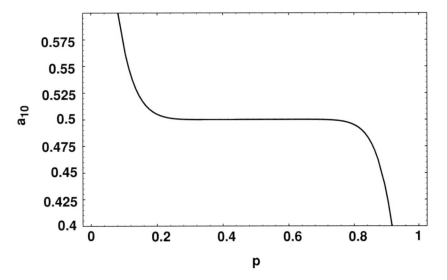

Figure 6.1 a_{10} for the quality control inspector.

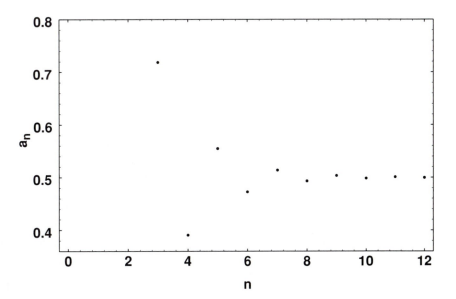

Figure 6.2 a_n for $p = \frac{1}{4}$.

of a_{10} in the early stages only. Figure 6.2 shows a graph of a_n for $p = \frac{1}{4}$, which indicates the very rapid convergence to $\frac{1}{2}$.

This evidence also suggests writing the solutions for Recursion 6.3 in terms of $(p - \frac{1}{2})$. Doing that we find

$$a_1 = 0$$

$$a_2 = 1 - p = \frac{1}{2} - \left(p - \frac{1}{2}\right)$$

$$a_3 = 2p(1 - p) = \frac{1}{2} - 2\left(p - \frac{1}{2}\right)^2$$

$$a_4 = 1 - 3p + 6p^2 - 4p^3 = \frac{1}{2} - 2^2\left(p - \frac{1}{2}\right)^3.$$

This strongly suggests that the general solution for recursion 6.3 is

$$a_n = \frac{1}{2} - 2^{n-2}\left(p - \frac{1}{2}\right)^{n-1}.$$

Direct substitution in Recursion 6.3 will verify that this conjecture is correct. Since $\mid p - \frac{1}{2} \mid \, < 1$, we see that $a_n \to \frac{1}{2}$ as $n \to \infty$. This could also have been predicted from Recursion 6.3, because, if $a_n \to L$, then $a_{n-1} \to L$ also; so, from Recursion 6.3,

$$L = pL + (1 - p)(1 - L),$$

whose solution is $L = \frac{1}{2}$.

Our solution used the fact that the first item was not inspected and so it may be thought that the long-range behavior of a_n may be dependent upon that assumption. This, however, is not true; in an exercise, the reader will be asked to show that $a_n \to \frac{1}{2}$ as $n \to \infty$ if $a_1 = 1$, that is, if the first item is inspected.

We used graphs and several specific values of a_n to conjecture the solution of the recursion as well as its long-term behavior. While it is often possible to find exact solutions for recursions, often the behavior of the solution for large values of n is of most interest; this behavior can frequently be predicted when exact solutions are not available. However, an exact solution for Recursion 6.3 can be constructed, and since the solution is typical of that of many kinds of recursions, it is shown now. ◀

Solution of the Recursion 6.3

While many computer algebra systems solve recursions, we give here some indication of the algebraic steps involved in this example as well as others in this chapter. We begin with the recursion

$$a_n = pa_{n-1} + (1 - p) \cdot (1 - a_{n-1}) \text{ for } n \geq 2$$

and write it as

$$a_n - (2p - 1)a_{n-1} = 1 - p, \quad n \geq 2.$$

Note first that the recursion has coefficients that are constants—they are not dependent upon the variable n. Note also that the right side of the equation is constant as well. We will consider here recursions having constant coefficients for the variables, but possibly having functions of n on the right side.

The solution of these equations is known to be composed of two parts: the solution of the *homogeneous* equation, in this case

$$a_{n,h} - (2p - 1)a_{n-1,h} = 0,$$

and a *particular solution,* some specific solution of

$$a_{n,p} - (2p - 1)a_{n-1,p} = 1 - p.$$

The general solution is known to be the sum of $a_{n,h}$ and $a_{n,p}$:

$$a_n = a_{n,h} + a_{n,p}.$$

We now show how to determine these two components of the solution. The homogeneous equation may be written as

$$a_{n,h} = (2p - 1)a_{n-1,h}.$$

This suggests that a value of the function $a_{n,h}$ is a constant multiple of the previous value $a_{n-1,h}$. Suppose that

$$a_{n,h} = r^n \text{ for some constant } r.$$

It follows that

$$r^n - (2p - 1)r^{n-1} = 0,$$

from which we conclude that $r = 2p - 1$. Since the equation is homogeneous, a constant multiple of a solution is also a solution, so

$$a_{n,h} = c(2p - 1)^n \text{ where } c \text{ is some constant.}$$

The equation $r^n - (2p-1)r^{n-1} = 0$ is often called the *characteristic equation.* The solutions for r are called *characteristic roots.*

The particular solution is some specific solution of

$$a_{n,p} - (2p-1)a_{n-1,p} = 1 - p.$$

Since the right side is a constant, we try a constant, say k, for a_n. The situation when the right side is a function of n is considerably more complex; we will encounter some of these equations later in this chapter. Substituting the constant k into Recursion 6.3 gives

$$k - (2p-1)k = 1 - p$$

whose solution is $k = \frac{1}{2}$. The complete solution for the recursion is the sum of the homogeneous solution and a particular solution so

$$a_n = \frac{1}{2} + c(2p-1)^n.$$

Now $a_1 = 0$, giving $c = -\frac{1}{2(2p-1)}$. Writing $2p - 1 = 2(p - \frac{1}{2})$, we find that

$$a_n = \frac{1}{2} - 2^{n-2}\left(p - \frac{1}{2}\right)^{n-1}, \text{ for } n \geq 1,$$

which is the solution previously found.

The reader who wants to learn more about the solution of recursions is urged to read Grimaldi [11] or Goldberg [8]. We proceed now with some more difficult examples.

Example 6.2.2

In Chapter 1 we considered the sample space when a loaded coin is flipped until two heads in a row occur. The reader may recall that if the event HH occurs for the first time at the nth toss, that the number of points in the sample space for n tosses can be predicted from that of $n - 1$ tosses and $n - 2$ tosses by the Fibonacci sequence. We did not, at that time, discover a formula for the probability that the event will occur for the first time at the nth toss; we will do so now.

Let b_n denote the probability that HH appears for the first time at the nth trial. Now consider a sequence of n trials for which HH appears for the first time at the nth trial. Since such a sequence can begin with either a tail or a head followed immediately by a tail (so that HH does not appear on the second trial),

$$b_n = qb_{n-1} + pqb_{n-2}, \; n \geq 3 \qquad (6.4)$$

is a recursion describing the problem. We also take $b_1 = 0$ and $b_2 = p^2$.

This recursion gives the following values for some small values of n:

$$b_3 = qp^2$$
$$b_4 = qp^2$$
$$b_5 = q^2p^2(1 + p)$$
$$b_6 = q^2p^2(1 + qp)$$
$$b_7 = p^2q^3(1 + pq + p + p^2).$$

It is now very difficult to detect a pattern in the results (although there is one). The behavior of b_n can be seen in the graphs in Figures 6.3 and 6.4 where the values of a_n are shown for a fair coin and then a loaded one.

We proceed with the solution of $b_n = qb_{n-1} + pqb_{n-2}$, $n \geq 3$, $b_1 = 0, b_2 = p^2$. Here the equation is homogeneous, and we write it as

$$b_n - qb_{n-1} - pqb_{n-2} = 0.$$

The solution in this case is similar to that of Example 6.2.1. If we presume that $b_n = r^n$, the characteristic equation becomes

$$r^2 - qr - pq = 0,$$

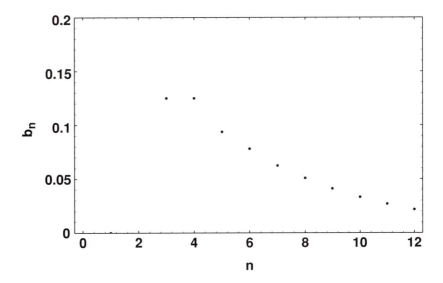

Figure 6.3 b_n for a fair coin.

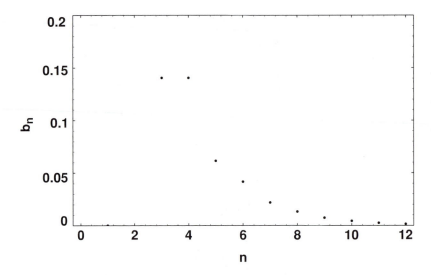

Figure 6.4 b_n for a coin with $p = \frac{3}{4}$.

giving two distinct characteristic roots, $r = \frac{q \pm \sqrt{q^2 + 4pq}}{2}$. Since the sum of solutions for a linear homogeneous equation must also be a solution, and because a constant multiple of a solution is also a solution, the general solution is

$$b_n = c_1 \left(\frac{q + \sqrt{q^2 + 4pq}}{2} \right)^n + c_2 \left(\frac{q - \sqrt{q^2 + 4pq}}{2} \right)^n .$$

The constants c_1 and c_2 can be determined from the boundary conditions $b_1 = 0, b_2 = p^2$, giving

$$b_n = \frac{2p^2}{q\sqrt{q^2 + 4pq} + q^2 + 4pq} \left(\frac{q + \sqrt{q^2 + 4pq}}{2} \right)^n$$

$$- \frac{2p^2}{q\sqrt{q^2 + 4pq} - q^2 - 4pq} \left(\frac{q - \sqrt{q^2 + 4pq}}{2} \right)^n . \qquad (6.5)$$

Computer algebra systems often solve recursions. Such a system solves the recursion 6.4 in the case where $p = q = \frac{1}{2}$ as

$$b_n = \left(\frac{1 + \sqrt{5}}{4} \right)^n \cdot \left(\frac{5 - \sqrt{5}}{10} \right) - \left(\frac{1 - \sqrt{5}}{4} \right)^n \cdot \left(\frac{5 + \sqrt{5}}{10} \right) .$$

This result can also be found by substituting $p = q = \frac{1}{2}$ in Formula 6.5. Other values for p and q make the solution much more complex.   ◀

Mean and Variance

The recursion $b_n = qb_{n-1} + pqb_{n-2}$, $n \geq 3$, can be used to determine the mean and variance of N, the number of tosses necessary to achieve two heads in a row. We employ a technique that is very similar to the one we used in Chapter 2 to find means and variances of random variables. Multiplying the recursion through by n and summing from 3 to infinity gives

$$\sum_{n=3}^{\infty} nb_n = \sum_{n=3}^{\infty} qnb_{n-1} + \sum_{n=3}^{\infty} pqnb_{n-2}.$$

This becomes

$$\sum_{n=2}^{\infty} nb_n - 2b_2 = q\sum_{n=3}^{\infty}[(n-1)+1]b_{n-1} + pq\sum_{n=3}^{\infty}[(n-2)+2]b_{n-2}.$$

Expanding and simplifying gives

$$E(N) - 2b_2 = qE(N) + q + pqE(N) + 2pq$$

from which it follows that

$$E(N) = \frac{1+p}{p^2}.$$

If $p = \frac{1}{2}$, this gives an average waiting time of six tosses to achieve two heads in a row. The result differs a bit from $\frac{1}{p^2}$, a result that might be anticipated from the geometric distribution, but we note that the variable is *not* geometric here.

The variance is calculated in much the same way. We begin with

$$\sum_{n=3}^{\infty} n(n-1)b_n = q\sum_{n=3}^{\infty}[(n-1)(n-2)+2(n-1)]b_{n-1}$$

$$+ pq\sum_{n=3}^{\infty}[(n-2)(n-3)+4(n-2)+2]b_{n-2}.$$

Expanding and simplifying gives

$$E[N(N-1)] - 2b_2 = qE[N(N-1)] + 2qE[N] + pqE[N(N-1)]$$

$$+ 4pqE[N] + 2pq.$$

It follows that $Var(N) = \dfrac{1 + 2p - 2p^2 - p^3}{p^4}$. If $p = \frac{1}{2}$, $Var(N) = 22$ tosses. We turn now to other examples.

Example 6.2.3

Consider again a sequence of Bernoulli trials with p the probability of success at a single trial. In a series of n trials, what is the probability that the sequence SF never appears?

Here a few points in the sample space will assist in seeing a recursion for the probability.

$$n = 2 \quad \begin{matrix} SS \\ FS \\ FF \end{matrix}$$

$$n = 3 \quad \begin{matrix} SSS \\ FSS \\ FFS \\ FFF \end{matrix}$$

$$n = 4 \quad \begin{matrix} SSSS \\ FSSS \\ FFSS \\ FFFS \\ FFFF \end{matrix}$$

It is now evident that a sequence in which SF never appears can arise in one of two mutually exclusive ways: Either the sequence is all F's, or when an S appears, it must be followed by all S's. The latter sequences end in S and are preceded by a sequence of $n - 1$ trials in which no sequence SF appears. So, if u_n denotes the probability that a sequence of n trials never contains the sequence SF, then

$$u_n = q^n + pu_{n-1}, \ n \geq 2, \ u_1 = 1$$

or

$$u_n - pu_{n-1} = q^n, n \geq 2, \ u_1 = 1. \tag{6.6}$$

A few values of u_n are as follows:

$$u_1 = 1$$
$$u_2 = q^2 + p$$
$$u_3 = q^3 + pq^2 + p^2.$$

These can be rewritten as

$$u_1 = \frac{q^2 - p^2}{q - p} \text{ if } p \neq q$$

$$u_2 = \frac{q^3 - p^3}{q - p} \text{ if } p \neq q$$

$$u_3 = \frac{q^4 - p^4}{q - p} \text{ if } p \neq q.$$

This leads to the conjecture that $u_n = \frac{q^{n+1}-p^{n+1}}{q-p}$, $n \geq 1$, $p \neq q$. The validity of this can be seen by substituting u_n into the original Recursion 6.6. Substituting in the right side of Recursion 6.6 we find, provided that $p \neq q$,

$$q^n + p \cdot \frac{q^n - p^n}{q - p}$$

which simplifies to

$$\frac{q^{n+1} - p^{n+1}}{q - p},$$

verifying the solution.

The solution of the Recursion 6.6 is also easy to construct directly. The characteristic equation is

$$r^n - p \cdot r^{n-1} = 0,$$

giving the characteristic root $r = p$. Therefore the homogeneous solution is

$$u_{n,h} = c \cdot p^n.$$

Now we seek a particular solution. Since the right side of $u_n - pu_{n-1} = q^n$ is q^n, suppose that we try

$$u_{n,p} = k \cdot q^n.$$

Substituting in the recursion, we have

$$k \cdot q^n - k \cdot p \cdot q^{n-1} = q^n,$$

and we find that

$$k = \frac{q}{q - p}, \text{ provided that } q \neq p.$$

So $u_{n,p} = \dfrac{q^{n+1}}{q - p}$, $q \neq p$, and the general solution is

$$u_n = cp^n + \frac{q^{n+1}}{q - p}, \text{ provided that } q \neq p.$$

By imposing the condition $u_1 = 1$ and simplifying, we find, as before, that

$$u_n = \frac{q^{n+1} - p^{n+1}}{q - p}, \quad n \geq 1, \ q \neq p. \tag{6.7}$$

We now investigate the case when $p = q$. In that situation, Recursion 6.6 becomes

$$u_n - \frac{1}{2}u_{n-1} = \left(\frac{1}{2}\right)^n. \tag{6.8}$$

Now the homogeneous solution is

$$u_{n,h} = c \cdot \left(\frac{1}{2}\right)^n,$$

so a particular solution $u_{n,p} = \left(\frac{1}{2}\right)^n$, a natural choice for $u_{n,p}$, will only produce 0 when substituted into the left side of Recursion 6.8. We try the function

$$u_{n,p} = k \cdot n \cdot \left(\frac{1}{2}\right)^n$$

for the particular solution. Then the left side of 6.8 becomes

$$k \cdot n \cdot \left(\frac{1}{2}\right)^n - \frac{1}{2} \cdot k \cdot (n - 1) \cdot \left(\frac{1}{2}\right)^{n-1}.$$

This simplifies to $k \cdot \left(\frac{1}{2}\right)^n$, so $k = 1$ and we have found the general solution,

$$u_n = c \cdot \left(\frac{1}{2}\right)^n + n \cdot \left(\frac{1}{2}\right)^n.$$

The boundary condition $u_1 = 1$ gives $c = 1$. The general solution in this case is

$$u_n = (n + 1)\left(\frac{1}{2}\right)^n.$$

This solution can also be found by using L'Hôspital's Rule in Formula 6.7. In this case we have

$$\lim_{p \to 1/2} \frac{q^{n+1} - p^{n+1}}{q - p}$$

$$= \lim_{p \to 1/2} \frac{(1 - p)^{n+1} - p^{n+1}}{1 - 2p}$$

$$= \lim_{p \to 1/2} \frac{-(n + 1)(1 - p)^n - (n + 1)p^n}{-2}$$

$$= (n + 1)(\frac{1}{2})^n. \qquad \blacktriangleleft$$

Exercises 6.2

Exercises 1–3 refer to a sequence of Bernoulli trials, where p is the probability of an event, and $p + q = 1$.

1. Describe a problem for which the recursion $a_n = qa_{n-1}$, $n > 1$, where $a_1 = 1 - q$ is appropriate. Then solve the recursion verifying that it does in fact describe the problem.

2. Let a_n denote the probability that a sequence of Bernoulli trials with probability of success p has an odd number of successes.

   a. Show that $a_n = p(1 - a_{n-1}) + qa_{n-1}$, for $n \geq 1$ if $a_0 = 0$. (Hint: Condition on the result of the first toss.)

   b. Solve the recursion in part **a**.

3. **a.** Find a recursion for the probability a_n that at least 2 successive successes occur in n Bernoulli trials.

   b. In part **a**, let $p = \frac{1}{2}$. Show that $a_n = 1 - \frac{f_{n+2}}{2^n}$, where f_n is the nth term in the Fibonacci sequence 1, 1, 2, 3, 5, 8, ...

4. Two machines in a manufacturing plant produce items that are either good (g) or unacceptable (u). Machine 1 has produced g good and u unacceptable items, while

the situation with machine 2 is exactly the reverse; it has produced u good items and g unacceptable items. An inspector is required to sample the production of the machines. To achieve a random order of items from each of the machines, the following plan is devised:

1. The first item is drawn from the output of machine 1.
2. Drawn items are returned to the output of the machine from which they were drawn.
3. If a sampled item is good, the next item is drawn from the first machine. If the sampled item is unacceptable, the next item is drawn from the second machine.

What is the probability that the nth sampled item is good?

5. A basketball player makes a series of free throw attempts. If he makes a shot, he makes the next one also with probability p_1. However, if he misses a shot, he makes the next one with probability p_2 where $p_1 \neq p_2$. If he makes the first shot, what is the probability that he makes the nth shot?

6. A coin, loaded to come up heads with probability p, is tossed until the sequence TH occurs for the first time. Let a_n denote the probability that the sequence TH occurs for the first time at the nth toss.

   a. Show that $a_n = pa_{n-1} + pq^{n-1}$, if $n \geq 3$ where $a_1 = 0$ and $a_2 = pq$.
   b. Show that the average waiting time for the first occurrence of the sequence TH is $\frac{1}{pq}$.
   c. If $p = q = \frac{1}{2}$, show that $a_n = \frac{n-1}{2^n}, n \geq 2$.

7. A party game starts with the host saying "yes" to the first person. This message is passed to the other guests in this way: If a person hears "yes," that is passed on with probability $\frac{3}{4}$; however, if a person hears "no," that is passed on with probability $\frac{2}{3}$.

   a. What is the probability that the nth person hears "yes"?
   b. Suppose we want the probability that the seventh person hears "yes" to be about $\frac{1}{2}$. What should the probability be that a "no" response is correctly passed on?

8. A coin has a 1 marked on one side and a 2 on the other. It is tossed repeatedly and the cumulative sum that has occurred is recorded. Let p_n denote the probability that the sum is n at some time during the course of the game.

   a. Show a sample space for several possible sums. Explain why the number of points in the sample space follows the Fibonacci sequence.
   b. Find an expression for p_n assuming the coin is fair.
   c. Show that $p_n \to \frac{2}{3}$ as $n \to \infty$.

    d. How should the coin be loaded so as to make p_{17}, the probability that the sum becomes 17 at some time, as large as possible?

9. Find the mean for the waiting time for the pattern HHH in tossing a coin loaded to come up heads with probability p.

▮ 6.3 ▶ **Random Walk and Ruin**

We now show an interesting application of recursions and their solutions, namely that of *random walk* problems.

    The theory of random walk problems is applicable to problems in many fields. We begin with a gambling situation to illustrate one approach to these problems. (Another avenue of approach will be shown in the sections on Markov chains later in this chapter.)

    Suppose a gambler is playing a game against an opponent where, at each trial, the gambler wins \$1 from or loses \$1 to the opponent. If, in the course of play, the gambler loses all his or her money, then the gambler is ruined; on the other hand, if he or she wins all the opponent's money, the gambler wins the game. We want to find the probability a_g that the player (the *gambler*) wins the game with an initial fortune of \$$g$ when the opponent (the *house*) initially has \$$h$.

    While the probability of winning at any particular trial is of obvious importance, we will see, in addition to this, the probability that the gambler wins the game is highly dependent upon the amount of money with which he or she starts, as well as the amount of money the house has.

    The game can be won with a fortune of \$$n$ under two mutually exclusive circumstances: The gambler wins the next trial (say with probability p) and goes on to win the game with a fortune of \$$(n + 1)$; or loses the next trial with probability $1 - p = q$ and subsequently wins the game with a fortune of \$$(n - 1)$. This leads to the recursion

$$a_n = pa_{n+1} + qa_{n-1}, \ a_0 = 0, \ a_{g+h} = 1.$$

The characteristic equation is $pr^2 - r + q = 0$, which has roots of 1 and q/p.

    Assuming that $p \neq q$, the solution is of the form

$$a_n = A + B \cdot \left(\frac{q}{p}\right)^n .$$

Using the fact that $a_0 = 0$ gives $0 = A + B$, so $a_n = A[1 - (\frac{q}{p})^n]$.

    Using the fact that $a_{g+h} = 1$ produces the solution

$$a_n = \frac{1 - \left(\dfrac{q}{p}\right)^n}{1 - \left(\dfrac{q}{p}\right)^{(g+h)}}, \ q \neq p.$$

So, in particular,

$$a_g = \frac{1 - \left(\dfrac{q}{p}\right)^g}{1 - \left(\dfrac{q}{p}\right)^{(g+h)}}, \quad q \neq p. \tag{6.9}$$

Formula 6.9 gives some interesting numerical results. Suppose that $p = 0.49$ so that the game is slightly unfavorable to the gambler, and that the gambler initially has $g = \$10$. The table below shows the probability that the gambler wins the game for various fortunes ($\$h$) that the opponent has:

$\$h$	Probability of winning
$10	0.401300
$14	0.305146
$18	0.238174
$22	0.189394
$26	0.152694
$30	0.124404

One conclusion to be drawn from the table is that the probability that the gambler wins drops rapidly as the opponent's fortune increases. Figure 6.5 shows graphs of the probability of winning with $\$g$ as a function of the opponent's fortune $\$h$.

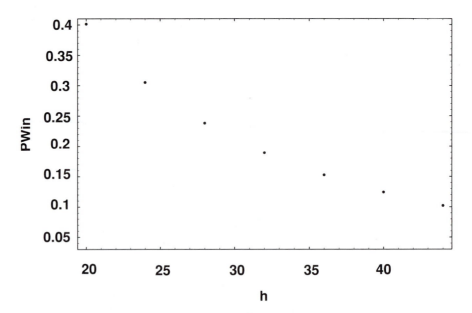

Figure 6.5   Probability of winning the gambler's ruin with an initial fortune of $10 against an opponent with an initial fortune of $\$h$ with $p = 0.49$.

Although the game is slightly (and only slightly) adverse for the gambler, the gambler still has, under some combinations of fortunes, a remarkably large probability of winning the game. If the opponent's fortune increases, however, that probability becomes very small very quickly. The table below shows, for $p = 0.49$, the probability that a gambler with an initial fortune of $10 wins the game over an opponent with an initial fortune of $$h$:

$h	Probability of winning
$90	0.00917265
$98	0.00662658
$106	0.00479399
$114	0.00347174
$122	0.00251603
$130	0.00182438

Now the gambler has little chance of winning the game, but the best chance occurs when the opponent has the least money, or, equivalently, when the ratio of the gambler's fortune to that of the opponent is as large as possible. It is interesting to examine the surface generated by various initial fortunes and values of p. This surface, for $h = \$30$, is shown in Figure 6.6. The contours of this surface appear in Figure 6.7.

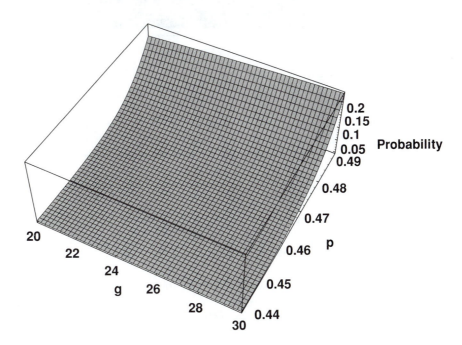

Figure 6.6 Probability of winning the gambler's ruin. The player has an initial fortune of $$g$. p is the probability that an individual game is won. The opponent initially has $30.

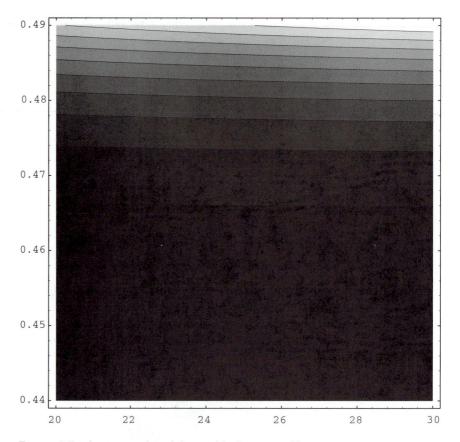

Figure 6.7   A contour plot of the gambler's ruin problem.

The chance of winning the game then is very slim, because the gambler must exhaust the opponent's fortune and has little hope of doing this. While the game is slightly unfavorable to the gambler, the real problem lies in the fact that the ratio of the gambler's fortune to that of the opponent is small. When this ratio increases, so does the gambler's chance of winning. These observations suggest two courses of action.

1. The gambler revises his or her plans and quits playing the game upon achieving a certain fortune (not necessarily that of the opponent).
2. The gambler bets larger amounts on each play of the game.

Either strategy will increase the player's chance of meeting his or her goals. For example, given the game where $p = 0.49$ and an initial fortune of \$10, the gambler's chance of doubling his or her money to \$20 is 0.4013, regardless of the opponent's fortune. The probability that a player with \$50 will reach \$60 in a game where he or she has probability 0.49 of winning on any play is about 0.64. Clearly the lesson here is that modest goals have a fair chance of being achieved, but the gambler must stop playing upon reaching them.

The following table shows the probability of winning a game against an opponent with initial fortune of \$100 when the bet on each play is \$25. The player's initial fortune is \$g. Again, $p = 0.49$.

\$g	Probability of winning
\$25	0.184326
\$50	0.307047
\$75	0.394564

Betting as much as possible in gambling games unfavorable to the gambler can be combined with alternative strategies as the gambler's fortune increases (if, in fact, it does!) to increase the gambler's chance of winning the game.

We turn now to the expected duration of the game.

Expected Duration of the Game

Let E_n denote the expected duration of the game if the gambler has \$n. Winning or losing the next trial increases E_n by 1, so

$$E_n = pE_{n+1} + qE_{n-1} + 1, \quad E_0 = 0, \quad E_{g+h} = 0.$$

This recursion is very similar to that for a_n, differing only in the boundary conditions and in the appearance of the constant 1. The characteristic roots are 1 and q/p again, and so the solution is of the form

$$E_n = A + B\left(\frac{q}{p}\right)^n + C \cdot n, \quad q \neq p.$$

Here, the term $C \cdot n$ represents the particular solution because a constant is a solution to the homogeneous equation.

The constant C must satisfy the equation $Cn - pC(n+1) - qC(n-1) = 1$, so $C = \frac{1}{q-p}$. The boundary conditions are then imposed, giving the result

$$E_n = \frac{n}{q-p} - \frac{g+h}{q-p}\frac{1-\left(\frac{q}{p}\right)^n}{1-\left(\frac{q}{p}\right)^{g+h}}, \quad q \neq p.$$

In particular, because the gambler starts with \$g,

$$E_g = \frac{g}{q-p} - \frac{g+h}{q-p}\frac{1-\left(\frac{q}{p}\right)^g}{1-\left(\frac{q}{p}\right)^{g+h}}, \quad q \neq p.$$

Some particular results from this formula may be of interest. Assume again that the game is slightly unfavorable to the gambler, so that $p = 0.49$, and assume that $N = \$100$. The game then has expected length 454 trials if the gambler starts with $10.

With how much money should the gambler start in order to maximize the expected length of the game? If the gambler and the house have a combined fortune of $100, a computer algebra system shows that the maximum expected length of the series is about 2088 games, occurring when $g = \$65$ and $h = \$35$. A graph of the expected duration of the game if $g + h = \$100$ is shown in Figure 6.8.

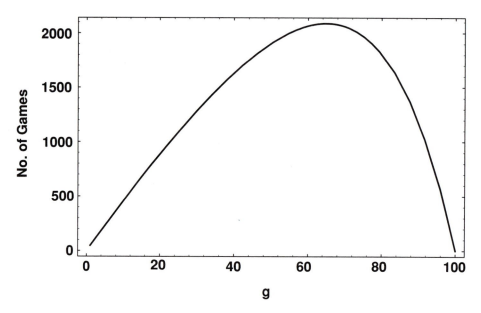

Figure 6.8   Expected duration of the gambler's ruin when the gambler initially has $g and the house has $(100 − g)$.

Exercises 6.3

1. Find the solution to the random walk and ruin problem if $p = \frac{1}{2}$.

2. Find the expected duration of the gambler's ruin game if $p = \frac{1}{2}$.

3. Show a graph of the probability of winning the gambler's ruin game if the game is favorable to the gambler with $p = 0.51$.

4. Show a graph of the expected duration of the gambler's ruin game when the game is favorable to the gambler with $p = 0.51$.

5. Show that if $p = 0.49$ and $h = \$116$, the probability that the gambler will be ruined is at least 0.99, regardless of the amount of money the gambler has.

6.4 ▶ Waiting Times for Patterns in Bernoulli Trials

In Chapter 1 we considered a game in which a fair coin is thrown until one of two patterns occurs: TH or HH. We concluded that the game is unfair because TH occurs before HH with probability $\frac{3}{4}$. We also considered other problems in Chapter 2, which we call *waiting time problems*, such as waiting for a single Bernoulli success. In section 6.2 we considered waiting for two successes in a row. We now want to consider more general waiting time problems, seeking probabilities as well as average waiting times.

Some of these problems may reveal solutions that are counter-intuitive. We have noted that the average waiting time for HH with a fair coin is six tosses, and the average waiting time for TH is four tosses, results that can easily be verified by simulation. This is a very surprising result for a fair coin; many people suspect that the average waiting time for these patterns for a fair coin should be the same, but they actually differ.

We now show how to determine probability generating functions from recursions. Since the direct construction of recursions for first-occurrence times involving complex patterns is difficult, we show how to arrive at the recursions by first creating a generating function for occurrence times. Then we use this generating function to find a recursion for first-occurrence times. The general technique will be illustrated by an example.

Example 6.4.1

A fair coin is thrown until the pattern THT occurs for the first time. On average, how many throws will this take?

First, let's look at the sample space. Let n be the number of tosses necessary to observe the pattern THT for the first time.

$$n = 3 \quad T\,H\,T$$

$$n = 4 \quad \begin{matrix} T\,T\,H\,T \\ H\,T\,H\,T \end{matrix}$$

$$n = 5 \quad \begin{matrix} T\,T\,T\,H\,T \\ H\,H\,T\,H\,T \\ H\,T\,T\,H\,T \end{matrix}$$

$$n = 6 \quad \begin{matrix} T\,T\,T\,T\,H\,T \\ H\,H\,H\,T\,H\,T \\ H\,H\,T\,T\,H\,T \\ T\,H\,H\,T\,H\,T \\ H\,T\,T\,T\,H\,T \end{matrix}$$

For $n = 7$, there are 9 sequences; if $n = 10$, there are 16 sequences. The pattern in the sequence $1, 2, 3, 5, 9, 16, \ldots$ may prove difficult for the reader to discover; we will find it later in this section. In any event, the task of identifying points for $n = 28$, say, is very difficult to say the least (there are 1,221,537 points to enumerate!).

Before solving the problem, let us first consider a definition of when the pattern THT *occurs*. In a sequence of throws, we examine the sequence from the beginning and note when we see the pattern THT, we say the pattern *occurs;* the examination of the sequence then begins again starting with the next throw. For example, in the sequence

$$HHTHTHTHTHHTHT,$$

the pattern THT occurs on the 5th and 12th throws; it does not occur on the 7th throw. This agreement, which may strike the reader as strange, is necessary for some simple results that follow.

Let us suppose that the pattern THT *occurs* on the nth throw. (This is not necessarily the *first* occurrence of the pattern.) Let u_n denote the probability that the pattern THT occurs at the nth throw; consider a sequence in which THT occurs at the nth throw. THT then occurs either at the nth trial or at the $n - 2$ trial, followed by HT. Since any sequence ending in THT has probability pq^2, it follows that

$$u_n + pq\, u_{n-2} = pq^2, \quad n \geq 3. \tag{6.10}$$

We take $u_0 = 1$ (again, so that a subsequent result is simple) and, of course, $u_1 = 0$ and $u_2 = 0$.

Results from the recursion are interesting. We find, for example, that

$$u_{15} = pq^2(1 - pq + p^2q^2 - p^3q^3 + p^4q^4 - p^5q^5 + p^6q^6)$$

$$= \frac{pq^2[1 + (pq)^7]}{1 + pq}.$$

In general, an examination of special values using a computer algebra system leads to the conjecture that

$$u_{2n} = u_{2n-1} = \frac{pq^2[1 - (-pq)^{n-1}]}{1 + pq}, \quad n = 1, 2, 3, \ldots.$$

This in fact will satisfy Formula 6.10 and is its general solution.

It is easy to see from the preceding that

$$u_{2n} \to \frac{pq^2}{1 + pq}$$

because $pq < 1$, and so $\lim_{n \to \infty} (pq)^{n-1} = 0$.

This result can also easily be found from Formula 6.10 because, if $u_n \to L$, say, then, $u_{n-2} \to L$ as well. So in this case we have

$$L + p \cdot q \cdot L = p \cdot q^2$$

whose solution is

$$L = \frac{pq^2}{1 + pq}.$$ ◀

Generating Functions

Now we seek a generating function for the sequence of u_n's.

Let $U(s) = \sum_{n=0}^{\infty} u_n \, s^n$ be the generating function for the u_n's. Multiplying both sides of Formula 6.10 by s^n and summing, we have

$$\sum_{n=3}^{\infty} u_n \, s^n + \sum_{n=3}^{\infty} u_{n-2} pq \, s^n = pq^2 \sum_{n=3}^{\infty} s^n.$$

Expanding and simplifying gives

$$U(s) - u_2 s^2 - u_1 s - u_0 + s^2 pq[U(s) - u_0] = \frac{pq^2 s^3}{1 - s},$$

from which it follows that

$$U(s) = 1 + \frac{pq^2 s^3}{(1 - s)(1 + pqs^2)}.$$

Now let $F(s)$ denote a generating function for f_n, the probability that THT occurs for the first time at the nth trial. It can be shown (see Feller [6]) that

$$F(s) = 1 - \frac{1}{U(s)}.$$

In this case we have

$$F(s) = \frac{pq^2 s^3}{1 - s + pqs^2 - p^2 qs^3}.$$

A power series expansion of $F(s)$, found using a computer algebra system, gives the following:

$$f_3 = pq^2$$

$$f_4 = pq^2$$

$$f_5 = pq^2(1 - pq)$$

$$f_6 = pq^2(1 - 2pq + qp^2).$$

Now we have a generating function whose coefficients give the probabilities of first-occurrence times. One could continue to find probabilities from this generating function using a computer algebra system, but we show now how to construct a recurrence from the generating function for $F(s)$.

Let

$$F(s) = \frac{pq^2 s^3}{1 - s + pqs^2 - p^2qs^3} = f_0 + f_1 s + f_2 s^2 + f_3 s^3 + f_4 s^4 + \cdots.$$

It follows that

$$pq^2 s^3 = (1 - s + pqs^2 - p^2qs^3)(f_0 + f_1 s + f_2 s^2 + f_3 s^3 + f_4 s^4 + \cdots).$$

By equating coefficients we have

$$f_0 = 0$$

$$f_1 = 0$$

$$f_2 = 0$$

$$f_3 = pq^2$$

and for $n \geq 4$,

$$f_n = f_{n-1} - pqf_{n-2} + p^2qf_{n-3} \tag{6.11}$$

We have succeeded in finding a recursion for first-occurrence times. The reader with access to a computer algebra system will discover some interesting patterns in the formulas for the f_n. Numerical results are also easy to find. For the case $p = q$, f_{20}, the probability that THT will be seen for the first time at the twentieth trial, is only 0.01295.

Finally, we find the pattern in the number of points in the sample space for first-time occurrences. Let w_n denote the number of ways THT can occur for the first time in n trials. Since, when $p = q = \frac{1}{2}$,

$$f_n = \frac{w_n}{2^n},$$

then

$$w_n = 2w_{n-1} - w_{n-2} + w_{n-3}, \ n \geq 6,$$

where $w_3 = 1$, $w_4 = 2$, and $w_5 = 3$, using Formula 6.11.

Average Waiting Times

The recurrence 6.11 can be used to find the mean waiting time for the first occurrence of the pattern THT. By multiplying through by n and summing, we find that

$$\sum_{n=4}^{\infty} n \, f_n = \sum_{n=4}^{\infty} [(n-1)+1] f_{n-1} - pq \sum_{n=4}^{\infty} [(n-2)+2] f_{n-2}$$

$$+ \, p^2 q \sum_{n=4}^{\infty} [(n-3)+3] f_{n-3}.$$

This can be simplified as

$$E(N) - 3pq^2 = E(N) + 1 - pq[E(N)+2] + p^2 q[E(N)+3],$$

from which we find that

$$E(N) = \frac{1+pq}{pq^2}.$$

For a fair coin, the average waiting time for THT to occur for the first time is ten trials.

Means and Variances by Generating Functions

The technique used in the previous section to find the average waiting time can also be used to find the variance of the waiting times. This was illustrated in Example 6.2.2. Now, however, we have the probability generating function for first-occurrence times, so it can be used to determine means and variances.

A computer algebra system will show that $F'(1) = \dfrac{1+pq}{pq^2}$ and that

$$F''(1) = \frac{2(q + 4p^2 - 2p^3 - 2p^4 + p^5)}{p^2 q^4}.$$

and since

$$\sigma^2 = F''(1) + F'(1) - [F'(1)]^2,$$

we find that

$$\sigma^2 = \frac{1 + 2pq - 5pq^2 + p^2q^2 - p^2q^3}{p^2q^4}.$$

With a fair coin, then, the average number of throws to see THT is 10 throws with variance 58.

Exercises 6.4

1. In Bernoulli trials with p the probability of success at any trial, consider the event SSS. Let u_n denote the probability that SSS occurs at the nth trial.

   a. Show that $u_n + pu_{n-1} + p^2u_{n-2} = p^3$, $n \geq 4$ and establish the boundary conditions.

   b. Find the generating function $U(s)$ from the recursion in part **a**. Use $U(s)$ to determine the probability that SSS occurs at the 20th trial if $p = \frac{1}{3}$.

   c. Find $F(s)$, the generating function for first-occurrence times of the pattern SSS. Again, if $p = \frac{1}{3}$, find the probability that SSS occurs for the first time at the 20th trial.

   d. Establish that w_n, the number of ways the pattern SSS can occur in n trials for the first time with a fair coin, is given by

   $$w_n = w_{n-1} + w_{n-2} + w_{n-3},$$

   for an appropriate range of values of n. This apparently would establish a "super-Fibonacci" sequence.

2. In waiting for the first occurrence of the pattern HTH in tossing a fair coin, we wish to create a fair game. We would like the probability that the event occurs for the first time in n or fewer trials to be $\frac{1}{2}$. Find n.

3. Find the variance of N, the waiting time for the first occurrence of THT, with a loaded coin.

4. Suppose we wait for the pattern $TTHTH$ in Bernoulli trials.

   a. Find a recursion for the probability of the occurrence of the pattern at the nth trial.

   b. Find a generating function for occurrence times and, from that, a recurrence for first-occurrence times.

   c. Find the mean and variance of first-occurrence times of the pattern.

6.5 ▶ Markov Chains

In practical discrete-probability situations, Bernoulli trials probably occur with the greatest frequency. Bernoulli trials are assumed to be independent with constant probability of success from trial to trial; these assumptions lead to the binomial random variable. Can the binomial be generalized in any way? One way to do this is to relax the assumption of independence, so that the outcome of a particular trial is dependent on the outcomes of the previous trials. The simplest situation assumes that the outcome of a particular trial is dependent only on the outcome of the immediately preceding trial. Such trials form what are called, in honor of the Russian mathematician, a *Markov chain*.

While relaxing the assumption of independence and, in addition, making only the outcome of the previous trial an influence on the next trial seem simple enough, they lead to a very complex though beautiful theory. We will consider only some of the simpler elements of that theory here and will frequently make statements without proof, although we will make them plausible.

Example 6.5.1

A gambler plays on one of four slot machines, each of which has probability $\frac{1}{10}$ of paying off with some reward. If the player wins on a particular machine, she continues to play on that machine; however, if she loses on any machine, she chooses one of the other machines with equal probability.

For convenience, number the machines 1, 2, 3, and 4. We are interested in the machine being played at the moment, and that is a function of whether or not the player won on the last play. Now let p_{ij} denote the probability that machine j is played immediately after machine i is played. We call this a *transition probability* because it gives the probability of going from machine i to machine j.

Now we find some of these transition probabilities. For example, $p_{12} = \frac{9}{10} \cdot \frac{1}{3} = \frac{3}{10}$, because the player must lose with machine 1 and then switch to machine 2 with probability $\frac{1}{3}$.

Also $p_{33} = \frac{1}{10}$ because the player must win with machine 3 and then stay with that machine, while $p_{42} = \frac{3}{10}$, the calculation being exactly the same as that for p_{12}.

The remaining transition probabilities are equally easy to calculate in this case. It is most convenient to display these transition probabilities as a matrix, T, whose entries are p_{ij}:

$$T = T[p_{ij}] = \begin{array}{c} \\ 1 \\ 2 \\ 3 \\ 4 \end{array} \overset{\begin{array}{cccc} 1 & 2 & 3 & 4 \end{array}}{\begin{pmatrix} \frac{1}{10} & \frac{3}{10} & \frac{3}{10} & \frac{3}{10} \\ \frac{3}{10} & \frac{1}{10} & \frac{3}{10} & \frac{3}{10} \\ \frac{3}{10} & \frac{3}{10} & \frac{1}{10} & \frac{3}{10} \\ \frac{3}{10} & \frac{3}{10} & \frac{3}{10} & \frac{1}{10} \end{pmatrix}}$$

This transition matrix is called *stochastic* because the row sums are each 1. Of course that has to be true because the player either stays with the machine currently being played or moves to some other machine for the next play. Actually, T is *doubly stochastic* because its column sums are 1 as well, but such matrices will not be of great interest to us. Matrices describing Markov chains, however, must be stochastic.

The course of the play is of interest, so we might ask, "What is the probability that the player moves from machine 2 to machine 4 after two plays?" We denote this probability by $p_{24}^{(2)}$.

Since the transition from machine 2 to machine 4 involves two plays, the player goes to machine 2 from machine 1 or machine 2 or machine 3 or machine 4, and on the second play moves to machine 4. Thus we see that

$$p_{24}^{(2)} = p_{21}p_{14} + p_{22}p_{24} + p_{23}p_{34} + p_{24}p_{44}$$

$$= \frac{3}{10}\frac{3}{10} + \frac{1}{10}\frac{3}{10} + \frac{3}{10}\frac{3}{10} + \frac{3}{10}\frac{1}{10} = \frac{6}{25}.$$

But this product is simply the dot product of the second row of T with the fourth column of T—an entry of the matrix product of T with itself. Hence $T^2 = [p_{ij}^{(2)}]$ where T^2 denotes the usual matrix product of T with itself. The reader should check that the remaining entries of T^2 give the proper two-step transition probabilities. We have

$$T^2 = \begin{pmatrix} \frac{7}{25} & \frac{6}{25} & \frac{6}{25} & \frac{6}{25} \\ \frac{6}{25} & \frac{7}{25} & \frac{6}{25} & \frac{6}{25} \\ \frac{6}{25} & \frac{6}{25} & \frac{7}{25} & \frac{6}{25} \\ \frac{6}{25} & \frac{6}{25} & \frac{6}{25} & \frac{7}{25} \end{pmatrix}.$$

The entries of T^n then represent transition probabilities in n steps. A computer algebra system is handy in finding these powers. In this case we find that

$$T^4 = \begin{pmatrix} \frac{157}{625} & \frac{156}{625} & \frac{156}{625} & \frac{156}{625} \\ \frac{156}{625} & \frac{157}{625} & \frac{156}{625} & \frac{156}{625} \\ \frac{156}{625} & \frac{156}{625} & \frac{157}{625} & \frac{156}{625} \\ \frac{156}{625} & \frac{156}{625} & \frac{156}{625} & \frac{157}{625} \end{pmatrix}.$$

Each of the entries in T^4 is now very close to $\frac{1}{4}$ and we conjecture that

$$
T^n \to
\begin{pmatrix}
\frac{1}{4} & \frac{1}{4} & \frac{1}{4} & \frac{1}{4} \\
\frac{1}{4} & \frac{1}{4} & \frac{1}{4} & \frac{1}{4} \\
\frac{1}{4} & \frac{1}{4} & \frac{1}{4} & \frac{1}{4} \\
\frac{1}{4} & \frac{1}{4} & \frac{1}{4} & \frac{1}{4}
\end{pmatrix}.
$$

This shows, provided that our conjecture is correct, that the player will play each of the machines about $\frac{1}{4}$ of the time as time goes on.

The vector $(\frac{1}{4}, \frac{1}{4}, \frac{1}{4}, \frac{1}{4})$ is called a *fixed vector* for the matrix T because

$$
\left(\frac{1}{4}, \frac{1}{4}, \frac{1}{4}, \frac{1}{4}\right)
\begin{pmatrix}
\frac{1}{10} & \frac{3}{10} & \frac{3}{10} & \frac{3}{10} \\
\frac{3}{10} & \frac{1}{10} & \frac{3}{10} & \frac{3}{10} \\
\frac{3}{10} & \frac{3}{10} & \frac{1}{10} & \frac{3}{10} \\
\frac{3}{10} & \frac{3}{10} & \frac{3}{10} & \frac{1}{10}
\end{pmatrix}
= \left(\frac{1}{4}, \frac{1}{4}, \frac{1}{4}, \frac{1}{4}\right).
$$

We say that a non-zero vector, w, is a *fixed vector* for the matrix T if

$$
wT = w.
$$

Many (but not all) transition matrices have fixed vectors, and when they do have fixed vectors the components are rarely equal, unlike the case with the matrix T. The fixed vector, if there is one, shows the *steady state* of the process under consideration.

Is the constant vector with each entry $\frac{1}{4}$ a function of the probability, $\frac{1}{10}$, of staying with a winning machine? To answer this, suppose p is the probability that the player stays with a winning machine and then switches with probability $\frac{1-p}{3}$ to each of the other machines. The transition matrix is then

$$
P =
\begin{pmatrix}
p & \dfrac{1-p}{3} & \dfrac{1-p}{3} & \dfrac{1-p}{3} \\[2mm]
\dfrac{1-p}{3} & p & \dfrac{1-p}{3} & \dfrac{1-p}{3} \\[2mm]
\dfrac{1-p}{3} & \dfrac{1-p}{3} & p & \dfrac{1-p}{3} \\[2mm]
\dfrac{1-p}{3} & \dfrac{1-p}{3} & \dfrac{1-p}{3} & p
\end{pmatrix}.
$$

We find that the vector $(\frac{1}{4}, \frac{1}{4}, \frac{1}{4}, \frac{1}{4})$ is a fixed point for the matrix P for any $0 < p < 1$, so the player will use each of the machines about $\frac{1}{4}$ of the time, regardless of the value for p!

It is common to refer to the possible positions in a Markov chain as *states*. In this example, the machines are the states, and the transition probabilities are probabilities of moving from state to state. The states in Markov chains often represent the outcomes of the experiment under consideration.   ◄

Example 6.5.2

Consider the transition matrix with three states

$$R = \begin{pmatrix} \frac{1}{2} & \frac{1}{4} & \frac{1}{4} \\ \frac{1}{8} & \frac{3}{4} & \frac{1}{8} \\ \frac{2}{3} & \frac{1}{6} & \frac{1}{6} \end{pmatrix}.$$

Calculation will show that powers of R approach the matrix

$$\begin{pmatrix} \frac{6}{17} & \frac{8}{17} & \frac{3}{17} \\ \frac{6}{17} & \frac{8}{17} & \frac{3}{17} \\ \frac{6}{17} & \frac{8}{17} & \frac{3}{17} \end{pmatrix}.$$

It will also be found that the solution of

$$(a, b, c) \begin{pmatrix} \frac{1}{2} & \frac{1}{4} & \frac{1}{4} \\ \frac{1}{8} & \frac{3}{4} & \frac{1}{8} \\ \frac{2}{3} & \frac{1}{6} & \frac{1}{6} \end{pmatrix} = (a, b, c),$$

with the restriction that $a + b + c = 1$, has the solution $(\frac{6}{17}, \frac{8}{17}, \frac{3}{17})$, so R has a fixed vector also.   ◄

These examples illustrate the remarkable fact that the powers of some matrices approach a matrix with equal rows. We can be more specific now about the conditions under which this happens.

Definition:
We call a matrix, T, *regular* if, for some n, the entries of T^n are all positive (no zeroes are allowed).

In the preceding examples, T and R are regular. We now state a theorem without proof.

THEOREM: If T is a regular transition matrix, then the powers of T, T^n, $n \geq 1$, each approach a matrix all of whose rows are the same probability vector w.   ■

Readers interested in the proof of this are referred to Isaacson and Madsen [17], or Kemeny et al. [20].

Example 6.5.3

The matrix

$$K = \begin{pmatrix} \frac{1}{2} & 0 & \frac{1}{2} \\ 0 & 1 & 0 \\ 0 & \frac{1}{4} & \frac{3}{4} \end{pmatrix}$$

is not regular, because the center row remains fixed, regardless of the power of K being considered. But

$$(0, 1, 0) \cdot \begin{pmatrix} \frac{1}{2} & 0 & \frac{1}{2} \\ 0 & 1 & 0 \\ 0 & \frac{1}{4} & \frac{3}{4} \end{pmatrix} = (0, 1, 0),$$

showing that the vector $(0, 1, 0)$ is a fixed vector for K.   ◀

Example 6.5.4

The matrix

$$A = \begin{pmatrix} 1 & 0 & 0 \\ a & 0 & 1-a \\ 0 & 0 & 1 \end{pmatrix}$$

for $0 < a < 1$, is not regular because $A^n = A$, for $n \geq 2$, and each row of A is a fixed vector. This shows that a nonregular matrix can have more than one fixed vector. If, however, a regular matrix has a unique fixed vector, then each row of the matrix T^n approaches that fixed vector.

To see this, suppose that the vector $w = (w_1, w_2, w_3)$ is a fixed probability vector for some 3-by-3 matrix T. Then $wT = w$, so $wT^2 = (wT)T = wT = w$ and so on. But if $T^n \to K$, where K is a matrix with identical fixed rows, say

$$K = \begin{pmatrix} a & b & c \\ a & b & c \\ a & b & c \end{pmatrix},$$

then $wK = w$, or

$$(w_1, w_2, w_3) \begin{pmatrix} a & b & c \\ a & b & c \\ a & b & c \end{pmatrix} = (w_1, w_2, w_3).$$

So $w_1 a + w_2 a + w_3 a = w$, and, because $w_1 + w_2 + w_3 = 1$, $w_1 = a$. A similar argument shows that $w_2 = b$ and $w_3 = c$, establishing the result. It is easy to reproduce the argument for any size matrix T.

What influence does the initial position in the chain have? We might conjecture that after a large number of transitions, the initial state has little, if any, influence on the long-term result. To see that this is indeed the case, suppose that there is a probability vector $P^0 = (p_1^0, p_2^0, p_3^0)$ whose components give the probability of the process starting in any of the three states, and where $\sum_{i=1}^3 p_i^0 = 1$. (We again argue, without any loss of generality, for a 3-by-3 matrix T.)

$P^0 T$ is a vector, say P^1, whose components give the probability that the process is in any of the three states after one step. Then $P^1 T = P^0 T \cdot T = P^0 T^2$ and so on. But if $T^n \to K$, and if the fixed vector for the matrix K is, say, (k_1, k_2, k_3), then

$$P^0 K = (k_1, k_2, k_3)$$

because the components of P^0 add up to 1, showing that the probability that the process is in any of the given states after a large number of transitions is independent of P^0.

Now we discuss a number of Markov chains and some of their properties. ◄

Example 6.5.5

The random walk and ruin problem of section 6.2 can be considered to be a Markov chain, the states representing the fortunes of the player. For simplicity, suppose that $1 is exchanged at each play; that the player has probability p of winning each game and probability $1 - p$ of losing each game, that the player's fortune is n and the opponent begins with $4; and that the boundary conditions are $P_0 = 0$ and $P_4 = 1$. There are five states representing fortunes of $0, $1, $2, $3 and $4. The transition matrix is

$$T = \begin{array}{c} \\ 0 \\ 1 \\ 2 \\ 3 \\ 4 \end{array} \begin{array}{ccccc} 0 & 1 & 2 & 3 & 4 \\ \begin{pmatrix} 1 & 0 & 0 & 0 & 0 \\ q & 0 & p & 0 & 0 \\ 0 & q & 0 & p & 0 \\ 0 & 0 & q & 0 & p \\ 0 & 0 & 0 & 0 & 1 \end{pmatrix} \end{array}$$

reflecting the fact that the game is over when the state $n = 0$ or $n = 4$ is reached. These states are called *absorbing states* because, once entered, they are impossible to leave. We call a Markov chain *absorbing* if it has at least one absorbing state, and if it is possible to move to some absorbing state from any nonabsorbing state in a finite number of moves. The matrix T describes an absorbing Markov chain.

It will be useful to reorder the states in an absorbing chain so that the absorbing states come first and the nonabsorbing states follow. If we do this for the matrix T, we find that

$$T = \begin{pmatrix} 1 & 0 & 0 & 0 & 0 \\ 0 & 1 & 0 & 0 & 0 \\ q & 0 & 0 & p & 0 \\ 0 & 0 & q & 0 & p \\ 0 & p & 0 & q & 0 \end{pmatrix}.$$

We can also write the matrix in block form as

$$T = \begin{pmatrix} I & O \\ R & Q \end{pmatrix}$$

where I is an identity matrix, O is a matrix each of whose entries is 0, and Q is a matrix whose entries are the transition probabilities from one nonabsorbing state to another. Any transition matrix with absorbing states can be written in the block form shown.

In addition, matrix multiplication shows that

$$T^n = \begin{pmatrix} I & O \\ R^n & Q^n \end{pmatrix}.$$

The entries of Q^n then give the probabilities of going from one nonabsorbing state to another in n steps.

It is a fact that if a chain has absorbing states, then eventually one of the absorbing states will be reached. The central reason for this is that any path avoiding the absorbing states has a probability that tends to 0 as the number of steps in the path increases.

The possible paths taken in a Markov chain are of some interest and one might consider how many times on average a nonabsorbing state is reached. Consider a particular nonabsorbing state, say j.

The entries of Q give the probabilities of reaching j from any other non-absorbing state, say i, in one step. The entries of Q^2 give the probabilities of reaching state j from state i in two steps, and in general the entries of Q^n give the probabilities of reaching state j from state i in n steps. Now define a sequence of indicator random variables:

$$X_k = \begin{cases} 1 & \text{if the chain is in state } j \text{ in } k \text{ steps} \\ 0 & \text{otherwise.} \end{cases}$$

Then X, the total number of times the process is in state j, is

$$X = \sum_k X_k,$$

and so the expected value of X – the expected number of times the chain is in state j – is

$$E(X) = I_j + \sum_i 1 \cdot q_{ij}^n$$

where q_{ij}^n is the (i, j) entry in the matrix Q^n, and where $I_j = 1$ or 0, depending on whether or not the chain starts in state j.

This shows that $E(X) = I + Q + Q^2 + Q^3 + \ldots$ where I is the n-by-n identity matrix.

It can be shown in this circumstance that $(I - Q)^{-1} = I + Q + Q^2 + Q^3 + \ldots$, and so the entries of $(I - Q)^{-1}$ give the expected number of times the process is in state j, given that it starts in state i.

The matrix $(I - Q)^{-1}$ is called the *fundamental matrix* for the absorbing Markov chain.   ◄

Example 6.5.6

Consider the transition matrix

$$P = \begin{pmatrix} 1 & 0 & 0 & 0 & 0 \\ 0 & 1 & 0 & 0 & 0 \\ \frac{1}{4} & 0 & 0 & \frac{3}{4} & 0 \\ 0 & 0 & \frac{1}{4} & 0 & \frac{3}{4} \\ 0 & \frac{3}{4} & 0 & \frac{1}{4} & 0 \end{pmatrix}$$

representing a random walk. The matrix Q here is

$$Q = \begin{pmatrix} 0 & \frac{3}{4} & 0 \\ \frac{1}{4} & 0 & \frac{3}{4} \\ 0 & \frac{1}{4} & 0 \end{pmatrix} \text{ and } I - Q = \begin{pmatrix} 1 & -\frac{3}{4} & 0 \\ -\frac{1}{4} & 1 & -\frac{3}{4} \\ 0 & -\frac{1}{4} & 1 \end{pmatrix},$$

so

$$(I - Q)^{-1} = \begin{pmatrix} \frac{13}{10} & \frac{6}{5} & \frac{9}{10} \\ \frac{2}{5} & \frac{8}{5} & \frac{6}{5} \\ \frac{1}{10} & \frac{2}{5} & \frac{13}{10} \end{pmatrix}$$

If the chain starts in state 1, it spends on average $\frac{13}{10}$ times in state 1, $\frac{6}{5}$ times in state 2, and $\frac{13}{10}$ times in state 3. This means that, starting in state 1, the total number of times in various states is $\frac{13}{10} + \frac{6}{5} + \frac{13}{10} = \frac{19}{5}$. So the average number of turns before absorption must be $\frac{19}{5}$ if the process begins in state 1.

Similar calculations can be made for the other beginning states. If we let V be a column vector each of whose entries is 1, then $(I - Q)^{-1}V$ represents the average number of times the process is in each state before being absorbed. Here,

$$(I - Q)^{-1}V = \begin{pmatrix} \frac{13}{10} & \frac{6}{5} & \frac{13}{10} \\ \frac{2}{5} & \frac{8}{5} & \frac{6}{5} \\ \frac{1}{10} & \frac{2}{5} & \frac{13}{10} \end{pmatrix} \begin{pmatrix} 1 \\ 1 \\ 1 \end{pmatrix} = \begin{pmatrix} \frac{19}{5} \\ \frac{16}{5} \\ \frac{9}{5} \end{pmatrix}.$$

We continue with further examples of Markov chains.   ◀

Example 6.5.7

In Example 6.5.5, we considered a game in which $1 was exchanged at each play, and where the game ended if either player was ruined. Now consider players who prefer that the game never end. They agree that if either player is ruined, the other player gives the ruined player $1 so that the game can continue. This creates a random walk with *reflecting barriers*.

Here's an example of such a random walk. p is the probability that a player wins at any play, $q = 1 - p$, and there are four possible states. The transition matrix is

$$M = \begin{pmatrix} 0 & 1 & 0 & 0 \\ q & 0 & p & 0 \\ 0 & q & 0 & p \\ 0 & 0 & 1 & 0 \end{pmatrix}.$$

It is probably not surprising to learn that M is not a regular transition matrix. Powers of M do, however, approach a matrix having, in this case, two sets of identical rows. We find, for example, that if $p = \frac{2}{3}$, then

$$M^n \rightarrow \begin{pmatrix} 0 & \frac{3}{7} & 0 & \frac{4}{7} \\ \frac{1}{7} & 0 & \frac{6}{7} & 0 \\ 0 & \frac{3}{7} & 0 & \frac{4}{7} \\ \frac{1}{7} & 0 & \frac{6}{7} & 0 \end{pmatrix}. \qquad \blacktriangleleft$$

Example 6.5.8

A private grade school offers instruction in grades $K, 1, 2,$ and 3. At the end of each academic year, a student can be promoted (with probability p), asked to repeat the grade (with probability r), or asked to return to the previous grade (with probability $1 - p - r = q$). The transition matrix is

$$G = \begin{array}{c} \\ K \\ 1 \\ 2 \\ 3 \end{array} \begin{array}{cccc} K & 1 & 2 & 3 \\ \begin{pmatrix} 1-p & p & 0 & 0 \\ q & r & p & 0 \\ 0 & q & r & p \\ 0 & 0 & q & 1-q \end{pmatrix} \end{array}$$

For the particular matrix

$$
\begin{array}{c c c c}
 & K & 1 & 2 & 3 \\
\end{array}
$$

$$
G = \begin{array}{c}
1 \\
2 \\
3
\end{array}
\begin{pmatrix}
0.1 & 0.2 & 0.7 & 0 \\
0 & 0.1 & 0.2 & 0.7 \\
0 & 0 & 0.1 & 0.9
\end{pmatrix}
$$

we find the fixed vector to be $(0.0025, 0.0175, 0.1225, 0.8575)$.

◀

Exercises 6.5

1. Show that an n-by-n doubly stochastic transition matrix has a fixed vector each of whose entries is $\frac{1}{n}$.

2. A family on vacation either goes camping or to a theme park. If the family camped one year, it goes camping again the next year with probability 0.7; if it went to a theme park one year, it goes camping the next year with probability 0.4. Show that the process is a Markov chain. In the long run how often does the family go camping?

3. Voters often change their party affiliation in subsequent elections. In a certain district, Republicans remain Republicans for the next election with probability 0.8. Democrats stay with their party with probability 0.9. Show that the process is a Markov chain and find the fixed vector.

4. A small manufacturing company has 2 boxes of parts. Box I has 5 good parts in it while box II has 6 good parts in it. There is 1 defective part, which initially is in the first box. A part is drawn from box I and put into box II; on the second draw, a part is drawn from box II and put into box I. After 5 draws, what is the probability that the defective part is in the first box?

5. Electrical usage during a summer month can be classified as normal, high, or low. Weather conditions often make this level of usage change according to the following matrix:

$$
\begin{array}{c c c c}
 & N & H & L \\
\end{array}
$$

$$
\begin{array}{c}
N \\
H \\
L
\end{array}
\begin{pmatrix}
\frac{3}{4} & \frac{1}{6} & \frac{1}{12} \\
\frac{2}{5} & \frac{1}{3} & \frac{4}{15} \\
\frac{1}{2} & \frac{2}{5} & \frac{1}{10}
\end{pmatrix}
$$

Find the fixed vector for this Markov chain and interpret its meaning.

6. A local stock either gains value (+), remains the same (0), or loses value (−) during a trading day according to the following matrix:

$$
\begin{array}{c c}
 & \begin{array}{ccc} + & 0 & - \end{array} \\
\begin{array}{c} + \\ 0 \\ - \end{array} &
\left(\begin{array}{ccc}
\frac{1}{3} & \frac{1}{3} & \frac{1}{3} \\
\frac{1}{2} & 0 & \frac{1}{2} \\
\frac{1}{1} & \frac{1}{4} & \frac{1}{2}
\end{array}\right)
\end{array}
$$

If you were to bet on the stock's performance tomorrow, how would you bet?

7. Show that the fixed vector for the transition matrix

$$
\begin{pmatrix} p & 1-p \\ r & 1-r \end{pmatrix}
$$

where $0 < p < 1$, $0 < r < 1$ and $1 - p + r \neq 0$, is

$$
\left(\frac{r}{1-p+r}, \frac{1-p}{1-p+r}\right).
$$

8. Alter the gambler's ruin situation in Example 6.5.5 as follows: Suppose that if a gambler is ruined, the opposing player returns \$2 to him so that the game can go on (in fact it can now go on forever!). Show the transition matrix if the probability of a gambler winning a game is $\frac{2}{3}$. If the matrix has a fixed point, find it.

9. The states in a Markov chain that is called *cyclical* are 0, 1, 2, 3, 4, and 5. If the chain is in state 0, the next state can be 5 or 1, and if the chain is in state 5, it can go to state 0 or 4. Show the transition matrix for this chain with probability $\frac{1}{2}$ of moving from one state to another possible state. If the matrix approaches a steady state, find it.

Chapter Review

This chapter considered two primary topics, recursions and Markov chains.

Recursions are used when it is possible to express one probability as a function of some variable, say n, in terms of other probabilities as functions of that same variable n. In Example 6.1.2, we tossed a loaded coin until it came up heads twice in a row. If a_n represents the probability that HH occurs for the first time at the nth toss, then $a_n = qa_{n-1} + pqa_{n-2}$, $n \geq 3$ with $a_1 = 0$, and $a_2 = p^2$. Values of a_n can easily be found using a computer algebra system. Such systems also frequently will solve recursions, producing formulas for the variable as a function of n. We showed an algebraic technique for solving recursions involving a *characteristic equation,* and *homogeneous* and *particular* solutions.

Generating functions associated with a recursion, such as $G(s) = \sum_{n=0}^{\infty} a_n \cdot s^n$, were also considered. These are often useful when recursions are not easily found directly. We illustrated how to find a recursion for the event "THT occurs at the nth trial," and a generating function, $U(s)$, for the probability that THT occurs at the nth trial. The generating function for first-time occurrences, $F(s)$, is simply related to that of $U(s)$:

$$F(s) = 1 - \frac{1}{U(s)}.$$

We then showed how to find a recursion for first-time occurrences, given $F(s)$.

When the events in question form a probability distribution, means and variances can be determined from recursions. For example, if the recursion is

$$a_n = f \cdot a_{n-1} + g \cdot a_{n-2}, \ n \geq 2$$

with initial values a_0 and a_1, and where f and g are constants, then

$$\sum_{n=2}^{\infty} na_n = f \cdot \sum_{n=2}^{\infty} [(n-1) + 1)]a_{n-1} + g \cdot \sum_{n=2}^{\infty} [(n-2) + 2]a_{n-2},$$

from which it follows that

$$E(N) = \frac{a_1 + f(1 - a_0) + 2g}{1 - f - g}.$$

Variances can also be determined from the recursion.

Markov chains arise when a process, consisting of a series of trials, can be regarded at any time as being in a particular *state*. Commonly, the matrix $T = [p_{ij}]$, where the p_{ij} represents the probability that the process goes from state i to state j, is called a *transition matrix*. The transition matrix clearly gives all the information needed about the process.

Entries of 1 in T indicate *absorbing states*, – states that cannot be left once they are entered. It is possible to partition the transition matrix for an absorbing chain as

$$T = \begin{pmatrix} I & O \\ R & Q \end{pmatrix}.$$

The matrix $I - Q$ is called the *fundamental matrix* for the Markov chain. The entries in $(I - Q)^{-1}$ give the average number of times the process is in state j, given that it started in state i.

Problems for Review

Exercises 6.2 # 2, 3, 4, 6, 9
Exercises 6.3 #1, 2
Exercises 6.4 #1, 3
Exercises 6.5 #2, 4, 6

Supplementary Exercises for Chapter 6

1. Consider a sequence of Bernoulli trials with a probability p of success.

   a. Find a recursion giving the probability u_n that the number of successes in n trials is divisible by 3.

   b. Find a recursion giving the probability that when the number of successes in n trials is divided by 3, the remainder is 1. (Hint: Write a system of three recursions involving u_n, the probability that the number of successes is divisible by 3; v_n, the probability that the number of successes leaves a remainder of 1 when divided by 3; and w_n, the probability that the number of successes leaves a remainder of 2 when divided by 3.)

2. Find a recursion for the probability q_n that there is no run of three successes in n Bernoulli trials where the probability of success at any trial is $\frac{1}{2}$.

3. Find the probability of an even number of successes in n Bernoulli trials where p is the probability of success at a single trial.

4. Find the probability that no two successive heads occur when a coin, loaded to come up heads with probability p, is tossed 12 times.

5. A loaded coin, whose probability of coming up heads at a single toss is p, is tossed and a running count of the heads and tails is kept. Show that if $u_n = P$ (heads and tails are equal at toss $2n$) then $u_n = \binom{2n}{n} p^n q^n$. Then find the probability that the heads and tails counts are equal for the first time at trial $2n$. [The binomial expansion of $(1 - 4pqs)^{-1/2}$ will help.]

6. Automobile buyers of brands A, B, and C, stay with or change brands according to the following matrix:

$$
\begin{array}{c} \\ A \\ B \\ C \end{array}
\begin{array}{ccc} A & B & C \\ \left(\begin{array}{ccc} \frac{4}{5} & \frac{1}{10} & \frac{1}{10} \\ \frac{1}{6} & \frac{2}{3} & \frac{1}{6} \\ \frac{1}{8} & \frac{1}{8} & \frac{3}{4} \end{array} \right) \end{array}
$$

   After several years, what share of the market does brand C have?

7. A baseball pitcher throws curves (C), sliders (S), and fastballs (F). She changes pitches with the following probabilities:

$$
\begin{array}{c} \\ C \\ S \\ F \end{array}
\begin{array}{ccc} C & S & F \\ \left(\begin{array}{ccc} \frac{1}{2} & \frac{1}{4} & \frac{1}{4} \\ 0 & \frac{2}{3} & \frac{1}{3} \\ \frac{1}{10} & \frac{3}{10} & \frac{3}{5} \end{array} \right) \end{array}
$$

The next batter hits fastballs well. In the long run, what proportion of the pitches will be fastballs?

8. A small town has two supermarkets, K and C. A shopper who last shopped at K is as likely as not to return there on the next shopping trip. However, if a shopper shops at C, the probability is $\frac{2}{3}$ that K will be chosen for the next shopping trip. What proportion of the time does the shopper shop at K?

9. Two players, A and B, play chess according to the following rule: The winner of a game plays the white pieces on the next game. If the probability of winning with the white pieces or the black pieces is p for either player, and if A plays the white pieces on the first game, what is the probability that A wins the nth game?

Bibliography

Where to Learn More

There is now a vast literature on the theory of probability. Many of the following references are cited in the text; other titles that may be useful to the student are included here as well.

1) Blom, Gunnar, Lars Holst, and Dennis Sandell, *Problems and Snapshots from the World of Probability*, Springer-Verlag, 1994.
2) David, F. N., and D. E. Barton, *Combinatorial Chance*, Charles Griffin & Company Limited, 1962.
3) Drane, J. Wanzer, Suhua Cao, Lixia Wang, and T. Postelnicu, Limiting Forms of Probability Mass Functions via Recurrence Formulas, *The American Statistician*, vol. 47, no. 4, November 1993, pp. 269–274.
4) Draper, N. R., and H. Smith, *Applied Regression Analysis*, Second Edition, John Wiley & Sons, 1981.
5) Duncan, Acheson J., *Quality Control and Industrial Statistics*, Fifth Edition, Richard D. Irwin, Inc., 1986.
6) Feller, William, *An Introduction to Probability and Its Applications*, Volumes I and II, John Wiley & Sons, 1968.
7) Gnedenko, B. V., *The Theory of Probability*, Chelsea Publishing Company, Fifth Edition, 1989.
8) Goldberg, Samuel, *Probability, An Introduction*, Prentice-Hall, Inc., 1960.
9) Goldberg, Samuel, *Introduction to Difference Equations*, Dover Publications, 1986.
10) Grant, Eugene L., and Richard S. Leavenworth, *Statistical Quality Control*, Sixth Edition, McGraw-Hill, 1988.
11) Grimaldi, Ralph P., *Discrete and Combinatorial Mathematics*, Third Edition, Addison-Wesley Publishing Co., 1994.
12) Hald, Anders, *A History of Probability and Statistics and Their Applications Before 1750*, John Wiley & Sons, 1990.

13) Hallinan, Arthur J., Jr., A Review of the Weibull Distribution, *Journal of Quality Technology*, vol. 25, no. 2, April 1993, pp. 85–93.

14) Hogg, Robert V., and Allen T. Craig, *Introduction to Mathematical Statistics*, Fourth Edition, Macmillan Publishing Company, 1986.

15) Huff, Barthel W., Another Definition of Independence, *Mathematics Magazine*, September–October, 1971, pp. 196–197.

16) Hunter, Jeffrey J., *Mathematical Techniques of Applied Probability*, Volumes 1 and 2, Academic Press, 1983.

17) Isaacson, Dean L., and Richard W. Madsen, *Markov Chains, Theory and Applications*, John Wiley & Sons, 1976.

18) Johnson, Norman L., Samuel Kotz and Adrienne W. Kemp, *Univariate Discrete Distributions*, Second Edition, John Wiley & Sons, 1992.

19) Johnson, Norman L., Samuel Kotz, and N. Balakrishnan, *Continuous Univariate Distributions*, Volumes 1 and 2, Second Edition, John Wiley and Sons, 1994

20) Kemeny, John G., and J. Laurie Snell, *Finite Markov Chains*, Springer-Verlag, 1976.

21) Kinney, John J., Tossing Coins Until All Are Heads, *Mathematics Magazine*, May, 1978, pp. 184–186.

22) Mosteller, Frederick, *Fifty Challenging Problems in Probability*, Addison-Wesley Publishing Co., 1965. Reprinted by Dover Publications.

23) Rao, C. Radhakrishna, *Linear Statistical Inference and Its Applications*, Second Edition, John Wiley & Sons, 1973.

24) Riordan, John, *Combinatorial Identities*, John Wiley & Sons, 1968.

25) Thompson, W. A., Jr., On the Foundations of Reliability, *Technometrics*, vol. 23, no. 1, February 1981, pp. 1–13.

26) Uspensky, J. V., *Introduction to Mathematical Probability*, McGraw-Hill, Inc., 1937.

27) Welch, B. L., The Significance of the Difference Between Two Means When the Population Variances are Unequal, *Biometrika* vol. 29, 1938, pp. 350–362.

28) *When AIDS Tests are Wrong*, The New York Times, September 5, 1987, p. 22.

29) Whitworth, William Allen, *Choice and Chance*, Fifth Edition, Hafner Publishing Company, 1965.

30) Wickham-Jones, Tom, *Mathematica Graphics*, Springer-Verlag, 1994.

31) Wolfram, Stephen, *Mathematica: A System for Doing Mathematics by Computer*, Addison-Wesley Publishing Co., 1991.

▶ *Appendix 1*

Use of Mathematica in Probability and Statistics

Reference is often made throughout the text to a computer algebra system. *Mathematica* was used for most of the work in this book, but many other computer algebra systems, such as *Maple* or *Derive*, are also capable of doing the tasks we need. We give examples of the use of all of the commands in Mathematica that have been used in the examples and graphs in this text, but we do not show the creation of every such graph. We make no attempt in this brief account to give a detailed explanation for any command. The reader is urged to consult Wolfram [31] and Wickham–Jones [30] for more complete explanations of the commands used. In addition, no attempt is made to carry out tasks required in the most efficient manner; the reader will find that Mathematica, like other computer algebra systems, often offers many alternative paths to the same result; the reader is urged to explore these systems.

The material here is referred to by chapter in the text and by examples within that chapter. We often do not repeat the conditions of the examples, so the reader should refer to the text before studying the solutions.

Entries in Mathematica appear in **bold-face** type; the responses follow in ordinary type.

The Mathematica kernel must first be loaded, usually by asking for a simple calculation. Once that result is returned, all the commands described here will work as shown; no other knowledge or experience with Mathematica is necessary. This appendix is in actuality a Mathematica notebook and will run on a computer exactly as presented here, with the exception of examples which use random samples. They will vary each time the program is run.

Chapter One

Section 1.1   Discrete Sample Spaces

The Fibonacci recursion is $a_n = a_{n-1} + a_{n-2}$, $a_2 = 1$, $a_3 = 1$. Values for a_n can be found directly using the recursion.

In[2]:=

```
a[n_]:=a[n-1]+a[n-2]
```

In[3]:=

```
a[2]=1
```

Out[3]:=

```
1
```

In[4]:=

```
a[3]=1
```

Out[4]:=

```
1
```

In[5]:=

```
Table[a[n],{n,2,15}]
```

Out[5]:=

```
{1, 1, 2, 3, 5, 8, 13, 21, 34, 55, 89, 144, 233, 377}
```

In[6]:=

```
a[25]
```

Out[6]:=

```
46368
```

Section 1.4   Conditional Probability and Independence

Example 1.4.4

Figure 1.9 shows the graph of $P(A|T^+)$ as a function of $P(A)$. It was drawn as follows:

In[7]:=

```
f[p_]:=(20000p)/(1+19999p)
```

In[8]:=

```
Plot[f[p],{p,0,1},Frame->True,AxesOrigin->{0,0.9875},
FrameLabel->{"P(A)","P(A|T+)"},
DefaultFont->{"Helvetica-Bold",12}]
```

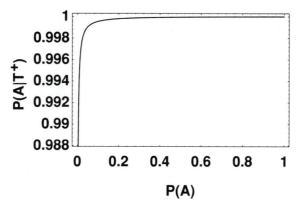

Out[8]:=

```
-Graphics-
```

This section also shows a three-dimensional graph of $P(A|T^+)$ as a function of both the incidence rate of the disease, r, as well as $p = P(T^+|A)$. This was done as follows:

In[9]:=

```
f[r_,p_]:=r*p/(r*p+(1-r)*(1-p))
```

In[10]:=

```
Plot3D[f[r,p],{r,0,1},{p,0,1},
 AxesLabel->{"r=P(A)","P(T+|A)","P(A|T+)"},
 ViewPoint->{0.965,-2.553,2.000},
 DefaultFont->{"Helvetica-Bold",12},PlotPoints->50]
```

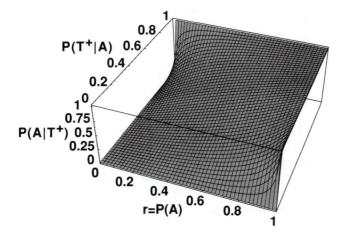

Out[10]:=

```
-SurfaceGraphics-
```

Example 1.5.1 (The Birthday Problem)

The table with exact values of $P(A)$ was constructed with this instruction:

In[11]:=

```
probs=Table[{i,1-Product[(366-r)/365,{r,i}]},
{i,1,40}]//N
```

Out[11]:=

```
{{1., 0}, {2., 0.00273973}, {3., 0.00820417},
  {4., 0.0163559}, {5., 0.0271356}, {6., 0.0404625},
  {7., 0.0562357}, {8., 0.0743353}, {9., 0.0946238},
  {10., 0.116948}, {11., 0.141141}, {12., 0.167025},
  {13., 0.19441}, {14., 0.223103}, {15., 0.252901},
  {16., 0.283604}, {17., 0.315008}, {18., 0.346911},
  {19., 0.379119}, {20., 0.411438}, {21., 0.443688},
  {22., 0.475695}, {23., 0.507297}, {24., 0.538344},
  {25., 0.5687}, {26., 0.598241}, {27., 0.626859},
  {28., 0.654461}, {29., 0.680969}, {30., 0.706316},
  {31., 0.730455}, {32., 0.753348}, {33., 0.774972},
  {34., 0.795317}, {35., 0.814383}, {36., 0.832182},
  {37., 0.848734}, {38., 0.864068}, {39., 0.87822},
  {40., 0.891232}}
```

Then the graph in Figure 1.13 was drawn with these commands:

In[12]:=

```
values=Table[i,{i,1,40}]
```

Out[12]:=

```
{1, 2, 3, 4, 5, 6, 7, 8, 9, 10, 11, 12, 13, 14, 15,
  16, 17, 18, 19, 20, 21, 22, 23, 24, 25, 26, 27,
  28, 29, 30, 31, 32, 33, 34, 35, 36, 37, 38, 39, 40}
```

In[13]:=

```
ListPlot[probs,Frame->True,
FrameLabel->{"n","Probability"},
Ticks->{values, Automatic},
PlotLabel->FontForm["Birthday Problem",
{"Helvetica-Bold",12}],
DefaultFont->{"Helvetica-Bold",12}]
```

Birthday Problem

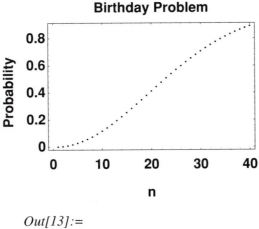

Out[13]:=

    -Graphics-

Section 1.7   Counting Techniques

Mathematica does exact arithmetic:

In[14]:=

  52!

Out[14]:=

    806581751709438785716606368564037669752895054408832771\
    824000000000000

The number of permutations of *r* objects chosen from *n* distinct objects is
$n!/(n − r)!$. For example, the number of distinct arrangements of 30 objects
chosen from 56 objects is 56!/26!. Mathematica will evaluate this exactly.

In[15]:=

  56!/26!

Out[15]:=

    176298944147904746509797704376907575853056000000000

Here are some more examples of permutations.

In[16]:=

  Permutations[{a,b,c}]

Out[16]:=

    {{a, b, c}, {a, c, b}, {b, a, c}, {b, c, a},
      {c, a, b}, {c, b, a}}

If some of the objects are alike, only the distinct permutations are returned.

In[17]:=

perms=Permutations[{a,a,b,b,c}]

Out[17]:=

```
{{a, a, b, b, c}, {a, a, b, c, b}, {a, a, c, b, b},
  {a, b, a, b, c}, {a, b, a, c, b}, {a, b, b, a, c},
  {a, b, b, c, a}, {a, b, c, a, b}, {a, b, c, b, a},
  {a, c, a, b, b}, {a, c, b, a, b}, {a, c, b, b, a},
  {b, a, a, b, c}, {b, a, a, c, b}, {b, a, b, a, c},
  {b, a, b, c, a}, {b, a, c, a, b}, {b, a, c, b, a},
  {b, b, a, a, c}, {b, b, a, c, a}, {b, b, c, a, a},
  {b, c, a, a, b}, {b, c, a, b, a}, {b, c, b, a, a},
  {c, a, a, b, b}, {c, a, b, a, b}, {c, a, b, b, a},
  {c, b, a, a, b}, {c, b, a, b, a}, {c, b, b, a, a}}
```

We can check that the number of these permutations is 5!/2!2!1!.

In[18]:=

5!/(2!*2!*1!)

Out[18]:=

30

In[19]:=

Length[perms]

Out[19]:=

30

Combinations are found using the Binomial[n, r] function. The number of distinct poker hands is:

In[20]:=

Binomial[52,5]

Out[20]:=

2598960

Mathematica can also evaluate binomial coefficients when n is negative:

In[21]:=

Binomial[-7,3]

Out[21]:=

-84

Example 1.7.2

In[22]:=

Binomial[3,1]*Binomial[8,4]/Binomial[11,5]

Out[22]:=

$$\frac{5}{11}$$

Example 1.7.3 *The Matching Problem*

A simulation of the matching problem can be done using the RandomPermutation command in the Combinatorica package. The package must be loaded first.

In[23]:=

<<DiscreteMath`Combinatorica`

If there are 9 numbers to be permuted, we can generate 20 random permutations of those 9 digits as follows:

In[24]:=

Table[RandomPermutation[9],{20}]

Out[24]:=

```
{{4, 6, 1, 8, 2, 9, 7, 3, 5},
 {5, 7, 1, 9, 4, 8, 2, 6, 3},
 {7, 5, 4, 1, 3, 2, 9, 6, 8},
 {5, 2, 9, 7, 6, 1, 4, 8, 3},
 {4, 2, 7, 6, 9, 8, 1, 3, 5},
 {3, 2, 8, 1, 5, 7, 4, 9, 6},
 {4, 5, 2, 9, 3, 8, 7, 6, 1},
 {7, 9, 5, 3, 6, 8, 4, 1, 2},
 {6, 5, 3, 7, 2, 8, 9, 1, 4},
 {1, 5, 6, 2, 3, 7, 9, 4, 8},
 {8, 6, 5, 2, 3, 9, 1, 7, 4},
 {2, 1, 6, 8, 5, 7, 9, 3, 4},
 {3, 7, 1, 8, 5, 2, 6, 9, 4},
 {2, 9, 4, 3, 6, 5, 7, 1, 8},
 {9, 5, 7, 4, 2, 1, 6, 3, 8},
 {6, 3, 7, 5, 1, 8, 4, 2, 9},
 {4, 6, 2, 9, 3, 8, 7, 1, 5},
 {8, 3, 7, 6, 1, 2, 4, 9, 5},
 {3, 5, 7, 2, 6, 9, 8, 1, 4},
 {6, 8, 1, 2, 9, 5, 3, 4, 7}}
```

▌Chapter Two

Each of the discrete probability distribution functions in Chapter Two is contained in Mathematica as a defined function. We give some examples of the use of these functions, of drawing graphs, and of drawing random samples from these distributions.

Probability distributions and some statistical analysis routines are contained in a number of subpackages in Mathematica. It is easiest to access all of these by loading a Master file.

In[25]:=

```
<<Statistics`Master`
```

Section 2.1   Random Variables

Example 2.1.1

We show a random sample of 100 drawn from the discrete uniform distribution $P(X = x) = 1/n, \ x = 1, 2, 3, \ldots, n$. The value of n must be specified. We take $n = 6$ here to simulate tosses of a fair die.

In[26]:=

```
data=Table[Random[DiscreteUniformDistribution[6]],
{100}]
```

Out[26]:=

```
{6, 5, 3, 2, 1, 5, 2, 1, 5, 5, 6, 5, 6, 3, 4, 6, 2,
 5, 4, 5, 3, 1, 4, 6, 5, 3, 5, 5, 2, 4, 2, 3, 6, 4,
 2, 4, 1, 6, 6, 4, 1, 5, 3, 4, 6, 1, 2, 4, 2, 2, 4,
 5, 2, 4, 1, 6, 3, 6, 6, 3, 2, 6, 1, 5, 3, 3, 5, 3,
 1, 3, 2, 1, 4, 2, 5, 5, 2, 1, 2, 3, 6, 4, 3, 2, 4,
 1, 1, 2, 4, 3, 5, 3, 6, 2, 3, 5, 2, 6, 6, 6}
```

These data can then be organized by counting the frequency with which each integer occurs. A subpackage to produce graphics must be loaded first.

In[27]:=

```
<<Graphics`Graphics`
```

In[28]:=

```
freq=BinCounts[data,{0,6,1}]
```

Out[28]:=

```
{13, 19, 17, 15, 18, 18}
```

In[29]:=

```
values={1,2,3,4,5,6}
```

Out[29]:=

{1, 2, 3, 4, 5, 6}
Now a histogram of the data can be shown.

In[30]:=

BarChart[Transpose[{freq,values}]]

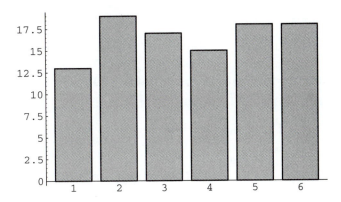

Out[30]:=

-Graphics-

Example 2.1.2

Sampling from the die loaded so that the probability that a face appears is proportional to the face is a bit more complex than sampling from a fair die. We sample from a discrete uniform distribution with values $1, 2, 3, \ldots, 21$; the value 1 becomes a one on the die; the next two values, namely 2 and 3, become a two on the die; the next three values are 4, 5, and 6, and they become a three on the die; and so on.

In[31]:=

**data=Table[Random[DiscreteUniformDistribution[21]],
{100}];**

(The ";" suppresses output. The next command gives the frequencies of 1's, 2's, $\ldots$, 21's.)

In[32]:=

freq=BinCounts[data,{0,21,1}]

Out[32]:=

{6, 2, 2, 5, 5, 3, 5, 8, 4, 5, 6, 7, 5, 5, 6, 6, 4,
  2, 7, 5, 2}

Now we collate the data as follows: the 1's; then the 2's and 3's; then the 4's, 5's, and 6's; and so on.

In[33]:=

```
orgdata=Table[Take[freq,{(1/2)*(2-m+m^2),m*(m+1)/2}],
{m,1,6}]
```

Out[33]:=

```
{{6}, {2, 2}, {5, 5, 3}, {5, 8, 4, 5},
   {6, 7, 5, 5, 6}, {6, 4, 2, 7, 5, 2}}
```

In[34]:=

```
orgfreq=Apply[Plus,orgdata,{1}]
```

Out[34]:=

```
{6, 4, 13, 22, 29, 26}
```

In[35]:=

```
BarChart[Transpose[{orgfreq,values}]]
```

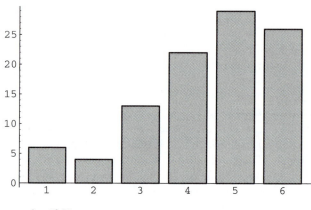

Out[35]:=

```
-Graphics-
```

Example 2.1.3

The probability distribution function when two dice are thrown is

$$P(X = x) = P(X = 14 - x) = (x - 1)/36$$
$$\text{for } x = 1, 2, 3, 4, 5, 6, 7.$$

A graph of this function is shown in Figure 2.2 and can be generated by the following commands:

In[36]:=

```
prob1=Table[(x-1)/36,{x,2,7}]
```

Out[36]:=

$$\{\frac{1}{36}, \frac{1}{18}, \frac{1}{12}, \frac{1}{9}, \frac{5}{36}, \frac{1}{6}\}$$

In[37]:=

```
prob2=Table[(14-x)/36,{x,9,13}]
```

Out[37]:=

$$\{\frac{5}{36}, \frac{1}{9}, \frac{1}{12}, \frac{1}{18}, \frac{1}{36}\}$$

In[38]:=

```
ts=Table[{i,i+1},{i,1,11}]
```

Out[38]:=

```
{{1, 2}, {2, 3}, {3, 4}, {4, 5}, {5, 6}, {6, 7},
 {7, 8}, {8, 9}, {9, 10}, {10, 11}, {11, 12}}
```

In[39]:=

```
ListPlot[Flatten[{prob1,prob2},1],Frame->True,
AxesOrigin->{1,0},FrameLabel->{"Sum","Probability"},
PlotRange->{0,0.18},FrameTicks->{ts,Automatic},
DefaultFont->{"Helvetica-Bold",12}]
```

Out[39]:=

```
-Graphics-
```

Sums on Three Fair Dice

First, all the possible permutations for the dice are generated, but we do not exhibit these here.

In[40]:=

```
perms=Flatten[Table[{i,j,k},{i,1,6},{j,1,6},
{k,1,6}],2];
```

Now we find all the possible sums and collect these values in a vector.

In[41]:=

```
values=Apply[Plus,perms,1]
```

Out[41]:=

```
{3, 4, 5, 6, 7, 8, 4, 5, 6, 7, 8, 9, 5, 6, 7, 8, 9,
  10, 6, 7, 8, 9, 10, 11, 7, 8, 9, 10, 11, 12, 8, 9,
  10, 11, 12, 13, 4, 5, 6, 7, 8, 9, 5, 6, 7, 8, 9,
  10, 6, 7, 8, 9, 10, 11, 7, 8, 9, 10, 11, 12, 8, 9,
  10, 11, 12, 13, 9, 10, 11, 12, 13, 14, 5, 6, 7, 8,
  9, 10, 6, 7, 8, 9, 10, 11, 7, 8, 9, 10, 11, 12, 8,
  9, 10, 11, 12, 13, 9, 10, 11, 12, 13, 14, 10, 11,
  12, 13, 14, 15, 6, 7, 8, 9, 10, 11, 7, 8, 9, 10,
  11, 12, 8, 9, 10, 11, 12, 13, 9, 10, 11, 12, 13,
  14, 10, 11, 12, 13, 14, 15, 11, 12, 13, 14, 15,
  16, 7, 8, 9, 10, 11, 12, 8, 9, 10, 11, 12, 13, 9,
  10, 11, 12, 13, 14, 10, 11, 12, 13, 14, 15, 11,
  12, 13, 14, 15, 16, 12, 13, 14, 15, 16, 17, 8, 9,
  10, 11, 12, 13, 9, 10, 11, 12, 13, 14, 10, 11, 12,
  13, 14, 15, 11, 12, 13, 14, 15, 16, 12, 13, 14,
  15, 16, 17, 13, 14, 15, 16, 17, 18}
```

In[42]:=

```
freq=BinCounts[values,{2,18,1}]
```

Out[42]:=

```
{1, 3, 6, 10, 15, 21, 25, 27, 27, 25, 21, 15, 10, 6,
  3, 1}
```

Figure 2.3 is a ListPlot of these values. The probabilities are the frequencies divided by 216.

Section 2.3   Expected Values of Discrete Random Variables

We find the mean and variance for the sums on three fair dice in the preceding example.

In[43]:=

```
probs=freq/216
```

Out[43]:=

$$\{\frac{1}{216},\ \frac{1}{72},\ \frac{1}{36},\ \frac{5}{108},\ \frac{5}{72},\ \frac{7}{72},\ \frac{25}{216},\ \frac{1}{8},\ \frac{1}{8},\ \frac{25}{216},\ \frac{7}{72},\ \frac{5}{72},$$

$$\frac{5}{108},\ \frac{1}{36},\ \frac{1}{72},\ \frac{1}{216}\}$$

In[44]:=

```
xs=Table[i,{i,3,18}]
```

Out[44]:=

```
{3, 4, 5, 6, 7, 8, 9, 10, 11, 12, 13, 14, 15, 16, 17,
   18}
```

In[45]:=

```
mean=Apply[Plus,xs*probs]
```

Out[45]:=

$$\frac{21}{2}$$

In[46]:=

```
variance=Apply[Plus,(xs^2)*probs]-mean^2
```

Out[46]:=

$$\frac{35}{4}$$

Section 2.4   Binomial Distribution

The binomial distribution is one of many probability distributions available in Mathematica. We show how to take a random sample of 1000 observations from a binomial distribution with $n = 20$ and $p = 3/4$ and then plot a histogram of the results.

In[47]:=

```
bindata=Table[Random[BinomialDistribution[20,3/4]],
{1000}];
```

In[48]:=

```
binfreq=BinCounts[bindata,{-1,20,1}]
```

Out[48]:=

```
{0, 0, 0, 0, 0, 0, 0, 0, 0, 2, 8, 26, 69, 105, 167,
   192, 208, 130, 70, 20, 3}
```

In[49]:=

```
binpoints=Table[i,{i,0,20,1}]
```

Out[49]:=

{0, 1, 2, 3, 4, 5, 6, 7, 8, 9, 10, 11, 12, 13, 14, 15, 16, 17, 18, 19, 20}

In[50]:=

```
BarChart[Transpose[{binfreq,binpoints}],
PlotRange->{Min[binfreq],Max[binfreq]},
AxesLabel->{x,frequency}]
```

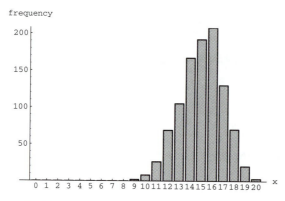

Out[50]:=

-Graphics-

Figure 2.9 is produced as follows:

In[51]:=

```
vals29=Table[PDF[BinomialDistribution[10,1/2],x],
{x,0,10}]
```

Out[51]:=

$$\{\frac{1}{1024}, \frac{5}{512}, \frac{45}{1024}, \frac{15}{128}, \frac{105}{512}, \frac{63}{256}, \frac{105}{512}, \frac{15}{128}, \frac{45}{1024}, \frac{5}{512}, \frac{1}{1024}\}$$

In[52]:=

```
xs29=Table[{i,i-1},{i,1,13}]
```

Out[52]:=

```
{{1, 0}, {2, 1}, {3, 2}, {4, 3}, {5, 4}, {6, 5},
   {7, 6}, {8, 7}, {9, 8}, {10, 9}, {11, 10},
   {12, 11}, {13, 12}}
```

In[53]:=

```
ListPlot[vals29,Frame->True,AxesOrigin->{0,0},
PlotRange->{0,0.3},FrameTicks->{xs29,Automatic},
FrameLabel->{"X","Probability"},
DefaultFont->{"Helvetica-Bold",12}]
```

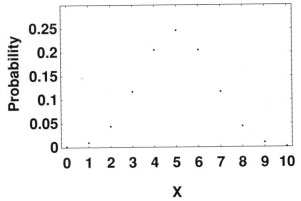

Out[53]:=

```
-Graphics-
```

Example   Flipping a Loaded Coin

Here we simulate 1000 flips of a coin loaded to come up heads with probability 2/5.

In[54]:=

```
biasdata=Table[Random[BinomialDistribution[1,2/5]],
{1000}];
```

In[55]:=

```
biasfreq=BinCounts[biasdata,{-1,1,1}]
```

Out[55]:=

```
{615, 385}
```

In[56]:=

```
biasvalues={0,1}
```

Out[56]:=

```
{0, 1}
```

In[57]:=

```
BarChart[Transpose[{biasfreq,biasvalues}],
AxesLabel->{face,frequency}]
```

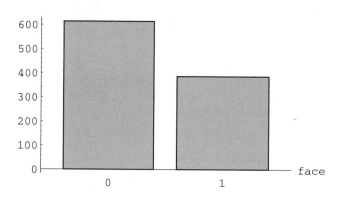

Out[57]:=

```
-Graphics-
```

Mathematica knows the mean and variance of the binomial distribution:

In[58]:=

```
Mean[BinomialDistribution[n,p]]
```

Out[58]:=

```
n p
```

In[59]:=

```
Variance[BinomialDistribution[n,p]]
```

Out[59]:=

```
n (1 - p) p
```


Section 2.6   Some Statistical Considerations

The confidence intervals shown in Figure 2.11 were generated and plotted as follows:

In[60]:=

```
sln=Simplify[Expand[Solve[100*p==X+2*
Sqrt[100*p*(1-p)],p]]]
```

Out[60]:=

$$\{\{p \;\text{->}\; \frac{10 \;+\; 5X \;-\; \text{Sqrt}[100 \;+\; 100\;X \;-\; X^2]}{520}\},$$

$$\{p \;\text{->}\; \frac{10 \;+\; 5X \;+\; \text{Sqrt}[100 \;+\; 100\;X \;-\; X^2]}{520}\}\}$$

In[61]:=

```
leftend=p/.sln[[1]]
```

Out[61]:=

$$\frac{10 \;+\; 5X \;-\; \text{Sqrt}[100 \;+\; 100\;X \;-\; X^2]}{520}$$

In[62]:=

```
rightend=p/.sln[[2]]
```

Out[62]:=

$$\frac{10 \;+\; 5X \;+\; \text{Sqrt}[100 \;+\; 100\;X \;-\; X^2]}{520}$$

In[63]:=

```
data={40,44,29,43,43,42,39,40,43,42,36,44,35,39,42}
```

Out[63]:=

```
{40, 44, 29, 43, 43, 42, 39, 40, 43, 42, 36, 44, 35,
   39, 42}
```

In[64]:=

```
endpts=Table[{leftend,rightend}/.X->data[[i]],
{i,1,Length[data]}]//N
```

Out[64]:=

```
{{0.307692, 0.5}, {0.344931, 0.539685},
   {0.208721, 0.387433}, {0.335563, 0.529822},
   {0.335563, 0.529822}, {0.326233, 0.519921},
   {0.298482, 0.48998}, {0.307692, 0.5},
   {0.335563, 0.529822}, {0.326233, 0.519921},
   {0.271095, 0.459674}, {0.344931, 0.539685},
   {0.26205, 0.449488}, {0.298482, 0.48998},
   {0.326233, 0.519921}}
```

In[65]:=

```
vert=Show[Graphics[Line[{{.4,0},{.4,16}}]]]
```

In[66]:=

```
Show[Graphics[Table[Line[{{endpts[[i]][[1]],i},
{endpts[[i]][[2]],i}}],{i,1,Length[endpts]}]],vert,
Frame->True]
```

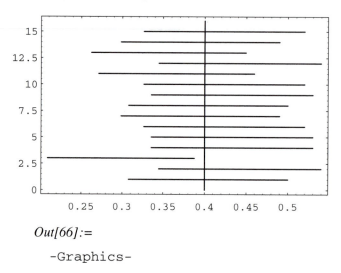

Out[66]:=

```
-Graphics-
```

Section 2.7   Hypothesis Tests

Alpha and beta errors in this section are sums of binomial probabilities:

In[67]:=

```
alpha=Sum[PDF[BinomialDistribution[20,0.2],x],{x,9,20}]
```

Out[67]:=

```
0.00998179
```

In[68]:=

```
beta = Sum[PDF[BinomialDistribution[20,0.3],x],
{x,0,8}]
```

Out[68]:=

```
0.886669
```

Section 2.10   Hypergeometric Distribution

Section 2.10.3

Figure 2.18 shows a hypergeometric distribution with $n = 30, D = 400$ and $N = 1000$. This distribution can be plotted with the following commands. First the hypergeometric function is called; Mathematica responds with the definition of the function.

In[69]:=

```
hyperfcn=PDF[HypergeometricDistribution[30,400,
1000],x]
```

Out[69]:=

```
If[Max[0, 30 - 1000 + 400] <= x <= Min[30, 400],
   If[IntegerQ[x],
```

$$\frac{\text{Binomial}[400,\ x]\ \text{Binomial}[1000 - 400,\ 30 - x]}{\text{Binomial}[1000,\ 30]}, 0], 0]$$

Now we generate a table of values of the function and then plot these values.

In[70]:=

```
xshyper=Table[{i,i-1},{i,1,24,2}]
```

Out[70]:=

```
{{1, 0}, {3, 2}, {5, 4}, {7, 6}, {9, 8}, {11, 10},
   {13, 12}, {15, 14}, {17, 16}, {19, 18}, {21, 20},
   {23, 22}}
```

In[71]:=

```
ListPlot[Table[hyperfcn,{x,0,24}],Frame->True,
FrameLabel->{"x","Probability"},PlotRange->{0,0.155},
FrameTicks->{xshyper,Automatic},AxesOrigin->{734,0},
DefaultFont->{"Helvetica-Bold",12}]
```

Out[71]:=

    -Graphics-

Section 2.13   Poisson Variable

The Poisson variable is called by specifying its parameter. Here is the way Figure 2.26 was drawn.

In[72]:=

```
ListPlot[Table[PDF[PoissonDistribution[4],x],
{x,0,15}],
Frame->True,DefaultFont->{"Helvetica-Bold",12}]
```

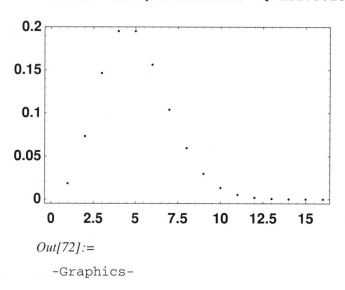

Out[72]:=

    -Graphics-

Chapter Three

The standard probability density functions, such as the uniform, exponential, normal, chi-squared, gamma, and Weibull distributions (as well as many others), are included in Mathematica. Some examples of their use are given here.

Section 3.1   Continuous Random Variables

Means and variances for continuous distributions can be found directly by integration. Here we use the probability density function $3x^2$, for x in the interval (0, 1).

Example 3.1.1

In[73]:=

```
mean=Integrate[3x^2 x,{x,0,1}]
```

Out[73]:=

$$\frac{3}{4}$$

In[74]:=

```
variance=Integrate[(3x^2)*(x^2),{x,0,1}]-mean^2
```

Out[74]:=

$$\frac{3}{80}$$

Section 3.3   Exponential Distribution

The probability density function for the exponential distribution can be found as follows. The value for λ must be specified.

In[75]:=

```
expdist=PDF[ExponentialDistribution[a],x]
```

Out[75]:=

$$\frac{a}{E^{ax}}$$

Note that the mean is then $1/a$. Probabilities are then found by integration.

Example 3.3.2

The probability that X exceeds 2 is given by the following:

In[76]:=

```
Integrate[PDF[ExponentialDistribution[1],x],
{x,2,Infinity}]
```

Out[76]:=

$$E^{-2}$$

Section 3.5   Normal Distribution

The probability density function of the normal distribution is found by specifying the mean and the standard deviation.

In[77]:=

```
normdist=PDF[NormalDistribution[a,b],x]
```

Out[77]:=

$$\frac{1}{b \, E^{(-a \, + \, x)^2/(2b^2)} \, Sqrt[2 \, Pi]}$$

Example 3.5.1

Here we seek the conditional probability that a score exceeds 600, given that it exceeds 500. There is no need to transform the scores or to consult a table.

In[78]:=

```
satdist=PDF[NormalDistribution[500,100],x];
```

In[79]:=

```
Integrate[satdist,{x,600,Infinity}]/
Integrate[satdist,{x,500,Infinity}]//N
```

Out[79]:=

```
0.317311
```

Section 3.6   Normal Approximation to the Binomial

Direct comparisons with exact binomial probabilities and the approximations given by the normal distribution are easily done. We use a binomial distribution with $n = 500$ and $p = 0.10$. To compare the exact value of $P(X = 53)$ with the normal curve approximation, calculate as follows:

In[80]:=

```
PDF[BinomialDistribution[500,0.10],53]
```

Out[80]:=

```
0.0524484
```

In[81]:=

```
Integrate[PDF[NormalDistribution[500*0.10,
Sqrt[500*0.10*0.90]],x],{x,52.5,53.5}]//N
```

Out[81]:=

```
0.0537716
```

Section 3.7   Gamma and Chi-Squared Distributions

We show here the form for calling gamma and chi-squared distributions, and we show two graphs. The number of degrees of freedom must be specified in the chi-squared distribution while the gamma distribution is characterized by two parameters, r and λ. In this example, the parameter in the chi-squared distribution is 6.

In[82]:=

```
Plot[PDF[ChiSquareDistribution[6],x],{x,0,16},
AxesLabel->{x,f},Frame->True,FrameLabel->{"x","f"},
DefaultFont->{"Helvetica-Bold",12}]
```

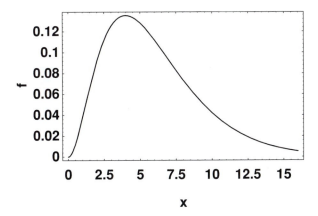

Out[82]:=

    -Graphics-

In[83]:=

```
gamdist=PDF[GammaDistribution[r,1/lambda],x]
```

Out[83]:=

$$\frac{x^{-1 + r}}{E^{lambda\ x}\ (\frac{1}{lambda})^{r}\ Gamma[r]}$$

Here r and λ are the parameters we have used. In the following graph $r = 6$ and $\lambda = \frac{1}{2}$.

In[84]:=

```
Plot[PDF[GammaDistribution[6,1/2],x],
{x,0,9},Frame->True,FrameLabel->{"x","f"},
DefaultFont->{"Helvetica-Bold",12}]
```

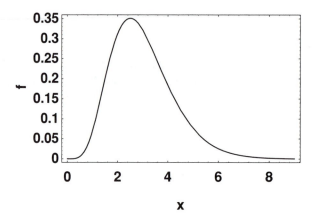

Out[84]:=

```
-Graphics-
```

Section 3.8   Weibull Distribution

The Weibull distribution is also an available probability density function in Mathematica. Parameters α and β must be specified.

In[85]:=

```
weibdist=PDF[WeibullDistribution[a,b],x]
```

Out[85]:=

$$\frac{a\ x^{-1+a}}{b^a\ E^{(x/b)^a}}$$

In[86]:=

```
Plot[PDF[WeibullDistribution[2,3],x],
{x,0,9},Frame->True,FrameLabel->{"x","f"},
DefaultFont->{"Helvetica-Bold",12}]
```

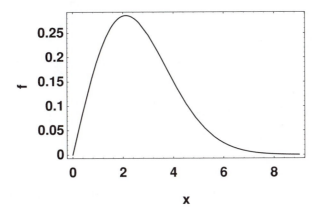

Out[86]:=

```
-Graphics-
```

Chapter Four

Section 4.5   Generating Functions

Generating functions are particularly easy to deal with in Mathematica because power series expansions are easily done; specific coefficients within these expansions can also be found.

Example 4.3.1

The generating function for the fair die is as follows:

In[87]:=

```
g[t_]:=(1/6)*(t+t^2+t^3+t^4+t^5+t^6)
```

Powers of $g[t]$ can be found.

In[88]:=

```
series=Expand[(g[t])^2]
```

Out[88]:=

$$\frac{t^2}{36} + \frac{t^3}{18} + \frac{t^4}{12} + \frac{t^5}{9} + \frac{5t^6}{36} + \frac{t^7}{6} + \frac{5t^8}{36} + \frac{t^9}{9} + \frac{t^{10}}{12}$$

$$+ \frac{t^{11}}{18} + \frac{t^{12}}{36}$$

If specific coefficients are needed, the following commands are useful:

In[89]:=

```
Coefficient[series,t,6]
```

Out[89]:=

$$\frac{5}{36}$$

Or the set of all coefficients beginning with the constant term can be found.

In[90]:=

```
CoefficientList[series,t]
```

Out[90]:=

$$\{0, \ 0, \ \frac{1}{36}, \ \frac{1}{18}, \ \frac{1}{12}, \ \frac{1}{9}, \ \frac{5}{36}, \ \frac{1}{6}, \ \frac{5}{36}, \ \frac{1}{9}, \ \frac{1}{12}, \ \frac{1}{18}, \ \frac{1}{36}\}$$

In this section we sought the probability that the sum is 17 when four fair dice are tossed. This can be found directly as follows:

In[91]:=

```
Coefficient[(g[t])^4,t,17]
```

Out[91]:=

$$\frac{13}{162}$$

Section 4.7   Probability Generating Functions for Some Specific Probability Distributions

The SymbolicSum command can be used to produce probability generating functions for discrete random variables.

In[92]:=

```
<<Algebra`SymbolicSum`
```

The probability generating function for the binomial distribution is

In[93]:=

```
SymbolicSum[(t^x)*Binomial[n,x]*(p^x)*(1-p)^(n-x),
{x,0,n}]
```

Out[93]:=

$$(1-p)^n \ (\frac{-1 \ + \ p \ - \ pt}{-1 \ + \ p})^n$$

This can be simplified to

Out[93]:=

$$(q \ + \ pt)^n$$

Section 4.8   Moment Generating Functions

Example 4.8.2

The moment generating function for the exponential distribution can be found as follows:

In[94]:=

```
expbinfcn=Integrate[Exp[t x]*Exp[-x],{x,0,Infinity}]
```

Out[94]:=

$$\frac{1}{1 \ - \ t}$$

This can be expanded in a power series.

In[95]:=

```
Series[expbinfcn,{t,0,10}]
```

Out[95]:=

$$1 \ + \ t \ + \ t^2 \ + \ t^3 \ + \ t^4 \ + \ t^5 \ + \ t^6 \ + \ t^7 \ + \ t^8 \ + \ t^9$$
$$+ \ t^{10} \ + \ O[t]^{11}$$

Section 4.10   Sums of Random Variables

Example 4.10.2   Sums of Exponential Random Variables

Here a limit must be calculated in order to show that the sum of exponential random variables becomes normal. We show the limit of $\log[M[Z;t]]$ as n becomes infinite.

In[96]:=

```
lmz=(-t*Sqrt[n])-n*Log[1-t/Sqrt[n]]
```

Out[96]:=

$$-(\text{Sqrt}[n] \ t) - n \ \text{Log}[1 - \frac{t}{\text{Sqrt}[n]}]$$

In[97]:=

```
Limit[lmz,n->Infinity]
```

Out[97]:=

$$\frac{t^2}{2}$$

Section 4.11   The Central Limit Theorem

We discussed sampling from the uniform distribution in this appendix in the material for Chapter Two. Here we want samples of size 3 from a uniform distribution on the integers from 1 to 20. We show how to draw 100 such samples and compute the sample mean of each one. We show a histogram of the sample means as an illustration of the central limit theorem.

In[98]:=

```
data=Table[Table[Random[DiscreteUniformDistribution
[20]],{3}],{100}];
```

In[99]:=

```
sums=Apply[Plus,data,1]
```

Out[99]:=

```
{24, 38, 33, 51, 16, 18, 29, 51, 33, 44, 41, 34, 16,
   39, 34, 46, 29, 26, 34, 29, 19, 26, 29, 20, 36, 22,
   32, 41, 34, 28, 34, 34, 23, 34, 45, 42, 10, 17, 35,
   50, 42, 30, 45, 38, 37, 23, 27, 12, 47, 24, 36, 19,
   39, 28, 35, 38, 37, 33, 49, 37, 33, 32, 25, 37, 37,
   33, 42, 24, 32, 12, 29, 35, 16, 51, 12, 12, 24, 22,
   23, 30, 37, 44, 46, 42, 42, 28, 19, 33, 43, 10, 23,
   34, 18, 21, 26, 37, 31, 34, 33, 55}
```

In[100]:=

```
means=sums/3;
```

In[101]:=

```
Length[means]
```

Out[101]:=

 100

In[102]:=

 Min[means]

Out[102]:=

 $\frac{10}{3}$

In[103]:=

 Max[means]

Out[103]:=

 $\frac{55}{3}$

In[104]:=

 freq=BinCounts[means,{3,59/3,1/3}]

Out[104]:=

 {2, 0, 4, 0, 0, 0, 3, 1, 2, 3, 1, 1, 2, 4, 4, 1, 3,
  1, 3, 5, 2, 1, 3, 7, 9, 3, 2, 7, 3, 2, 0, 2, 5, 1,
  2, 2, 2, 1, 0, 1, 1, 3, 0, 0, 0, 1, 0, 0, 0, 0}

In[105]:=

 Apply[Plus,freq]

Out[105]:=

 100

In[106]:=

 values=Table[i,{i,10/3,59/3,1/3}];

In[107]:=

 xticks=Table[{3i-2,i+2},{i,1,16}]

Out[107]:=

 {{1, 3}, {4, 4}, {7, 5}, {10, 6}, {13, 7}, {16, 8},
  {19, 9}, {22, 10}, {25, 11}, {28, 12}, {31, 13},
  {34, 14}, {37, 15}, {40, 16}, {43, 17}, {46, 18}}

In[108]:=

```
BarChart[Transpose[{freq,values}],
PlotRange->{0,Max[freq]},Ticks->{xticks,Automatic}]
```

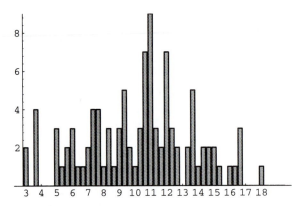

Out[108]:=

```
-Graphics-
```

This sample shows some departure from normality, but that is not unexpected with samples of size 3.

Section 4.12   Weak Law of Large Numbers

The following commands show the construction of Figure 4.10.

In[109]:=

```
vals=Table[Random[BinomialDistribution[100,0.38]],
{100}]
```

Out[109]:=

```
{39, 40, 41, 40, 33, 37, 36, 37, 38, 39, 37, 34, 45,
  37, 28, 36, 35, 39, 38, 44, 44, 39, 42, 32, 35, 44,
  30, 37, 32, 38, 41, 31, 39, 45, 36, 43, 40, 38, 36,
  38, 37, 45, 43, 39, 39, 36, 31, 38, 35, 38, 26, 28,
  34, 25, 35, 46, 39, 32, 45, 32, 44, 39, 40, 40, 37,
  40, 42, 32, 44, 35, 41, 40, 39, 42, 41, 39, 37, 42,
  37, 40, 43, 41, 38, 38, 47, 33, 44, 44, 38, 38, 32,
  42, 40, 34, 36, 33, 31, 37, 40, 39}
```

In[110]:=

```
newvals=Table[Sum[vals[[i]],{i,1,j}]/(100*j),
{j,1,100}]
```

Out[110]:=

$$\{\frac{39}{100}, \frac{79}{200}, \frac{2}{5}, \frac{2}{5}, \frac{193}{500}, \frac{23}{60}, \frac{19}{50}, \frac{303}{800}, \frac{341}{900}, \frac{19}{50}, \frac{417}{1100},$$

$$\frac{451}{1200}, \frac{124}{325}, \frac{533}{1400}, \frac{187}{500}, \frac{597}{1600}, \frac{158}{425}, \frac{671}{1800}, \frac{709}{1900}, \frac{753}{2000},$$

$$\frac{797}{2100}, \frac{19}{50}, \frac{439}{1150}, \frac{91}{240}, \frac{189}{500}, \frac{989}{2600}, \frac{1019}{2700}, \frac{66}{175}, \frac{272}{725},$$

$$\frac{563}{1500}, \frac{1167}{3100}, \frac{599}{1600}, \frac{1237}{3300}, \frac{641}{1700}, \frac{659}{1750}, \frac{1361}{3600}, \frac{1401}{3700},$$

$$\frac{1439}{3800}, \frac{59}{156}, \frac{1513}{4000}, \frac{31}{82}, \frac{319}{840}, \frac{819}{2150}, \frac{1677}{4400}, \frac{143}{375}, \frac{219}{575},$$

$$\frac{1783}{4700}, \frac{607}{1600}, \frac{464}{1225}, \frac{947}{2500}, \frac{32}{85}, \frac{487}{1300}, \frac{991}{2650}, \frac{223}{600}, \frac{1021}{2750},$$

$$\frac{261}{700}, \frac{709}{1900}, \frac{2159}{5800}, \frac{551}{1475}, \frac{559}{1500}, \frac{114}{305}, \frac{2319}{6200}, \frac{337}{900}, \frac{2399}{6400},$$

$$\frac{609}{1625}, \frac{619}{1650}, \frac{1259}{3350}, \frac{3}{8}, \frac{1297}{3450}, \frac{2629}{7000}, \frac{267}{710}, \frac{271}{720}, \frac{2749}{7300},$$

$$\frac{2791}{7400}, \frac{236}{625}, \frac{2871}{7600}, \frac{727}{1925}, \frac{59}{156}, \frac{2987}{7900}, \frac{3027}{8000}, \frac{307}{810}, \frac{3111}{8200},$$

$$\frac{3149}{8300}, \frac{3187}{8400}, \frac{1617}{4250}, \frac{3267}{8600}, \frac{3311}{8700}, \frac{61}{160}, \frac{3393}{8900}, \frac{3431}{9000},$$

$$\frac{3463}{9100}, \frac{701}{1840}, \frac{709}{1860}, \frac{3579}{9400}, \frac{723}{1900}, \frac{19}{50}, \frac{3679}{9700}, \frac{929}{2450}, \frac{313}{825},$$

$$\frac{759}{2000}\}$$

In[111]:=

```
ListPlot[newvals,AxesOrigin->{0,0.35},Frame->True,
FrameLabel->{"Number of Trials","Probability
of Success"},DefaultFont->{"Helvetica-Bold",12}]
```

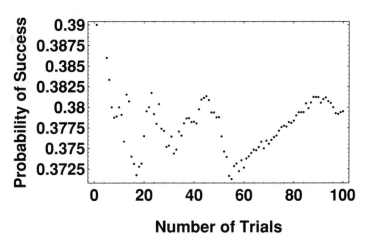

Out[111]:=

-Graphics-

Section 4.13 Distribution of the Sample Variance

Figure 4.11 shows the variances of samples of size 3 drawn without replacement from a uniform distribution on the set $\{1, 2, \ldots, 20\}$. The graph is drawn as follows:

In[112]:=

```
data=Flatten[Table[{i,j,k},{i,1,20},{j,i+1,20},
{k,j+1,20}],2];
```

In[113]:=

```
vars=Table[Variance[data[[i]]],{i,1,Length[data]}];
```

In[114]:=

```
sortvars=Sort[vars];
```

In[115]:=

```
valuesvars=Union[sortvars];
```

In[116]:=

```
freqvars=Table[Count[sortvars,valuesvars[[i]]],
{i,1,Length[valuesvars]}];
```

In[117]:=

```
ticks={{20,20},{40,40},{60,60},{80,80},{100,100}}
```

Out[117]:=

```
{{20, 20}, {40, 40}, {60, 60}, {80, 80}, {100, 100}}
```

In[118]:=

```
ListPlot[Transpose[{valuesvars,freqvars}],
FrameLabel->{"Variance","Frequency"},
PlotRange->{0,Max[freqvars]},Ticks->{ticks,Automatic},
DefaultFont->{"Helvetica-Bold",12},Frame->True]
```

Out[118]:=

```
-Graphics-
```

Section 4.14 Hypothesis Tests and Confidence Intervals for a Single Mean

Example 4.14.1

If the mean and standard deviation are known, which is the case in this example, a confidence interval can be found using the following:

In[119]:=

NormalCI[2200,4591.84/5]//N

Out[119]:=

```
{400.032, 3999.97}
```

Note that we used the standard deviation of the mean, based on a sample size of 25. Note also that the default is a 95% confidence interval. This can be changed to any probability.

In[120]:=

NormalCI[2200,4591.84/5,ConfidenceLevel->0.8436]//N

Out[120]:=

```
{898.412, 3501.59}
```

The Student t distribution is included in Mathematica. Here is the Student t distribution with 6 degrees of freedom:

In[121]:=

tdist=PDF[StudentTDistribution[6],x]

Out[121]:=

$$\frac{405 \left(\dfrac{1}{6+x^2}\right)^{7/2}}{2}$$

Note that Mathematica gives the probability density function.

In[122]:=

Plot[tdist,{x,-3,3},AxesLabel->{t,f}]

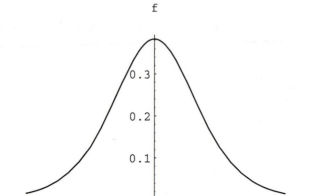

Out[122]:=

-Graphics-

Confidence intervals can be calculated.

Example 4.14.4

Suppose the data are as follows:

In[123]:=

data={8,7,3,5,9,4,10,2,6,7}

Out[123]:=

{8, 7, 3, 5, 9, 4, 10, 2, 6, 7}

In[124]:=

Variance[data]

Out[124]:=

$\frac{203}{30}$

In[125]:=

Mean[data]

Out[125]:=

$\frac{61}{10}$

In[126]:=

> **StudentTCI[61/10,Sqrt[(203/30)/10],9]//N**

Out[126]:=

> {4.23916, 7.96084}

Section 4.15 Tests on Two Samples

Mathematica can do hypothesis testing and find confidence intervals directly from sample data. As an example, suppose the following are samples from two normal populations.

In[127]:=

> **data1={7,4,5,6,7,3,4,2,5,8,9,12}**

Out[127]:=

> {7, 4, 5, 6, 7, 3, 4, 2, 5, 8, 9, 12}

In[128]:=

> **data2={8,9,5,6,7,8,3,4,5,6,3,4,5,7,6,3,4,9}**

Out[128]:=

> {8, 9, 5, 6, 7, 8, 3, 4, 5, 6, 3, 4, 5, 7, 6, 3, 4, 9}

A 95% confidence interval for the difference between the means can be found. We can assume the variances are equal or not. The default assumption is that the variances are unequal.

In[129]:=

> **MeanDifferenceCI[data1,data2,EqualVariances->True]**

Out[129]:=

> $\{\frac{1}{3} - 0.341401 \, \text{Sqrt}[\frac{55}{2}], \frac{1}{3} + 0.341401 \, \text{Sqrt} \, [\frac{55}{2}]\}$

In[130]:=

> **%//N**

Out[130]:=

> {-1.45699, 2.12366}

Now we assume that the variances are not equal.

In[131]:=

> **MeanDifferenceCI[data1,data2]**

Out[131]:=

$\{\frac{1}{3} - 0.699221 \ \text{Sqrt}[\frac{173}{22}], \ \frac{1}{3} + 0.699221 \ \text{Sqrt} \ [\frac{173}{22}]\}$

In[132]:=

%//N

Out[132]:=

{-1.62744, 2.2941}

These give quite different confidence intervals.

We can also find a confidence interval for the ratio of the variances. We must specify the sample ratio of the variances and then the number of degrees of freedom for the numerator and the denominator, respectively.

In[133]:=

FRatioCI[Variance[data1]/Variance[data2],11,17]

Out[133]:=

{0.681112, 6.41411}

Since 1 is in this interval, we can presume that the true variances are equal.

p values are given for hypothesis tests. We must specify the data and the difference between the means that we wish to test. Here we choose 1 for this difference. Again, the true variances are presumed to be unequal.

In[134]:=

MeanDifferenceTest[data1,data2,1]

Out[134]:=

OneSidedPValue -> 0.242332

Undoubtedly we would not agree that the difference between the means is 1.

Section 4.16   Least Squares Linear Regression

We show the construction of Figure 4.18.

In[135]:=

rdata={{92,104},{86,91},{104,123},{109,102},{75,86},
{100,99},{91,92},{110,114},{128,99}}

Out[135]:=

{{92, 104}, {86, 91}, {104, 123}, {109, 102},
  {75, 86}, {100, 99}, {91, 92}, {110, 114},
  {128, 99}}

In[136]:=

```
scatter=ListPlot[rdata,Frame->True,AxesOrigin->
{70,80},
PlotStyle->{PointSize[0.015]},
FrameLabel->{"Math Score","IQ"},
DefaultFont->{"Helvetica-Bold",12}]
```

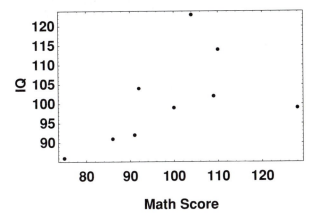

Out[136]:=

```
-Graphics-
```

To fit a straight line to the data we use the Regress command. The braces $\{1, x\}$ indicate that we fit a straight line with an intercept. Mathematica offers a wide range of functions that can be specified in addition to this simple linear function.

In[137]:=

```
Regress[rdata,{1,x},x]
```

Out[137]:=

```
{ParameterTable ->
        Estimate   SE         TStat      PValue,
    1   63.0792    24.2581    2.60034    0.0354077
    x   0.382444   0.241314   1.58484    0.157022
RSquared -> 0.264065, AdjustedRSquared -> 0.158931,
EstimatedVariance -> 113.217,
ANOVATable ->
            DoF SoS        MeanSS     FRatio     PValue
    Model   1   284.368    284.368    2.5117     0.157022
    Error   7   792.521    113.217
    Total   8   1076.89
```

In[138]:=

```
r[x_]:=63.0792+0.38244*x
```

In[139]:=

stline=Plot[r[x],{x,70,130}]

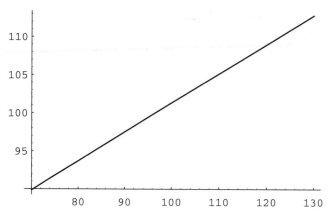

Out[139]:=

-Graphics-

In[140]:=

Show[scatter,stline]

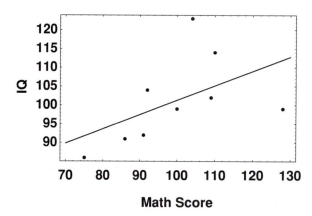

Out[140]:=

-Graphics-

Section 4.17   Control Chart for X Bar

Example 4.17.1

We show a plot of a quality control chart as in Figure 4.20.

In[141]:=

```
data={{3,-1,6,4},{9,0,3,-2},{-12,4,-9,-6},{11,9,4,1},
{-1,-2,4,-1},{8,1,-2,3},{-1,-4,-9,-3},{-8,-3,-6,-4},
{1,-2,-2,1},{-2,-2,-3,-2},{0,4,1,-2}}
```

Out[141]:=

```
{{3, -1, 6, 4}, {9, 0, 3, -2}, {-12, 4, -9, -6},
  {11, 9, 4, 1}, {-1, -2, 4, -1}, {8, 1, -2, 3},
  {-1, -4, -9, -3}, {-8, -3, -6, -4},
  {1, -2, -2, 1}, {-2, -2, -3, -2}, {0, 4, 1, -2}}
```

In[142]:=

```
means=Table[Mean[data[[i]]],{i,1,Length[data]}]
```

Out[142]:=

$$\{ 3, \frac{5}{2}, -(\frac{23}{4}), \frac{25}{4}, 0, \frac{5}{2}, -(\frac{17}{4}), -(\frac{1}{2}), -(\frac{9}{4}), \frac{3}{4}\}$$

In[143]:=

```
ucl=Show[Graphics[Line[{{0,5.0974},{11.25,
5.0974}}]]]
```

In[144]:=

```
lcl=Show[Graphics[Line[{{0,-6.188},{11.25,
-6.188}}]]]
```

In[145]:=

```
ltext=Show[Graphics[Text["LCL",{11.25,
-6.188}]]]
```

In[146]:=

```
utext=Show[Graphics[Text["UCL",{11.25,5.0974}]]]
```

In[147]:=

```
Show[ListPlot[means,AxesOrigin->{0,0}],ucl,lcl,ltext,
utext,AxesLabel->{"x","Mean"},
DefaultFont->{"Helvetica-Bold",12}]
```

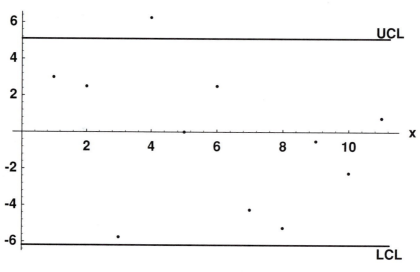

Out[147]:=

```
-Graphics-
```

Chapter Five

Section 5.2   Joint and Marginal Distributions

Example 5.2.1

Probabilies for the joint distribution of X and Y are found by the following:

In[148]:=

```
probs=Table[Binomial[5,x]*Binomial[5-x,y]*
(1/2)^(10-x),{x,0,5},{y,0,5-x}]
```

Out[148]:=

$$\{\{\frac{1}{1024},\ \frac{5}{1024},\ \frac{5}{512},\ \frac{5}{512},\ \frac{5}{1024},\ \frac{1}{1024}\},$$
$$\{\frac{5}{512},\ \frac{5}{128},\ \frac{15}{256},\ \frac{5}{128},\ \frac{5}{512}\},\ \{\frac{5}{128},\ \frac{15}{128},\ \frac{15}{128},\ \frac{5}{128}\},$$
$$\{\frac{5}{64},\ \frac{5}{32},\ \frac{5}{64}\},\ \{\frac{5}{64},\ \frac{5}{64}\},\ \{\frac{1}{32}\}\}$$

The marginal distribution for *X* can be found by using the following:

In[149]:=

```
Apply[Plus,probs,1]
```

Out[149]:=

$$\{\frac{1}{32},\ \frac{5}{32},\ \frac{5}{16},\ \frac{5}{16},\ \frac{5}{32},\ \frac{1}{32}\}$$

Example 5.2.2

The joint probability distribution function is plotted as follows:

In[150]:=

```
Clear[f,g]
```

In[151]:=

```
f[x_,y_]:=x^2+(8/3) x y
```

In[152]:=

```
Plot3D[f[x,y],{x,0,1},{y,0,1},ViewPoint ->
{1.3,-1.8,2},
AxesLabel->{"X","Y","f"},
DefaultFont->{"Helvetica-Bold",12}]
```

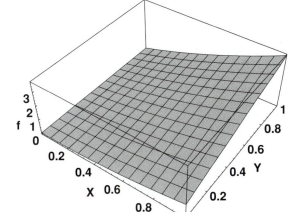

Out[152]:=

```
-SurfaceGraphics-
```

Marginal distributions are easily found.

In[153]:=

```
f[x_]:=Integrate[f[x,y],{y,0,1}]
```

In[154]:=

```
f[x]
```

Out[154]:=

$$\frac{4\,x}{3} + x^2$$

In[155]:=

```
g[y_]:=Integrate[f[x,y],{x,0,1}]
```

In[156]:=

```
g[y]
```

Out[156]:=

$$\frac{1}{3} + \frac{4\,y}{3}$$

Probabilities are volumes.

In[157]:=

```
Integrate[f[x,y],{x,1/2,1},{y,0,2/3}]
```

Out[157]:=

$$\frac{5}{12}$$

Section 5.3   Conditional Distributions and Densities

Conditional distributions are ratios of the distributions previously calculated.

In[158]:=

```
fxgivy[x_]:=f[x,y]/g[y]
```

In[159]:=

```
fxgivy[x]
```

Out[159]:=

$$\frac{x^2 + \dfrac{8\ x\ y}{3}}{\dfrac{1}{3} + \dfrac{4\ y}{3}}$$

In[160]:=

```
fygivx[y_]:=f[x,y]/f[x]
```

In[161]:=

```
fygivx[y]
```

Out[161]:=

$$\frac{x^2 + \dfrac{8\ x\ y}{3}}{\dfrac{4\ x}{3} + x^2}$$

Section 5.4   Expected Values

Conditional and unconditional expectations can now be found. We continue with Example 5.2.2.

The expected value of Y given X is

In[162]:=

```
eygivx=Integrate[y*fygivx[y],{y,0,1}]
```

Out[162]:=

$$\frac{8}{3\ (4\ +\ 3\ x)} + \frac{3\ x}{2\ (4\ +\ 3\ x)}$$

The integral of $E[Y|X = x] * f[x]$ gives the expected value of Y.

In[163]:=

```
ey=Integrate[eygivx*f[x],{x,0,1}]
```

Out[163]:=

$$\frac{11}{18}$$

Section 5.6   Bivariate Normal Densities

The bivariate normal density is defined as follows:

In[164]:=

```
k[r_]:=1/(2 Pi Sqrt[1-r^2])
```

In[165]:=

```
k[r]
```

Out[165]:=

$$\frac{1}{2 \text{ Pi Sqrt}[1 - r^2]}$$

In[166]:=

```
Clear[f]
```

In[167]:=

```
f[x_,y_]:=k[r]*Exp[(-1/(2(1-r^2)))*
(x^2-2 r x y +y^2)]
```

In[168]:=

```
f[x,y]
```

Out[168]:=

$$\frac{1}{2 \text{ E}^{(x^2 - 2 r x y + y^2)/(2 (1 - r^2))} \text{ Pi Sqrt}[1 - r^2]}$$

Now let the correlation coefficient be 1/2.

In[169]:=

```
r=1/2
```

Out[169]:=

$$\frac{1}{2}$$

Figure 5.9 is drawn as follows:

In[170]:=

```
Plot3D[f[x,y],{x,-3,3},{y,-3,3},
AxesLabel->{"x","y","f"},PlotPoints->40,
```

DefaultFont->{"Helvetica-Bold",12}]

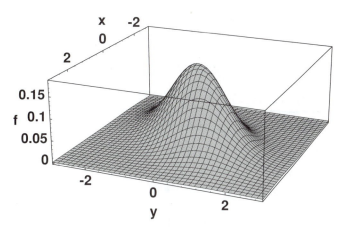

Out[170]:=

    -SurfaceGraphics-

A contour plot is found as follows:

In[171]:=

ContourPlot[f[x,y],{x,-3,3},{y,-3,3},PlotPoints->40]

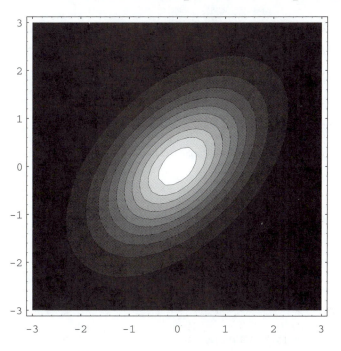

Out[171]:=

    -ContourGraphics-

Chapter Six

Recursions and Markov chains are easily done in Mathematica.

Section 6.2   Some Recursions and their Solutions

Example 6.2.1

The recursion is first defined.

In[172]:=

```
Clear[a]
```

In[173]:=

```
a[n_]:=a[n]=p a[n-1]+(1-p)*(1-a[n-1])
```

In[174]:=

```
a[1]=0
```

Out[174]:=

```
0
```

In[175]:=

```
a[2]=1-p
```

Out[175]:=

```
1 - p
```

Now we find some values for a_n.

In[176]:=

```
Table[Simplify[Expand[a[n]]],{n,1,7}]
```

Out[176]:=

$$\{0,\ 1-p,\ 2\ (1-p)\ p,\ 1-3\ p+6\ p^2-4\ p^3,$$

$$4\ p\ (1-3\ p+4\ p^2-2\ p^3),$$

$$1-5\ p+20\ p^2-40\ p^3+40\ p^4-16\ p^5,$$

$$2\ p\ (3-15\ p+40\ p^2-60\ p^3+48\ p^4-16\ p^5)\}$$

We now look at a particular value, a_4, and expand it about 1/2.

In[177]:=

```
exp=Normal[Series[a[4],{p,1/2,3}]]
```

Out[177]:=

$$\frac{1}{2} - 4 \left(-(\tfrac{1}{2}) + p\right)^3$$

This suggests a simple form for a_n.

Mathematica will also solve the recursion; we choose the special case where $p = 1/3$. The RSolve package is needed first.

In[178]:=

```
<<DiscreteMath`RSolve`
```

In[179]:=

```
RSolve[{c[n]==(1/3)*c[n-1]+(2/3)*(1-c[n-1])/;n>=3,
c[1]==0,c[2]==2/3},c[n],n]
```

Out[179]:=

$$\left\{\left\{c[n] \; \rightarrow \; \frac{1}{2} - \frac{(-(\tfrac{1}{3}))^{-1+n}}{2}\right\}\right\}$$

This shows that $c_n \rightarrow 1/2$ as n becomes large.

Section 6.5   Markov Chains

Example 6.5.1

The matrix in this example is entered as follows:

In[180]:=

```
m={{1/10,3/10,3/10,3/10},{3/10,1/10,3/10,3/10},
{3/10,3/10,1/10,3/10},{3/10,3/10,3/10,1/10}}
```

Out[180]:=

$$\left\{\left\{\frac{1}{10}, \frac{3}{10}, \frac{3}{10}, \frac{3}{10}\right\}, \left\{\frac{3}{10}, \frac{1}{10}, \frac{3}{10}, \frac{3}{10}\right\},\right.$$

$$\left.\left\{\frac{3}{10}, \frac{3}{10}, \frac{1}{10}, \frac{3}{10}\right\}, \left\{\frac{3}{10}, \frac{3}{10}, \frac{3}{10}, \frac{1}{10}\right\}\right\}$$

This can be written in the usual way as follows:

In[181]:=

```
MatrixForm[m]
```

Out[181]//MatrixForm=

$$
\frac{1}{10} \quad \frac{3}{10} \quad \frac{3}{10} \quad \frac{3}{10}
$$

$$
\frac{3}{10} \quad \frac{1}{10} \quad \frac{3}{10} \quad \frac{3}{10}
$$

$$
\frac{3}{10} \quad \frac{3}{10} \quad \frac{1}{10} \quad \frac{3}{10}
$$

$$
\frac{3}{10} \quad \frac{3}{10} \quad \frac{3}{10} \quad \frac{1}{10}
$$

Powers of *m* can be computed, indicating the limiting form of this matrix.

In[182]:=

```
MatrixForm[m.m]
```

Out[182]//MatrixForm=

$$
\frac{7}{25} \quad \frac{6}{25} \quad \frac{6}{25} \quad \frac{6}{25}
$$

$$
\frac{6}{25} \quad \frac{7}{25} \quad \frac{6}{25} \quad \frac{6}{25}
$$

$$
\frac{6}{25} \quad \frac{6}{25} \quad \frac{7}{25} \quad \frac{6}{25}
$$

$$
\frac{6}{25} \quad \frac{6}{25} \quad \frac{6}{25} \quad \frac{7}{25}
$$

For higher powers of *m*, it is best to use the MatrixPower command. Here is the 20th power of *m* written in the usual form for a matrix.

In[183]:=

```
MatrixForm[MatrixPower[m,20]]
```

Out[183]//MatrixForm=

$$\frac{23841857910157}{95367431640625} \quad \frac{23841857910156}{95367431640625} \quad \frac{23841857910156}{95367431640625}$$

$$\frac{23841857910156}{95367431640625}$$

$$\frac{23841857910156}{95367431640625} \quad \frac{23841857910157}{95367431640625} \quad \frac{23841857910156}{95367431640625}$$

$$\frac{23841857910156}{95367431640625}$$

$$\frac{23841857910156}{95367431640625} \quad \frac{23841857910156}{95367431640625} \quad \frac{23841857910157}{95367431640625}$$

$$\frac{23841857910156}{95367431640625}$$

$$\frac{23841857910156}{95367431640625} \quad \frac{23841857910156}{95367431640625} \quad \frac{23841857910156}{95367431640625}$$

$$\frac{23841857910157}{95367431640625}$$

Each of the entries here is very close to 1/4.
    The fixed point for *m* can also be found.

In[184]:=

```
Solve[{a,b,c,d}.m=={a,b,c,d},{a,b,c,d}]
```

Out[184]:=

```
{{a -> d, b -> d, c -> d}}
```
This indicates that the entries for the fixed vector are all equal.

Example 6.5.6

In this example we need various matrix products and an inverse. These are done
in the following way.

In[185]:=

```
q={{0,3/4,0},{1/4,0,3/4},{0,1/4,0}}
```

Out[185]:=

$$\{\{0, \ \frac{3}{4}, \ 0\}, \ \{\frac{1}{4}, \ 0, \ \frac{3}{4}\}, \ \{0, \ \frac{1}{4}, \ 0\}\}$$

In[186]:=

```
iminusq=IdentityMatrix[3]-q
```

Out[186]:=

$$\{\{1, \ -(\frac{3}{4}), \ 0\}, \ \{-(\frac{1}{4}), \ 1, \ -(\frac{3}{4})\}, \ \{0, \ -(\frac{1}{4}), \ 1\}\}$$

In[187]:=

inv=Inverse[iminusq]

Out[187]:=

$$\{\{\frac{13}{10}, \frac{6}{5}, \frac{9}{10}\}, \{\frac{2}{5}, \frac{8}{5}, \frac{6}{5}\}, \{\frac{1}{10}, \frac{2}{5}, \frac{13}{10}\}\}$$

In[188]:=

inv.{1,1,1}

Out[188]:=

$$\{\frac{17}{5}, \frac{16}{5}, \frac{9}{5}\}$$

► *Appendix 2*

Answers–
Odd Numbered Exercises

Chapter One

Exercises 1.1

1. a) $S = \{(x, y)|x \neq y,\ x, y \in \{1, 2 \ldots, 9\}\}$. S has
$9 \cdot 8 = 72$ points.

   b) $\frac{13}{16}$

3. $S = \{(x_1, x_2, x_3, x_4, x_5)|x_i \in \{G, N\}, i = 1, 2, \ldots, 5\}$
or $S = \{G, N\} \times \{G, N\} \times \{G, N\} \times \{G, N\} \times \{G, N\}$,
where $A \times B$ denotes the cross product of the sets A
and B. S has 32 points.

5. There are 15 sample points:

AAAA	NNAAAA
	NANAAA
NAAAA	NAANAA
ANAAA	NAAANA
AANAA	ANNAAA
AAANA	ANANAA
	ANAANA
	AANNAA
	AANANA
	AAANNA

7. $S = \{(x, y)|x, y \in \{1, 2 \ldots, 12\}, x \neq y\}$. S contains
$12 \cdot 11 = 132$ points.

9. There are 15 sample points.

<u>32</u>	62
43	63
<u>42</u>	64
52	<u>65</u>
53	72
<u>54</u>	73
	74
	75
	76

11.

HHHHH
HHHHT
HHHTH
HHTHT
HHHTT
HHTHH

Exercises 1.3

1. a) 5/16

   b) 13/16

3. a) $S = \{(x, y) | x, y \in \{1, 2, 3, 4, 5\}\}$,

  b) 10/25

  c) 6/25

5. $P(A \cap B) \leq \min\{P(A), P(B)\}$ so the given
probabilities are impossible.

7. 0.7

9. Consider $P(B) = P[B \cap (A \cup \overline{A})]$
$= P[(B \cap A) \cup (B \cap \overline{A})]$. Then use the addition law.

11. P (exactly one of A, B) $= P(A) + P(B)$
  $- 2P(A \cap B)$

Exercises 1.4

1. a) 58/135

  b) 15/29

3. a) 4.9%

  b) 35/98

5. Use the addition law.

7. a) 1/7

  b) 1/10

9. 0.994

11. a) 4

  b) $n \geq \dfrac{\log(1-r)}{\log(1-p)}$

13. 1/4. [Consider $P(1|1$ or even).]

15. a) No

  b) No

  c) 1/6

  d) 1/3

17. 2/3

19. a) 17/70

  b) 12/17

21. 1/2

23. a) $p_1 = p, p_2 = q + r(p - q)$,
   $p_3 = 2r(r - 1)(p - q) + p$, where $q = 1 - p$.
  b) No
  c) $r = 1/2$

25. a) 1/7

  c) 2/5

27. The estimate of p is $2\times$ sample proportion answering
"yes"$-1/2$.

Exercises 1.5

3.

r	2	3	4	5	6	7	8	9
p_r	$\dfrac{1}{12}$	$\dfrac{17}{72}$	$\dfrac{41}{96}$	$\dfrac{89}{144}$	$\dfrac{1,343}{1,728}$	$\dfrac{3,071}{3,456}$	$\dfrac{39,547}{41,472}$	$\dfrac{122,491}{124,416}$

r	10	11	12	13
p_r	$\dfrac{495,739}{497,664}$	$\dfrac{2,984,059}{2,985,984}$	$\dfrac{35,829,883}{35,831,808}$	1

5. a) 3/8

  b) 4/9

7. a) 96/15,625 = 0.006144

  b) 19/8,855 = 0.002146

Exercises 1.6

1. a) $1 - (1 - p_A)(1 - p_B)(1 - p_C)$

  b) $p_A p_B + p_A p_C + p_B p_C - 2p_A p_B p_C$

3. $p_A p_C (2 - p_A)(2 - p_C)$

Exercises 1.7

1. a) 4/9

  b) 1/9

  c) 2/9

3. There are more than $26^3 = 17,576$ people in
Indianapolis.

5. a) 280

  b) 431

7. 34,650

9. 7/24

11. a) At least one six in 4 rolls of a fair die.

13. a) 0.4035

  b) P(at least 1 duplicate) $= \dfrac{1,316,998,181}{1,334,062,100}$
   $= 0.9872091$

15. 1/70

17. 2,162/54,145 = 0.0399298

19. 201,600

21. 13,860

23. $(N - 2)/N$

25. 35

27. a) 11/16

    b) 7/8

Supplementary Exercises for Chapter 1

1. $\dfrac{77}{969} = 0.0794634$

3. 1/2

5. 3/4

7. 4

9. 5/33

11. a) 2/5

    b) 5/9

    c) 1/30

13. a) 2/5

    b) 5/6

    c) 2/3

15. 20% A's, 50% B's, and 30% C's

17. $\dfrac{217,006,443}{318,555,566} = 0.68122$

19. a) 7^{10}

    b) 5^{10}

    c) 207,446,400

21. $1 - \sqrt{2}/2$

25. a) $S = \{(x, y)\,|\,x, y \in \{1, 2, 3, 4, 5, 6\}, x < y\}$

    b) 3/5

    c) 13/15

27. $\dfrac{100}{429}$

29. $\dfrac{4\binom{39}{13} - 6\binom{26}{13} + 4}{\binom{52}{13}} = \dfrac{1,621,364,909}{31,750,677,980} = 0.0510655$

31. $\dfrac{2^{50}}{\binom{100}{50}} = 1.11595 \times 10^{-14}$

33. 4/5

35. a) 0.30

    b) 0.10

39. a) 1/208,012

    b) 12/13

43. a) 0.936

    b) 6

45. $\dfrac{(m^2 - m + 1)}{m^2(m - 1)}$

47. 0.4

49. $1 - \dfrac{N!}{(N - n)!N^n}$

51. a) 7,694,644,696,200

    b) 1,889,912,732,400

    c) 70,671,744

    d) 211,360,681,548

53. $2/n$

55. 21,349/22,407 = 0.952783

Chapter Two

Exercises 2.2

1. a) $S = \{(x_1, x_2, x_3, x_4)\,|\,x_i \in \{H, T\}, i = 1, 2, 3, 4\}$

    b) 3/16

    c) 7/15

3. a) $k = 6/31$

    b) 25/31

    c) $F(x) = \begin{cases} 0 & x \le 1 \\ 6/31 & 1 \le x < 2 \\ 18/31 & 2 \le x < 3 \\ 27/31 & 3 \le x < 4 \\ 1 & x \ge 4 \end{cases}$

7.

Sum	2	3	4	5	6	7	8	9	10	11	12
$441 \times$ Prob	36	60	73	76	70	56	35	20	4	1	

9. $f(x) = \begin{cases} 0 & y < 2 \\ 1/4 & 2 \le y < 3 \\ 1/2 & 3 \le y < 4 \\ 3/4 & 4 \le y < 5 \\ 1 & y \ge 5 \end{cases}$

11. $f(-1) = 1/3$, $f(0) = 1/2$, $f(2) = 1/6$

13. $F(x) = i/n, i \le x < i+1, i = 1, 2, \ldots, n$

Exercises 2.3

1. a) $E(X) = 1/2$, $\text{Var}(X) = 0.45$
   b) $3,900

3. $E(X) = \$12,900$, $\text{Var}(X) = \$31,900,000$,
   $\sigma = \$1786.06$

5. $E(X) = 3/5$, $\text{Var}(X) = 72/175$

7. $E(X) = \dfrac{n+1}{2}$, $\text{Var}(X) = \dfrac{n^2-1}{12}$

9. 9/25

11. a) $P(X = x) = \begin{cases} 1/6 & \text{if } x = 1, 2, 3 \\ 1/8 & \text{if } x = 4, 5, 6 \\ 1/16 & \text{if } x = 7, 8 \end{cases}$

   b) $1.00

13. $E(X)$

Exercises 2.5

1. a) 0.1091
   b) 0.999437

3. a) 0.10292
   b) 20
   c) 0.3445

5. a) 0.854134
   b) 0.0591414

7. a) 0.004629630
   b) 0.0006766
   c) 0.0791125

9. a) X is binomial with $n = 7$ and $p = 0.2$.
   b) 73/15,625
   c) 73/2,313

11. $-\$60,100$

13.

N	-3	-2	-1	0	1	2
$59{,}049 * \text{Prob}$	29,161	10	40	80	80	29,678

15. a) 0.966634
   b) 93

17. 5/36

19. 0.0973832

21. a) 0.128506
   b) 0.0318415

23. 65/256

Exercises 2.6

1. 10 to 20

5. 0.23598, 0.285551

7. 0.330775, 0.473463

9. A: 0.102589, 0.151481
   J: 0.0275727, 0.0495844
   No

Exercises 2.7

1. a) 0.0866925
   b) 0.60801

3. $c = 4$ and $\beta = 0.99992$

5. 16

7. H_a

9. Yes

Exercises 2.8

1. 0.413143, 0.527647

3. Yes

5. 400

7. 0.527975, 0.667947

9. 0.419418, 0.580582

Exercises 2.9

1. 3/4

3. a) 0.0064
   b) 0.04096

5. 8/65

7. 0.123959

9. 11/32

11. 0.0313811

13. a) 1/7
   b) 7

15. a) $21
   b) No

17. r/t

Exercises 2.10

1. 677,007/832,370

3. a)

X	3	4	5	6	7	8
56 * Prob	1	3	6	10	15	21

   b) $\mu = 27/4, \sigma^2 = 27/16$

5. a) 0.880527
   b) 79/3200

7. a) 0.956094
   b) 0.05629

9. 0.0573656

Exercises 2.11

1. 49/50

3. 0.68122

5. a) 27,683/32,340
   b) 97/2000

7. 0.104399

9. 10

11. a) 0.999616
   b) 0.00740206

13. a) 314.837

Exercises 2.14

1. 0.104137

3. a) 0.0916
   b) 0.18045
   c) 0.9380

5. a) 0.0469122
   b) 0.0661276

7. The Poisson approximation gives 0.067086.

9. a) 0.227975
   b) 30 seconds

11. 0.014388

13. a) 0.265026
   b) 9.1774×10^{-5}

15. a) 0.04979
   b) 0.03374
   c) 1.535 ft^3

17. 23

19. a) 0.180
   b) 0.143
   c)

X	0	1	2	3
Prob.	e^{-2}	$2e^{-2}$	$2e^{-2}$	$1 - 5e^{-2}$

   d) 0.218
   e) 4

21. a)

Y	0	1	2	3
Prob.	e^{-4}	$4e^{-4}$	$8e^{-4}$	$\frac{32}{3}e^{-4}$

Y	4	5	6 or more
Prob.	$\frac{32}{3}e^{-4}$	$\frac{128}{15}e^{-4}$	$1 - \frac{643}{15}e^{-4}$

   b) 3.80457

Supplementary Problems for Chapter 2

1. 0.938031

3. a) 0.0111603
   b) 0.0127952
   c) 29

5. 4/5

7. 1/4

9. a) 2,162/54,145
   b) 109

11. 5/273

13. 0.231639

15. a) $\displaystyle\sum_{y=1}^{\infty} \binom{2y-1}{1} p^2 (1-p)^{2y-2}$
   b) 1/3

17. 8/65

19. a) 0.013695
   b) 0.33282

21. 0.865618

23. 20

25. $p_A^3 + 3p_A^2(1 - p_A)(1 - p_B^2)$
   $+ 3p_A(1 - p_A)^2(1 - p_B)^2$

27. 0.74286

29. 0.707143

31. 0.206051

33. 0.290228, 0.379419

35. Yes

37. 1692

39. a)

W:	-1	0	1	2
$216 \times$ Prob:	125	75	15	1

   b) $-1/2$
   c) Make m as large as possible.

41. 0.112553

43. 1/12

Chapter Three

Exercises 3.1

1. b) 19/64
   c) 19/27
   d) 0.693361

3. b) 1/4
   c) $\arccos(1/3)$

5. a) $e^{-3/4}$
   b) $\mu = 1/3, \sigma^2 = 1/9$

7. a) 3/8
   b) 1/4

9. b) 3/4
   c) $\mu = 0, \sigma^2 = 2/3$

11. a) 3/2

   b) $F(y) = \begin{cases} 0 & y \le 0 \\ \dfrac{y^3}{2} + \dfrac{y^2}{2} & 0 \le y \le 1 \\ 1 & y \ge 1 \end{cases}$

   c) $P(Y > y) = \begin{cases} 1 & y \le 0 \\ 1 - \dfrac{y^3}{2} - \dfrac{y^2}{2} & 0 \le y \le 1 \\ 0 & y \ge 1 \end{cases}$

13. a) $\mu = 0, \sigma^2 = 27/5$

15. $\mu = 16/35, \sigma^2 = 201/4{,}900$

19. b) $\mu = 3/2, \sigma^2 = 1/4$

Exercises 3.2

1. 1/2

3. 1/3

5. 0.597126

7. The exact probability is 3/5.

9. $\dfrac{\sqrt{3} - 1}{2}$

11. $\ln 1.765 - 1$

Exercises 3.4

1. a) $e^{-2/3}$
   b) $e^{-1/3}$

3. a) $e^{-17/15}$
   b) 0.293681

5. A should be taken.

7. $\lambda = 1/2$

9. $e^{-6.25}$

13. 122,000 miles

15. b) $F(x) = 1 - e^{-3(x-2)}, x \ge 2$
   c) e^{-6}
   d) 1.69495×10^{-4}

17. a) λ
   b) $e^{-2/\lambda}$

Exercises 3.5

1. a) 0.0026
   b) 0.682689

3. III

5. 68.75%

7. Yes

9. $37.98

13. 17.01% will be outside the warning limits.

15. a) 0.213485
    b) 12 bags gives a probability of about 0.95

17. −0.427183855

19. 0.975412

21. 64.71

23. a) 0.308538
    b) 0.2113
    c) 0.0152

25. D_2

Exercises 3.6

Answers given are the exact result followed by the normal approximation.

1. 0.0795892 and 0.0666053 are the exact probabilities. The normal approximations are 0.07965579 and 0.0668073.

3. a) 0.219353, 0.214145
   b) 0.934849

5. 26, 26

7. 26, 26

9. 0.229707, 0.230906

11. 0.9851, 0.9854

13. 0.657949, 0.642834

Exercises 3.7

1. a) $\mu = 1/2, \sigma^2 = 1/20$
   b) 0.66229

3. b) $3e^{-2}$
   c) $5/4\sqrt{e}$

9. Yes

11. a) $e^{-y/12}$
    b) $\frac{1}{12}e^{-y/12}, y \geq 0$
    c) 0.434598

Exercises 3.8

1. b) 0.301305
   c) 0.67032

3. a) $e^{-t^2/20,000}$
   b) 125.331 hours
   c) 0.324652

5. b) 0.283468

7. $\beta(\ln 2)^{1/\alpha}$

9. 0.443559

Supplementary Exercises for Chapter 3

1. 0.0730169

3. 0.1686

5. 2.27789×10^{-4}

7. a) 0.367879
   b) 0.306432

9. b) 0.7875
   c) 62/15
   d) 729/8000

11. b) 112/243
    c) $F(x) = \begin{cases} 0 & \text{if } x < 0 \\ 5x^4 - 4x^5 & \text{if } 0 \leq x < 1 \\ 1 & \text{if } x \geq 1 \end{cases}$

13. $\sqrt{11}/5$

17. $\mu = 2, \sigma^2 = 1/5$

19. a) $k = 2$
    b) $F(x) = \begin{cases} 0 & \text{if } x < 0 \\ 1 - e^{-x^2} & \text{if } x \geq 0 \end{cases}$

21. 0.0220301

23. $F(x) = \begin{cases} 0 & \text{if } x < 0 \\ x^2 & \text{if } 0 \le x < 1/2 \\ 6x - 3x^2 - 2 & \text{if } 1/2 \le x < 1 \\ 1 & \text{if } x \ge 1 \end{cases}$

25. $\mu = 1, \sigma^2 = 1/2$

27. $P(x = -2) = 1/6, P(x = -1) = 1/3,$
$P(x = 2) = 1/2$

29. $2/3$

Chapter Four

Exercises 4.3

1. $g(y) = 1/9, 4 \le y \le 13$

3. $g(y) = 1/(2\sqrt{y}), 0 < y < 1$

5. a) $E(Y) = e^{\mu + 1/2\sigma^2}, \text{Var}(Y) = e^{2\mu+\sigma^2}(e^{\sigma^2} - 1)$

   b) $g(y) = \dfrac{1}{\sigma y \sqrt{2\pi}} e^{-1/2(\frac{\ln y - \mu}{\sigma})^2}, y > 0$

9. $g(y) = \dfrac{1}{3y^{2/3}} e^{-y^{1/3}}, y > 0$

11. $g(y) = \dfrac{1}{2\sqrt{1 - y^2}}, \sin(-1) \le y \le \sin(1)$

13. a) $g(y) = \dfrac{2y}{(1 + y^2)^2}, y \ge 0$

15. $602/3$

17. $g(y) = n\lambda e^{-n\lambda y}, y \ge 0$

19. $g(y) = \dfrac{1}{2\sqrt{y}} - \dfrac{1}{4}, 0 < y < 4$

21. No

23. a) $Y = a + (b - a)X$

   b) $Y = -\dfrac{1}{\lambda} \ln X.$

27. 35

Exercises 4.4

3.

Sum	3	4	5	6	7	8	9	10	11	12
$64 \times$ Prob	1	3	6	10	12	12	10	6	3	1

5. a) $P(X + Y = z) = (\frac{1}{2})^{z-2} - 6(\frac{1}{3})^z, z = 2, 3, 4, \ldots$

   b) $7/2$

Exercises 4.5

3. All the subsets of $\{a, b, c\}$.

5. $1 - (1 - \dfrac{1}{t}) \log(1 - t)$

7. a) $t^k P_X(t)$

   b) $P_X(t^k)$

9.

Sum	2	3	4	5	6	7	8	9	10	11	12
$126 \times$ Prob	1	3	6	10	15	21	20	18	15	11	6

Exercises 4.7

1. a) $e^{-\lambda(1-t)}$

3. a) $16/31$

   b) $\frac{1}{31}(16 + 8t + 4t^2 + 2t^3 + t^4)$

   c) $\mu = 26/31, \sigma^2 = 1, 122/961$

5. a) $\dfrac{71,393,157}{1,073,741,824} = 0.0664901$

   b) $\dfrac{2,046,448,125}{2,147,483,648} = 0.952952$

7. a) e^{t-1}

   b) $\mu = \sigma^2 = 1$

Exercises 4.8

3. a) $e^{-\lambda(1-e^t)}$

5. a) $\mu = -5/6, \sigma^2 = 65/36$

   b) $1 - \dfrac{5}{6}t + \dfrac{5}{4}t^2 - \dfrac{23}{36}t^3 + \dfrac{17}{48}t^4 + \cdots$

7. a) $\left(\dfrac{\lambda}{\lambda - t}\right)^r$

   b) $\mu = r/\lambda, \sigma^2 = r/\lambda^2$

9. $\dfrac{1}{3t}(e^{-t} - e^{-2t}) + \dfrac{2}{9t}(e^{4t} - e^t)$

11. a) 0.6826

   b) 0.8414

13. a) $\dfrac{1}{t}(e^{3t} - e^{2t})$

  b) $\mu = 5/2, \sigma^2 = 1/12$

15. a) $\mu = 5/3, \sigma^2 = 10/9$

  b) X is binomial with $n = 5$ and $p = 1/3$.

Exercises 4.10

  1. a) $\mu = 1/5, \sigma^2 = 1/25$
   b) $f(x) = 5e^{-5x}, x \geq 0$

  3. $\mu = 10, \sigma^2 = 20$

  5. 0.97725

  7. $\mu = \sigma^2 = 1$.

  9. M is the generating function for the variable $X - \lambda$ where X is Poisson with parameter λ.

15. d) $X_1 + X_2$ has probability distribution given by $g(y)$.

17. $\dfrac{1}{t^2}(e^t + e^{-t} - 2)$

19. 0.814453

Exercises 4.11

  3. 0.974085

  5. 166

  7. a) 0.3993
   b) 1008

  9. 0.008853

11. 0.0681571

13. 0.0227501

15. a) 0.382925
   b) 0.682689

17. The probability that $\overline{X}$ is in the range 3.11 to 3.61 is 0.95833.

Exercises 4.13

  1. 77.7867, 1789.36

  3. $\dfrac{(n-1)s^2}{\chi_U^2}, \dfrac{(n-1)s^2}{\chi_L^2}$ where χ_U^2 and χ_L^2 are upper and lower chi-squared values.

  5. 12

  7. 0.898904

  9. 2.55646, 13.1746

Exercises 4.14

  1. a) Yes
   b) 0.03338

  3. a) $\overline{x} < 0.248897$
   b) 0.41365
   c) $n = 49$

  5. a) 0.0228
   b) 0.158655

  7. a) 5.175, 10.825
   b) 3.8756, 45.282

  9. a) Reject if $\overline{x} > 4.02816$ or if $\overline{x} < 3.98179$.
   b) Reject H_0.
   c) 0.877193

11. Yes

13. a) 0.055115
   b) 0.21186

15. a) No
   b) 0.837945
   c) 28

17. a) $\overline{x} < 73.4538$
   b) Yes
   c) 0.939007
   d) 2.34×10^{-12}

19. a) 40,526
   b) 1,089,055 to 3,749,072
   c) 40,778

Exercises 4.15

  1. a) Accept H_0.
   b) Yes

  3. Accept H_0.

  5. 0.153189, 11.9492

  7. -0.223, 8.223

  9. a) Accept H_0.

b) 0.141803, 4.16881

c) Reject H_0.

11. a) -960.19, -309.81

b) -949.55, -320.55

c) 0.00242706

13. a) Accept H_0.

b) Reject H_0.

15. a) Reject the null hypothesis that painting does not reduce the top speed.

17. No, if $\alpha = 5\%$

19. $n = 110$

Exercises 4.16

1. b) $y = -0.273177 + 1.61719x$

c) $F(1, 4) = 99.041.$ $P[F(1, 4) > 99.041]$ $= 5.726 \times 10^{-4}$, so the fit is a very good one.

3. b) $y = 66.198 - 0.8682x$

c) $F(1.8) = 16.9988.$ $P[F(1, 8) > 16.9988]$ $= 3.33 \times 10^{-3}$, so the fit is a very good one.

5. b) $y = -.192719 + 1.22056x$
$x = 7.31915 + 0.485106y$

7. a) $y = 0.489784 + 0.272431x$

b) $r = 0.92$

9. a) $\hat{\beta} = \dfrac{\sum_{i=1}^{n} x_i^2 y_i}{\sum_{i=1}^{n} x_i^4}$

b) $\hat{\beta} = 0.953807$

11. a) $\hat{\beta} = \dfrac{\sum_{i=1}^{n} x_i y_i}{\sum_{i=1}^{n} x_i^2}$

Exercises 4.17

1. b) 8.8292, 11.2458

Supplementary Exercises for Chapter 4

1. a) $\dfrac{2}{a^2 t^2}(e^{at} - 1) - \dfrac{2}{at}$

b) $\mu = a/3,\ \sigma^2 = a^2/18$

3. a) $\dfrac{2^{1+t} - 1}{1 + t}$

b) $\mu = 0.3862,\ \sigma^2 = 0.03909$

5. $\beta^k \Gamma(1 + \frac{k}{\alpha})$

7. 4/81

9. Poisson $[\lambda p]$

11. 16

13. a) 0.23975

b) 0.0569231

15. a) 0.135666

b) 0.0139034

c) 0.8133

d) 0.0126737

17. a) $\dfrac{2}{t^2}(e^t - t - 1)$

b) $\dfrac{2}{(k + 1)(k + 2)}$

c) $\dfrac{8e^{t/2}}{t^2}(e^{t/2} - \dfrac{t}{2} - 1)$

19. a) $\mu = 5/3,\ \sigma^2 = 5/9$

b) $\dfrac{e^t}{2} + \dfrac{e^{2t}}{3} + \dfrac{e^{3t}}{6}$

21. a) $P(x = 1) = 2/5,\ P(x = 2) = 1/5,$ $P(x = 3) = 2/5$

b) $\mu = 2,\ \sigma^2 = 4/5$

23. $\dfrac{1}{k\beta} e^{\frac{-y}{k\beta}},\ y > 0$

25. a) $g(y) = y^{-1/2} - 1, 0 < y \leq 1$

27. $g(y) = \dfrac{y}{\sqrt{1 - y^2}}, 0 < y < 1$

29. $g(y) = 2/y^3,\ y > 1$ $E(Y) = 2$

31. $g(y) = \dfrac{3 - \sqrt{y + 1}}{9}, -1 \leq y \leq 8$

33. $g(y) = \begin{cases} 2/3 & \text{if } 0 < y < 1 \\ 1/3 & \text{if } 1 \leq y \leq 2 \end{cases}$

35. 0.017008

37. Yes. $P(x_{30}^2 > 56.4) = 0.002456$.

39. No

41. 5.88586, 8.17407

43. a) Accept H_0.

b) 9.44793, 11.1521

45. a) 70

   b) 0.617075

47. 0.965068

49. a) Reject H_0.

   b) 0.00925637

51. 658.88, 701.13

53. Yes

55. a) Accept H_0.

   b) 0.2610863

57. a) Reject H_0.

   b) Reject H_0.

59. a) $\bar{x} < 181.997$

   b) 0.857088

61. 1.48773, 3.97234

Chapter Five

Exercises 5.2

1. a)

x	2	2.5	3	3.5	4
$f(x)$	0.17	0.14	0.20	0.33	0.16

y	2	2.5	3	3.5	4
$g(y)$	.05	0.15	0.28	0.40	0.12

   b) $E(X) = 3.085$, $E(Y) = 3.195$

   c) 0.68

3. a) $f(x, y) = \binom{4}{x}\binom{4}{y}\binom{44}{2-x-y}/\binom{52}{2}$,
   $x = 0, 1, 2;\ y = 0, 1, 2;\ x + y \leq 2$

   b) $f(x) = \begin{cases} \frac{188}{221} & \text{if } x = 0 \\ \frac{32}{221} & \text{if } x = 1 \\ \frac{1}{221} & \text{if } x = 2. \end{cases}$

   $g(y)$ is similar.

5. a) 1/65

   b) $f(x) = \frac{x+1}{65}, x = 1, 2, 3, \ldots, 10$

   $g(y) = \begin{cases} \frac{10}{65} & \text{if } y = 9 \text{ or } 10 \\ \frac{y+1}{65} & \text{if } y = 0, 1, \ldots, 8 \end{cases}$

7. a) $P(X = x, Y = y)$
   $= \binom{n_1}{x}\binom{n_2}{y}p^{x+y}q^{n_1+n_2-x-y}$,
   $x = 0, 1, \cdots, n_1;\ y = 0, 1, \cdots, n_2$

   b) $X + Y$ is binomial with parameters $n_1 + n_2$ and p.

9. a) $f(x, y) = \frac{e^{-6}6^x}{x!}\binom{x}{y}\left(\frac{3}{4}\right)^y\left(\frac{1}{4}\right)^{x-y}$,
   $x = 0, 1, \ldots;\ y = 0, 1, \ldots, x$

   b) $g(y) = \frac{e^{-9/2}(9/2)^y}{y!}, \quad y = 0, 1, \ldots$

   c) $E(Y) = \frac{3}{4} \cdot 6$

11. $P(X = x, Y = y) = \frac{\binom{3}{x}\binom{2}{y}\binom{2}{3-x-y}}{\binom{7}{3}}$,
   $x = 0, 1, 2, 3;\ y = 0, 1, 2;\ x + y \leq 3$

13. b) $f(x) = 12x^3(1 - x^2), 0 < x < 1$
   $g(y) = 12y^2(1 - y), 0 < y < 1$

15. a) 3

   b) $f(x) = 6x(1 - x)$, $0 < x < 1$

   $g(y) = \begin{cases} \frac{3}{2}(1 - y)^2 & 0 < y < 1 \\ \frac{3}{2}(1 + y)^2 & -1 < y < 0 \end{cases}$

17. a)

$W \backslash Y$	0	1	2	$f(w)$
0	1/8	1/8	0	2/8
1	1/8	2/8	1/8	4/8
2	0	1/8	1/8	2/8
$g(y)$	2/8	4/8	2/8	

   b) $f(y) = \begin{cases} 1/4 & \text{if } y = 0 \text{ or } 2 \\ 1/2 & \text{if } y = 1 \end{cases}$

   $g(w)$ is similar.

19. b) 1/16

   c) 7/16

21. b) $f(x) = \frac{3}{2}(1 - x^2), 0 < x < 1$
   $g(y) = 3y^2, 0 < y < 1$

Exercises 5.3

1. a) $f(x|y) = \dfrac{2(x+y)}{2y+1}, 0 \le x \le 1$

$f(y|x) = \dfrac{2(x+y)}{2x+1}, 0 \le y \le 1$

b) $E(X|Y=y) = \dfrac{3y+2}{6y+3}$

$E(Y|X=x) = \dfrac{3x+2}{6x+3}$

3. $f(x|y) = \dfrac{2x}{1-y^2}, y < x < 1$

$f(y|x) = \dfrac{2y}{x^2}, 0 < y < x$

5. a) $k = 8$

b) $E(Y) = 8/15$

7. a) $E(Y|X) = \frac{2}{3}x\frac{(1-x^3)}{1-x^2}, 0 < x < 1 \quad E(X|Y)$

$= \frac{2}{3}y^{1/2}\frac{1-y^{3/2}}{1-y}, 0 < y < 1$

b) 3/8

9. a) $f(x) = 1 + 2x - 3x^2, \quad 0 < x < 1$

$g(y) = \frac{1}{2}(5 - 8y + 3y^2), \quad 0 < y < 1$

b) $f(y|x) = \dfrac{1}{1-x}, \quad 0 < y < 1 - x$

Exercises 5.5

11. b) $f(x|y) = 2x/y, \quad 0 < x < \sqrt{y}$

c) $\dfrac{2\sqrt{y}}{3}$

13. $f(x|y) = \dfrac{1}{2\sqrt{y}}, \quad -\sqrt{y} < x < \sqrt{y}$

$f(y|x) = \dfrac{1}{1-x^2}, \quad x^2 < y < 1$

15. a) $f(x) = \dfrac{4-x}{6}, \quad x = 1, 2, 3.$

$g(y)$ is similar.

b) $f(y|x) = \dfrac{1}{4-x}, \quad y = 1, 2, \ldots, 4 - x$

17. $-1/3$

19. b) $f(x) = x^3/4, \quad 0 < x < 2$

$g(y) = y - y^3/4, \quad 0 < y < 2$

c) No

d) 7/12

21. $220 - 30\sqrt{2}$

25. $14/\sqrt{247}$

Exercises 5.6

1. c) 0.355435

Exercises 5.7

1. $g(z) = \dfrac{2}{(z+2)^2}, \quad z > 0$

3. $g(z) = \begin{cases} \frac{2}{3}z^2(3-z), & 0 < z < 1 \\ \frac{2}{3}(z^3 - 3z^2 + 4), & 1 \le z < 2 \end{cases}$

Supplementary Exercises for Chapter 5

1. a) 3/128

b) No

c) $f(x) = \dfrac{4 + 3x^2}{32}, \quad -2 < x < 2$

$g(y) = \dfrac{4 + 3y^2}{32}, \quad -2 < y < 2$

$f(x|y) = \dfrac{3(x^2 + y^2)}{16 + 12y^2}, \quad -2 < x < 2$

$f(y|x) = \dfrac{3(x^2 + y^2)}{16 + 12x^2}, \quad -2 < y < 2$

3. a) $f(x, y) = \begin{cases} 1/36 & \text{if } x = y, \\ & \quad x = 1, 2, \ldots, 6, \\ & \quad y = 1, 2, \ldots; 6 \\ 2/36 & \text{if } x > y, \\ & \quad x = 1, 2, \ldots, 6, \\ & \quad y = 1, 2, \ldots, 6 \end{cases}$

b) $E(X) = 161/36$

c) $E(Y) = 91/36$

d) $E(X + Y) = E(X) + E(Y)$

5. 1/3

7. a) 3/16

  b) $\sqrt{y}/2$, $0 < x < \sqrt{y}$

9. $\dfrac{16 - 12\log 2}{9} = 0.853582$

11. $\dfrac{1}{9(1 - e^{-1/3})^2}$

13. a) $f(x, y) = 12x$, $0 < y < 2x, 0 < x < 1/2$

  b) $g(y) = \frac{3}{2}(1 - y^2)$, $0 < y < 1$

  c) No

15. a) 2

  b) No

  c) 76/81

17. a) 19

  b) -2

  c) $-1/5\sqrt{3}$

  d) 1

  e) 250

  f) 44

19. a) 12

  b) $f(x) = 6x^5$, $0 < x < 1$
      $g(y) = 6y(1 - y)$, $0 < y < 1$

c) No

d) 27/32

21. a)

$Y\backslash X$	0	1	2	3	4	$g(y)$
0	1/16	0	0	0	0	1/16
1	0	4/16	3/16	0	0	7/16
2	0	0	3/16	2/16	0	5/16
3	0	0	0	2/16	0	2/16
4	0	0	0	0	1/16	1/16
$f(x)$	1/16	4/16	6/16	4/16	1/16	

  b) 3/16

  c) 2

23. a) 1/2

  b) $f(x) = 1$, $0 < x < 1$

  $$g(y) = \frac{1}{\sqrt{y}} - 1,   0 < y < 1$$

  c) $1 - \sqrt{2}/2$

Chapter Six

Exercises 6.2

1. X is a geometric random variable. $a_n = pq^{n-1}, n \geq 1$.

3. a) $a_n = p^2 + qa_{n-1} + pqa_{n-2}, n \geq 2. a_0 = a_1 = 0$;
  $a_2 = p^2$

5. $\dfrac{(1 - p_1)(p_1 - p_2)^{n-1} + p_2}{1 - p_1 + p_2}, n \geq 1$

7. a) $\dfrac{3}{7}\left(\dfrac{5}{12}\right)^n + \dfrac{4}{7}, n = 0, 1, 2, \ldots$

  b) 0.754238

9. $\dfrac{1 + p + p^2}{p^3}$

Exercises 6.3

1. $p_n = n/N$

Exercises 6.4

1. a) $u_0 = 1, u_1 = u_2 = 0$

  b) $U(s) = 1 + \dfrac{(ps)^3}{(1 - s)(1 + ps + (ps)^2)}$;

  $\dfrac{29, 801, 576}{1, 162, 261, 467} = 0.025641$

  c) $F(s) = \dfrac{(ps)^3}{(ps)^3 + (1 - s)(1 + ps + (ps)^2)}$;

  $\dfrac{58, 333, 184}{3, 486, 784, 401} = 0.0167298$

3. $\dfrac{1 + 2pq - 5pq^2 + p^2q^2 - p^2q^3}{p^2q^4}$

Exercises 6.5

3. $(1/3, 2/3)$

5. $(444/699, 165/699, 90/699)$

9.
$$\begin{pmatrix} 1/3 & 0 & 1/3 & 0 & 1/3 & 0 \\ 0 & 1/3 & 0 & 1/3 & 0 & 1/3 \\ 1/3 & 0 & 1/3 & 0 & 1/3 & 0 \\ 0 & 1/3 & 0 & 1/3 & 0 & 1/3 \\ 1/3 & 0 & 1/3 & 0 & 1/3 & 0 \\ 0 & 1/3 & 0 & 1/3 & 0 & 1/3 \end{pmatrix}$$

Supplementary Exercises for Chapter 6

1. a) $u_n = q^n + \sum_{k=3}^{n} \binom{k-1}{2} p^3 q^{k-3} u_{n-k}$,

$n \geq 3$, $u_0 = 1$, $u_1 = q$, $u_2 = q^2$, $u_3 = p^3 + q^3$.

b) $u_n = p w_{n-1} + q u_{n-1}$; $v_n = p u_{n-1} + q v_{n-1}$;
$w_n = p v_{n-1} + q w_{n-1}$;
$v_n = 3q v_{n-1} - 3q^2 v_{n-2} + (p^3 + q^3) v_{n-3}$,
$n \geq 4$, $v_0 = 0$, $v_1 = p$, $v_2 = 2pq$, $v_3 = 3pq^2$.

3. $\frac{1}{2}[1 - (q - p)^n]$

5. $\binom{2n}{n} \dfrac{1}{2n - 1} p^n q^n$

7. No

9. $p_n = \frac{1}{2}\big[1 + (2p - 1)^n\big]$, $n \geq 1$

► *Appendix 3*

Tables

Standard Normal Distribution

The entries in this table give the areas under the standard normal curve from 0 to z.

z	.00	.01	.02	.03	.04	.05	.06	.07	.08	.09
0.0	.0000	.0040	.0080	.0120	.0160	.0199	.0239	.0279	.0319	.0359
0.1	.0398	.0438	.0478	.0517	.0557	.0596	.0636	.0675	.0714	.0753
0.2	.0793	.0832	.0871	.0910	.0948	.0987	.1026	.1064	.1103	.1141
0.3	.1179	.1217	.1255	.1293	.1331	.1368	.1406	.1443	.1480	.1517
0.4	.1554	.1591	.1628	.1664	.1700	.1736	.1772	.1808	.1844	.1879
0.5	.1915	.1950	.1985	.2019	.2054	.2088	.2123	.2157	.2190	.2224
0.6	.2257	.2291	.2324	.2357	.2389	.2422	.2454	.2486	.2517	.2549
0.7	.2580	.2611	.2642	.2673	.2704	.2734	.2764	.2794	.2823	.2852
0.8	.2881	.2910	.2939	.2967	.2995	.3023	.3051	.3078	.3106	.3133
0.9	.3159	.3186	.3212	.3238	.3264	.3289	.3315	.3340	.3365	.3389
1.0	.3413	.3438	.3461	.3485	.3508	.3531	.3554	.3577	.3599	.3621
1.1	.3643	.3665	.3686	.3708	.3729	.3749	.3770	.3790	.3810	.3830
1.2	.3849	.3869	.3888	.3907	.3925	.3944	.3962	.3980	.3997	.4015
1.3	.4032	.4049	.4066	.4082	.4099	.4115	.4131	.4147	.4162	.4177
1.4	.4192	.4207	.4222	.4236	.4251	.4265	.4279	.4292	.4306	.4319
1.5	.4332	.4345	.4357	.4370	.4382	.4394	.4406	.4418	.4429	.4441
1.6	.4452	.4463	.4474	.4484	.4495	.4505	.4515	.4525	.4535	.4545
1.7	.4554	.4564	.4573	.4582	.4591	.4599	.4608	.4616	.4625	.4633
1.8	.4641	.4649	.4656	.4664	.4671	.4678	.4686	.4693	.4699	.4706
1.9	.4713	.4719	.4726	.4732	.4738	.4744	.4750	.4756	.4761	.4767
2.0	.4772	.4778	.4783	.4788	.4793	.4798	.4803	.4808	.4812	.4817
2.1	.4821	.4826	.4830	.4834	.4838	.4842	.4846	.4850	.4854	.4857
2.2	.4861	.4864	.4868	.4871	.4875	.4878	.4881	.4884	.4887	.4890
2.3	.4893	.4896	.4898	.4901	.4904	.4906	.4909	.4911	.4913	.4916
2.4	.4918	.4920	.4922	.4925	.4927	.4929	.4931	.4932	.4934	.4936
2.5	.4938	.4940	.4941	.4943	.4945	.4946	.4948	.4949	.4951	.4952
2.6	.4953	.4955	.4956	.4957	.4959	.4960	.4961	.4962	.4963	.4964
2.7	.4965	.4966	.4967	.4968	.4969	.4970	.4971	.4972	.4973	.4974
2.8	.4974	.4975	.4976	.4977	.4977	.4978	.4979	.4979	.4980	.4981
2.9	.4981	.4982	.4982	.4983	.4984	.4984	.4985	.4985	.4986	.4986
3.0	.4987	.4987	.4987	.4988	.4988	.4989	.4989	.4989	.4990	.4990

The *t* Distribution Table

**The entries in the table give the critical values
of *t* for the specified number of degrees
of freedom and areas in the right tail.**

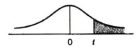

df	Area in the Right Tail under the *t* Distribution Curve					
	.10	.05	.025	.01	.005	.001
1	3.078	6.314	12.706	31.821	63.657	318.309
2	1.886	2.920	4.303	6.965	9.925	22.327
3	1.638	2.353	3.182	4.541	5.841	10.215
4	1.533	2.132	2.776	3.747	4.604	7.173
5	1.476	2.015	2.571	3.365	4.032	5.893
6	1.440	1.943	2.447	3.143	3.707	5.208
7	1.415	1.895	2.365	2.998	3.499	4.785
8	1.397	1.860	2.306	2.896	3.355	4.501
9	1.383	1.833	2.262	2.821	3.250	4.297
10	1.372	1.812	2.228	2.764	3.169	4.144
11	1.363	1.796	2.201	2.718	3.106	4.025
12	1.356	1.782	2.179	2.681	3.055	3.930
13	1.350	1.771	2.160	2.650	3.012	3.852
14	1.345	1.761	2.145	2.624	2.977	3.787
15	1.341	1.753	2.131	2.602	2.947	3.733
16	1.337	1.746	2.120	2.583	2.921	3.686
17	1.333	1.740	2.110	2.567	2.898	3.646
18	1.330	1.734	2.101	2.552	2.878	3.610
19	1.328	1.729	2.093	2.539	2.861	3.579
20	1.325	1.725	2.086	2.528	2.845	3.552
21	1.323	1.721	2.080	2.518	2.831	3.527
22	1.321	1.717	2.074	2.508	2.819	3.505
23	1.319	1.714	2.069	2.500	2.807	3.485
24	1.318	1.711	2.064	2.492	2.797	3.467
25	1.316	1.708	2.060	2.485	2.787	3.450
26	1.315	1.706	2.056	2.479	2.779	3.435
27	1.314	1.703	2.052	2.473	2.771	3.421
28	1.313	1.701	2.048	2.467	2.763	3.408
29	1.311	1.699	2.045	2.462	2.756	3.396
30	1.310	1.697	2.042	2.457	2.750	3.385
31	1.309	1.696	2.040	2.453	2.744	3.375
32	1.309	1.694	2.037	2.449	2.738	3.365
33	1.308	1.692	2.035	2.445	2.733	3.356
34	1.307	1.691	2.032	2.441	2.728	3.348
35	1.306	1.690	2.030	2.438	2.724	3.340
36	1.306	1.688	2.028	2.434	2.719	3.333
37	1.305	1.687	2.026	2.431	2.715	3.326
38	1.304	1.686	2.024	2.429	2.712	3.319
39	1.304	1.685	2.023	2.426	2.708	3.313
40	1.303	1.684	2.021	2.423	2.704	3.307
∞	1.282	1.645	1.960	2.326	2.576	3.090

Chi-Squared Distribution Table

The entries in this table give
the critical values of χ^2 for the
specified number of degrees of freedom
and areas in the right tail.

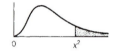

df	.995	.990	.975	.950	.900	.100	.050	.025	.010	.005
					Area in the Right Tail under the Chi-square Distribution Curve					
1	0.000	0.000	0.001	0.004	0.016	2.706	3.841	5.024	6.635	7.879
2	0.010	0.020	0.051	0.103	0.211	4.605	5.991	7.378	9.210	10.597
3	0.072	0.115	0.216	0.352	0.584	6.251	7.815	9.348	11.345	12.838
4	0.207	0.297	0.484	0.711	1.064	7.779	9.488	11.143	13.277	14.860
5	0.412	0.554	0.831	1.145	1.610	9.236	11.070	12.833	15.086	16.750
6	0.676	0.872	1.237	1.635	2.204	10.645	12.592	14.449	16.812	18.548
7	0.989	1.239	1.690	2.167	2.833	12.017	14.067	16.013	18.475	20.278
8	1.344	1.646	2.180	2.733	3.490	13.362	15.507	17.535	20.090	21.955
9	1.735	2.088	2.700	3.325	4.168	14.684	16.919	19.023	21.666	23.589
10	2.156	2.558	3.247	3.940	4.865	15.987	18.307	20.483	23.209	25.188
11	2.603	3.053	3.816	4.575	5.578	17.275	19.675	21.920	24.725	26.757
12	3.074	3.571	4.404	5.226	6.304	18.549	21.026	23.337	26.217	28.300
13	3.565	4.107	5.009	5.892	7.042	19.812	22.362	24.736	27.688	29.819
14	4.075	4.660	5.629	6.571	7.790	21.064	23.685	26.119	29.141	31.319
15	4.601	5.229	6.262	7.261	8.547	22.307	24.996	27.488	30.578	32.801
16	5.142	5.812	6.908	7.962	9.312	23.542	26.296	28.845	32.000	34.267
17	5.697	6.408	7.564	8.672	10.085	24.769	27.587	30.191	33.409	35.718
18	6.265	7.015	8.231	9.390	10.865	25.989	28.869	31.526	34.805	37.156
19	6.844	7.633	8.907	10.117	11.651	27.204	30.144	32.852	36.191	38.582
20	7.434	8.260	9.591	10.851	12.443	28.412	31.410	34.170	37.566	39.997
21	8.034	8.897	10.283	11.591	13.240	29.615	32.671	35.479	38.932	41.401
22	8.643	9.542	10.982	12.338	14.041	30.813	33.924	36.781	40.289	42.796
23	9.260	10.196	11.689	13.091	14.848	32.007	35.172	38.076	41.638	44.181
24	9.886	10.856	12.401	13.848	15.659	33.196	36.415	39.364	42.980	45.559
25	10.520	11.524	13.120	14.611	16.473	34.382	37.652	40.646	44.314	46.928
26	11.160	12.198	13.844	15.379	17.292	35.563	38.885	41.923	45.642	48.290
27	11.808	12.879	14.573	16.151	18.114	36.741	40.113	43.195	46.963	49.645
28	12.461	13.565	15.308	16.928	18.939	37.916	41.337	44.461	48.278	50.993
29	13.121	14.256	16.047	17.708	19.768	39.087	42.557	45.722	49.588	52.336
30	13.787	14.953	16.791	18.493	20.599	40.256	43.773	46.979	50.892	53.672
40	20.707	22.164	24.433	26.509	29.051	51.805	55.758	59.342	63.691	66.766
50	27.991	29.707	32.357	34.764	37.689	63.167	67.505	71.420	76.154	79.490
60	35.534	37.485	40.482	43.188	46.459	74.397	79.082	83.298	88.379	91.952
70	43.275	45.442	48.758	51.739	55.329	85.527	90.531	95.023	100.425	104.215
80	51.172	53.540	57.153	60.391	64.278	96.578	101.879	106.629	112.329	116.321
90	59.196	61.754	65.647	69.126	73.291	107.565	113.145	118.136	124.116	128.299
100	67.328	70.065	74.222	77.929	82.358	118.498	124.342	129.561	135.807	140.169

The *F* Distribution Table

.01

F

Area in the Right Tail under the *F* Distribution Curve = .01

								Degrees of Freedom for the Numerator										
	1	2	3	4	5	6	7	8	9	10	11	12	15	20	25	30	40	50	100
1	4052	5000	5403	5625	5764	5859	5928	5981	6022	6056	6083	6106	6157	6209	6240	6261	6287	6303	6334
2	98.50	99.00	99.17	99.25	99.30	99.33	99.36	99.37	99.39	99.40	99.41	99.42	99.43	99.45	99.46	99.47	99.47	99.48	99.49
3	34.12	30.82	29.46	28.71	28.24	27.91	27.67	27.49	27.35	27.23	27.13	27.05	26.87	26.69	26.58	26.50	26.41	26.35	26.24
4	21.20	18.00	16.69	15.98	15.52	15.21	14.98	14.80	14.66	14.55	14.45	14.37	14.20	14.02	13.91	13.84	13.75	13.69	13.58
5	16.26	13.27	12.06	11.39	10.97	10.67	10.46	10.29	10.16	10.05	9.96	9.89	9.72	9.55	9.45	9.38	9.29	9.24	9.13
6	13.75	10.92	9.78	9.15	8.75	8.47	8.26	8.10	7.98	7.87	7.79	7.72	7.56	7.40	7.30	7.23	7.14	7.09	6.99
7	12.25	9.55	8.45	7.85	7.46	7.19	6.99	6.84	6.72	6.62	6.54	6.47	6.31	6.16	6.06	5.99	5.91	5.86	5.75
8	11.26	8.65	7.59	7.01	6.63	6.37	6.18	6.03	5.91	5.81	5.73	5.67	5.52	5.36	5.26	5.20	5.12	5.07	4.96
9	10.56	8.02	6.99	6.42	6.06	5.80	5.61	5.47	5.35	5.26	5.18	5.11	4.96	4.81	4.71	4.65	4.57	4.52	4.41
10	10.04	7.56	6.55	5.99	5.64	5.39	5.20	5.06	4.94	4.85	4.77	4.71	4.56	4.41	4.31	4.25	4.17	4.12	4.01
11	9.65	7.21	6.22	5.67	5.32	5.07	4.89	4.74	4.63	4.54	4.46	4.40	4.25	4.10	4.01	3.94	3.86	3.81	3.71
12	9.33	6.93	5.95	5.41	5.06	4.82	4.64	4.50	4.39	4.30	4.22	4.16	4.01	3.86	3.76	3.70	3.62	3.57	3.47
13	9.07	6.70	5.74	5.21	4.86	4.62	4.44	4.30	4.19	4.10	4.02	3.96	3.82	3.66	3.57	3.51	3.43	3.38	3.27
14	8.86	6.51	5.56	5.04	4.69	4.46	4.28	4.14	4.03	3.94	3.86	3.80	3.66	3.51	3.41	3.35	3.27	3.22	3.11
15	8.68	6.36	5.42	4.89	4.56	4.32	4.14	4.00	3.89	3.80	3.73	3.67	3.52	3.37	3.28	3.21	3.13	3.08	2.98
16	8.53	6.23	5.29	4.77	4.44	4.20	4.03	3.89	3.78	3.69	3.62	3.55	3.41	3.26	3.16	3.10	3.02	2.97	2.86
17	8.40	6.11	5.18	4.67	4.34	4.10	3.93	3.79	3.68	3.59	3.52	3.46	3.31	3.16	3.07	3.00	2.92	2.87	2.76
18	8.29	6.01	5.09	4.58	4.25	4.01	3.84	3.71	3.60	3.51	3.43	3.37	3.23	3.08	2.98	2.92	2.84	2.78	2.68
19	8.18	5.93	5.01	4.50	4.17	3.94	3.77	3.63	3.52	3.43	3.36	3.30	3.15	3.00	2.91	2.84	2.76	2.71	2.60
20	8.10	5.85	4.94	4.43	4.10	3.87	3.70	3.56	3.46	3.37	3.29	3.23	3.09	2.94	2.84	2.78	2.69	2.64	2.54
21	8.02	5.78	4.87	4.37	4.04	3.81	3.64	3.51	3.40	3.31	3.24	3.17	3.03	2.88	2.79	2.72	2.64	2.58	2.48
22	7.95	5.72	4.82	4.31	3.99	3.76	3.59	3.45	3.35	3.26	3.18	3.12	2.98	2.83	2.73	2.67	2.58	2.53	2.42
23	7.88	5.66	4.76	4.26	3.94	3.71	3.54	3.41	3.30	3.21	3.14	3.07	2.93	2.78	2.69	2.62	2.54	2.48	2.37
24	7.82	5.61	4.72	4.22	3.90	3.67	3.50	3.36	3.26	3.17	3.09	3.03	2.89	2.74	2.64	2.58	2.49	2.44	2.33
25	7.77	5.57	4.68	4.18	3.85	3.63	3.46	3.32	3.22	3.13	3.06	2.99	2.85	2.70	2.60	2.54	2.45	2.40	2.29
30	7.56	5.39	4.51	4.02	3.70	3.47	3.30	3.17	3.07	2.98	2.91	2.84	2.70	2.55	2.45	2.39	2.30	2.25	2.13
40	7.31	5.18	4.31	3.83	3.51	3.29	3.12	2.99	2.89	2.80	2.73	2.66	2.52	2.37	2.27	2.20	2.11	2.06	1.94
50	7.17	5.06	4.20	3.72	3.41	3.19	3.02	2.89	2.78	2.70	2.63	2.56	2.42	2.27	2.17	2.10	2.01	1.95	1.82
100	6.90	4.82	3.98	3.51	3.21	2.99	2.82	2.69	2.59	2.50	2.43	2.37	2.22	2.07	1.97	1.89	1.80	1.74	1.60

Degrees of Freedom for the Denominator

The *F* Distribution Table (continued)

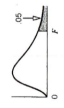

Area in the Right Tail under the *F* Distribution Curve = .05

Degrees of Freedom for the Numerator

	1	2	3	4	5	6	7	8	9	10	11	12	15	20	25	30	40	50	100
1	161.5	199.5	215.7	224.6	230.2	234.0	236.8	238.9	240.5	241.9	243.0	243.9	246.0	248.0	249.3	250.1	251.1	251.8	253.0
2	18.51	19.00	19.16	19.25	19.30	19.33	19.35	19.37	19.38	19.40	19.40	19.41	19.43	19.45	19.46	19.46	19.47	19.48	19.49
3	10.13	9.55	9.28	9.12	9.01	8.94	8.89	8.85	8.81	8.79	8.76	8.74	8.70	8.66	8.63	8.62	8.59	8.58	8.55
4	7.71	6.94	6.59	6.39	6.26	6.16	6.09	6.04	6.00	5.96	5.94	5.91	5.86	5.80	5.77	5.75	5.72	5.70	5.66
5	6.61	5.79	5.41	5.19	5.05	4.95	4.88	4.82	4.77	4.74	4.70	4.68	4.62	4.56	4.52	4.50	4.46	4.44	4.41
6	5.99	5.14	4.76	4.53	4.39	4.28	4.21	4.15	4.10	4.06	4.03	4.00	3.94	3.87	3.83	3.81	3.77	3.75	3.71
7	5.59	4.74	4.35	4.12	3.97	3.87	3.79	3.73	3.68	3.64	3.60	3.57	3.51	3.44	3.40	3.38	3.34	3.32	3.27
8	5.32	4.46	4.07	3.84	3.69	3.58	3.50	3.44	3.39	3.35	3.31	3.28	3.22	3.15	3.11	3.08	3.04	3.02	2.97
9	5.12	4.26	3.86	3.63	3.48	3.37	3.29	3.23	3.18	3.14	3.10	3.07	3.01	2.94	2.89	2.86	2.83	2.80	2.76
10	4.96	4.10	3.71	3.48	3.33	3.22	3.14	3.07	3.02	2.98	2.94	2.91	2.85	2.77	2.73	2.70	2.66	2.64	2.59
11	4.84	3.98	3.59	3.36	3.20	3.09	3.01	2.95	2.90	2.85	2.82	2.79	2.72	2.65	2.60	2.57	2.53	2.51	2.46
12	4.75	3.89	3.49	3.26	3.11	3.00	2.91	2.85	2.80	2.75	2.72	2.69	2.62	2.54	2.50	2.47	2.43	2.40	2.35
13	4.67	3.81	3.41	3.18	3.03	2.92	2.83	2.77	2.71	2.67	2.63	2.60	2.53	2.46	2.41	2.38	2.34	2.31	2.26
14	4.60	3.74	3.34	3.11	2.96	2.85	2.76	2.70	2.65	2.60	2.57	2.53	2.46	2.39	2.34	2.31	2.27	2.24	2.19
15	4.54	3.68	3.29	3.06	2.90	2.79	2.71	2.64	2.59	2.54	2.51	2.48	2.40	2.33	2.28	2.25	2.20	2.18	2.12
16	4.49	3.63	3.24	3.01	2.85	2.74	2.66	2.59	2.54	2.49	2.46	2.42	2.35	2.28	2.23	2.19	2.15	2.12	2.07
17	4.45	3.59	3.20	2.96	2.81	2.70	2.61	2.55	2.49	2.45	2.41	2.38	2.31	2.23	2.18	2.15	2.10	2.08	2.02
18	4.41	3.55	3.16	2.93	2.77	2.66	2.58	2.51	2.46	2.41	2.37	2.34	2.27	2.19	2.14	2.11	2.06	2.04	1.98
19	4.38	3.52	3.13	2.90	2.74	2.63	2.54	2.48	2.42	2.38	2.34	2.31	2.23	2.16	2.11	2.07	2.03	2.00	1.94
20	4.35	3.49	3.10	2.87	2.71	2.60	2.51	2.45	2.39	2.35	2.31	2.28	2.20	2.12	2.07	2.04	1.99	1.97	1.91
21	4.32	3.47	3.07	2.84	2.68	2.57	2.49	2.42	2.37	2.32	2.28	2.25	2.18	2.10	2.05	2.01	1.96	1.94	1.88
22	4.30	3.44	3.05	2.82	2.66	2.55	2.46	2.40	2.34	2.30	2.26	2.23	2.15	2.07	2.02	1.97	1.94	1.91	1.85
23	4.28	3.42	3.03	2.80	2.64	2.53	2.44	2.37	2.32	2.27	2.24	2.20	2.13	2.05	2.00	1.96	1.91	1.88	1.82
24	4.26	3.40	3.01	2.78	2.62	2.51	2.42	2.36	2.30	2.25	2.22	2.18	2.11	2.03	1.97	1.94	1.89	1.86	1.80
25	4.24	3.39	2.99	2.76	2.60	2.49	2.40	2.34	2.28	2.24	2.20	2.16	2.09	2.01	1.96	1.92	1.87	1.84	1.78
30	4.17	3.32	2.92	2.69	2.53	2.42	2.33	2.27	2.21	2.16	2.13	2.09	2.01	1.93	1.88	1.84	1.79	1.76	1.70
40	4.08	3.23	2.84	2.61	2.45	2.34	2.25	2.18	2.12	2.08	2.04	2.00	1.92	1.84	1.78	1.74	1.69	1.66	1.59
50	4.03	3.18	2.79	2.56	2.40	2.29	2.20	2.13	2.07	2.03	1.99	1.95	1.87	1.78	1.73	1.69	1.63	1.60	1.52
100	3.94	3.09	2.70	2.46	2.31	2.19	2.10	2.03	1.97	1.93	1.89	1.85	1.77	1.68	1.62	1.57	1.52	1.48	1.39

Degrees of Freedom for the Denominator

▶ *Index*